High T_c Superconductors and Related Transition Metal Oxides

Annette Bussmann-Holder
Hugo Keller (Eds.)

High T_c Superconductors and Related Transition Metal Oxides

Special Contributions in Honor of K. Alex Müller on the Occasion of his 80th Birthday

Springer

Dr. Annette Bussmann-Holder
Max-Planck-Institut für Festkörperforschung
Heisenbergstraße 1
70569 Stuttgart, Germany
a.bussmann-holder@fkf.mpg.de

Dr. Hugo Keller
Universität Zürich
Physik-Institut
Winterthurer Straße 190
8057 Zürich, Switzerland
keller@physik.unizh.ch

Library of Congress Control Number: 2007922939

ISBN 978-3-540-71022-6 Springer Berlin Heidelberg New York
DOI 10.1007/978-3-540-71023-3

Springer is a part of Springer Science+Business Media

springer.com

Cover Design: WMXDesign GmbH, Heidelberg, Germany
Typesetting and Production: LE-TEX Jelonek, Schmidt & Vöckler GbR, Leipzig

Printed on acid-free paper 2/3100 YL – 5 4 3 2 1 0

Um ein tadelloses Mitglied einer Schafherde sein zu können,
muss man vor allem ein Schaf sein.

In order to be an irreproachable member of a flock of sheep,
most notably one has to be a sheep.

Albert Einstein

K. Alex Müller

Preface

When initiating this book and inviting friends of K. Alex Müller to send contributions to it in order to honor his 80'th birthday, we frequently were told that it would be difficult to write anything related to high temperature superconductivity. It seemed that most of his friends and colleagues associate almost exclusively high temperature superconductivity with his name and have forgotten that K. Alex Müller's scientific career is more tightly bound to phase transitions, critical phenomena, electron paramagnetic resonance (EPR), and ferroelectricity than to superconductivity, even though he caused – together with Georg Bednorz – a revolution in this field. More than half of his research life concentrates on the above mentioned phenomena and major breakthroughs have been achieved by him in this area. Starting point was the investigation of the electron paramagnetic resonance of Mn^{IV} in $SrTiO_3$ where he observed that distorted regimes appear in the nominally cubic phase. A homogeneous state is only realized when the crystal is in the tetragonal phase. Even though $SrTiO_3$ remained in the center of his research interests early on, the Jahn–Teller effect was fascinating him. Experimentally and theoretically he started various projects in this field which has the unique property that a vibronic state dominates the physics, where electronic and ionic degrees of freedom are undistinguishable. This state is a consequence of a competition between enhancing the entropy and lowering the energy of the system and intimately related to ferrodistortive transitions. These have been investigated in detail by him concentrating on the rotational instability of the oxygen octahedral in perovskites where he showed that the rotation angle plays the role of an order parameter. Together with Harry Thomas he established a theory of phase transitions, where soft mode dynamics are predicted in coexistence with elastic instabilities stemming from the coupling to strain fields. However, in ferrodistortive transitions the Jahn–Teller effect is the important one, since here electron-lattice interactions play a crucial role. Since phase transitions are intimately related to critical exponents, EPR techniques have been used by him in order to determine these for the rotational instabilities in various perovskites. The essence of this work was that in the critical region physically measurable variables become independent of interatomic distances, since due to cooperative phenomena the correlation lengths are much larger than the range of forces between the individual particles. The phenomenon of phase transitions has been his focus for many years, where especially those related to ferroelectricity and zone boundary instabilities have attracted his interest. Besides of EPR also electron spin resonance (ESR) methods have been used by him. It should be emphasized here that K. Alex Müller's work always represented a mixture of theoretical considerations and experimental results. In the following years he concentrated on details related to structural phase transitions, where he was able to show that rotational fluctuations are present already above the instability, and a crossover from a fast to slow fluctuation regime sets in upon approaching the critical temperature . This notion is nowadays reinvented, since high resolution local probes enable to detect the onset of fluctuations far above T_c. Besides of perovskites, he also worked on hydrogen-bonded ferroelectrics, where for the first time evidence for the existence of locally polarized clusters was obtained which form far above the actual phase transition temperature and exhibit distinct life times. Importantly, these extra dynamics are present independent of the tunneling motion of the protons and are related to the heavy ion sublattice. We believe this work is the first to provide insight into the phase transition mechanism of hydrogen-bonded ferroelectrics which is beyond the concept of purely hydrogen ordering induced. Intimately related to this is the classification of ferroelectrics into either order/disorder or displacive ones. From the very beginning of the research in ferroelectricity this scheme has been used to classify this material class. It is based on the assumption that in an order/disorder dominated transition the involved ions tunnel between equivalent lattice sites in a random manner above T_c and order cooperatively into one of the minima at T_c. In a displacive transition, on the other hand, the motion of two sublattice ions against each other slows down with decreasing temperature to freeze out at T_c, and thus determines the low temperature polar structure. From his EPR measurements K. Alex Müller came to the conclusion that such a strict distinction of the two pictures is not possible, but that always a coexistence regime must be present. Interestingly, he resolved with this proposal a long standing controversy, since long wave length experiments showed – in contrast to local probes – displacive characteristics, whereas the local probes provided evidence for order/disorder dynamics. This experimental discrepancy has been shown to be a consequence of different time and length scales involved in the different experiments. The issue of time and length scales has been followed by him until now. An important point was addressed by K. Alex Müller within the context of ferroelectricity. While he concentrated for a while on the structural instability observed in $SrTiO_3$ at 104K, which is driven by the condensation of a zone boundary mode, he got more interested in the dielectric behavior at lower temperatures where the dielectric constant continuously grows with decreasing temperature. Such a feature is reminiscent of a polar instability. However, and opposite to e.g. $BaTiO_3$ and $KNbO_3$ no instability sets in, since quantum fluctuations suppress the phase transition. This led him to introduce the concept of quantum paraelectricity which is a rare phenomenon and only observed in one other perovskite, namely $KTaO_3$. Ferroelectricity can be induced in these systems by proper doping. Especially, Ca doping of $SrTiO_3$ shows interesting aspects, since a displacive type transition can be induced by a adding a small amount of Ca. The nature of the transition changes substantially with increasing doping from a XY quantum ferroelectric to a random-field induced domain state. Using photoelectron and inverse photoemission spectroscopy the band structure of $SrTiO_3$ has been investigated by him which shows the peculiar feature that the valence band states are not of purely oxygen 2p character, but have small admixtures of Ti d-states. This p-d hybridization has been shown then to be a basic ingredient for phase transition mechanism in perovksite oxides. Since lattice dynamics is a central issue to ferroelectrics as it can discriminate between order/disorder and displacive transitions, he used EPR in order to answer this question. Especially in the case of $BaTiO_3$ it remained unclear how to classify this compound, since again different experiments reported different dynamics. K. Alex Müller was able to measure the degree of anharmonicity which is a direct probe of the transition dynamics. He observed that the anharmonicity of $BaTiO_3$ is 34% larger than the one of $SrTiO_3$ and $KTaO_3$. Correspondingly, the double-well potential has a large potential barrier as is typical for order/disorder transitions. The same conclusions have been reached only recently by NMR techniques which probe local properties.

While K. Alex Müller is very well known for his extended research in phase transitions, critical phenomena, and ferroelectricity, it has frequently been overlooked that he worked also in superconductivity concentrating on now very popular aspects . He observed that the superconducting properties of granular aluminum are highly inhomogeneous and characterized by local transition temperatures. Using microwave measurements of granular aluminum sheets in a cavity he obtained direct information on the number and temperature dependence of superconducting Cooper pairs. Specifically. inhomogeneity has important influences on the resistive transitions which take place by the formation of superconducting regions below a certain critical temperature. But the locking of phases between the grains produces a continuous superconducting path only when a lower temperature is reached. The notion of inhomogeneity is nowadays an important issue in high temperature superconductivity where substantial evidence for multi-components exists.

The discovery of high temperature superconductivity in cuprates did not end his interests in phase transitions, but was actually a result of his former research in transition metal oxides where electron lattice interaction effects have been shown to be crucial. His detailed knowledge of the Jahn–Teller effect played here an important role, since from his collaboration with Harry Thomas in Basel and Heinz Bilz in Stuttgart the possibility seemed to be given that polaron formation could be a new glue in Cooper pair formation. Since this process is not tied to the energy scale of a typical lattice vibration, there was a chance of raising the superconducting transition temperature in an unexpected way. His success justified this assumption. However, the research in superconductivity took an unexpected way, since the antiferromagnetic properties of the parent compounds have frequently been taken as evidence that purely electronic mechanisms are at work. These proposals intensified K. Alex Müller's work in the field of high temperature superconductivity, since they were against his own conviction. His scientific focus after the discovery was consequently on exploring the microscopic origin of the pairing mechanism in cuprates. Since in conventional superconductors the isotope effect on T_c has proven that electron-lattice interactions are the source of Cooper pair formation, he initiated a vast and extremely successful research at the University of Zürich on isotope effects in cuprates. In order to identify the locus of pairing, carefully prepared site-selective oxygen isotope experiments have been performed showing that the largest effect on T_c stems from the planes, whereas only a negligible isotope effect can be attributed to the out of plane oxygen ions. Besides of this isotope effect, various others have been discovered which are mostly highly unconventional, unexpected and beyond a BCS framework. It has to be mentioned here, that a substantial and sign reversed isotope effect has been observed by different methods and in various cuprate families on the so-called pseudo gap temperature T^*. Furthermore the superfluid density has been shown to be isotope dependent, thereby showing that the conventional BCS pairing mechanism can be excluded, since within BCS theory the penetration depth shows no isotope effect. Using his preferred method of EPR he could show that the EPR line width in Mn^{2+} doped $La_{2-x}Sr_xCuO_4$ is larger for ^{18}O than for ^{16}O, being pronounced at low temperatures and decreasing with increasing Sr content. This effect has been explained in terms Cu S=1/2 relaxation coupled linearly to local Q_4/Q_5 tilting modes of the CuO_6 octahedra. Since these modes are coupled sterically to the Q_2 type Jahn–Teller mode, the results support the idea of polaron formation as being a relevant ingredient to superconductivity. An important issue has been raised by K. Alex Müller early on, namely, that the pairing state is not of d-wave symmetry alone, but carries a substantial s-wave component. This implies that cuprates are multi component systems where the dynamical interplay between the components gives rise to multi-band superconductivity. These ideas have recently been confirmed experimentally by muon spin rotation, which unambiguously demonstrates that besides of a leading d-wave component, also a pronounced s-wave component is present in cuprates. This observation provides further evidence that cuprates are inhomogeneous and a composite of at least two components, a hole rich polaronic distorted one and a hole poor undistorted matrix, as confirmed early on by his EPR experiments. Even though the origin about the pairing glue in cuprates is not generally agreed on, many facts support his original concept of Jahn–Teller polaron formation.

In addition to cuprates also manganites have been of extreme interest to him in the more recent past. Their structural similarity to cuprates and perovskite ferroelectrics suggests that electron-lattice interactions are important in understanding their rich phenomenology. That here again the Jahn–Teller effect is active and polaron formation a clue in understanding the physics, could be proven by showing that the ferromagnetic transition temperature carries a substantial isotope effect. In addition, an isotope effect on the magnetic exchange coupling constant has been detected via EPR measurements. By comparing these results to those obtained for iso-structural ruthenates, where no isotope effect has been observed, it became obvious that the orbital degeneracy of the Mn^{3+} state is lifted through a lattice distortion, i.e., the Jahn–Teller effect. The isotope effect observed via EPR, on the other hand, is a consequence of polaron formation. Both effects established that Jahn–Teller polaron formation is the origin of the giant magneto resistent effect.

Besides of K. Alex Müller's broad scientific contributions to solid state physics and in view of the numerous honors, including the Nobel prize, he received, his personal style and his attitude towards his friends and collaborators never changed during the years. Opposite to many of his colleagues with comparable merits, he never acted like a despot who takes his opinion as dogma to be adopted by everyone. He suggested many research fields which typically yielded extremely successful results. His intuition, his *nose*, has always led to new areas and influenced whole research fields. Even if there was disbelieve in his *nose*, he was patient enough to await the sometimes long way of others to share his opinion. *This might take even 10 years or longer!*

K. Alex Müller is still engaged in science and participates in the university life. He is regularly coming to his office at the University of Zürich and contributes actively to ongoing research fields. In particular, he attends all seminars held at the Physik-Institut, even if students are the speakers and actively participates in the discussions. His interest in the young research students' careers and scientific developments as well as their social life is intense and very supportive. Also he tries not to miss social activities of the Physik-Insitut as, e.g., the annual Christmas party, the annual excursion, birthday parties and personal celebrations, which is encouraging for all members of the Physik-Institut. Besides of visiting regularly the university he still attends many conferences and presents his new view points in science. Unlike many other people of his age, he even learned new technologies like power point presentation and manages to use these techniques in a very professional way. Again in contrast to many others, he typically attends conferences, symposia and schools for the full duration and contributes to all discussions.

In his private life he adopted similar attitudes as those that had emerged from his research. The strict classification scheme into classes, which had been applied to ferroelectrics for a long time and which he showed to be invalid, has been avoided by him, showing that life is not black or white, but that shades and many colors determine the human character. Time and length scales have been important for him in physics, and also in life where the family entity is the small scale of its own importance. On the other hand, social life requires to consider global schemes to be taken into account as well. Looking through a microscope on something does not tell you what the macroscopic shape looks like. Only both together will give you information about the whole object. Similarly, time scales can be measured on the scale of the individual life time. However, history and future influence this in a dramatic way, and larger time scales have to considered.

We both are grateful to share friendship with Alex. We profited in an enormous way from this, as well as from his scientific input on our work. What we would like to emphasize is, that however strongly he was convinced of something, he never applied any pressure on us in order to persuade us that he was right.

Finally, some personal interactions should be mentioned which emphasize his humor and friendly attitude. Alex has proven many times that he is a perfect gentleman. We will never forget the occasion that when leaving a restaurant he brought a nice fur to Annette belonging to another lady, even though this lady heavily protested. On another occasion he forgot the name of the hotel he stayed in during a conference. He was convinced it should be *Kodak* but we could not find such a place even in the telephone book. By chance we finally found the right hotel with a completely different name after walking for a long time in different directions. Since he insists to drive old cars instead of a nice fancy brand new Mercedes with all kinds of high tech electronic support, it occurred that he arrived with a broken starter and we had to push it to get it running. Instead of going to a repair place he prefers to do this himself. For this reason he has installed a garage in his house which is not always very much appreciated by his wife, since he touches unintentionally with his dirty hands everything which was just polished. Interestingly, he wrote a paper about this hobby, entitled "garagists and medical doctors", where he suggests many analogies between both. Another hobby of Alex is downhill skiing. He counts the number of skiing days per year and is highly disappointed if a certain critical number is not reached. He is a real expert in this field and skis so excellently that many young skiers have substantial difficulties to follow him. Sometimes it seems that he was born with skis on his feet. There are many more things to mention about Alex as a human being, but these are postponed to his next round birthday.

Dear Alex, we wish you and your family many more happy years together, many more winters with skiing in the Swiss mountains, many more oldtimers to be repaired, many more conferences to attend and many more years in physics such that we can continue to profit from your fruitful suggestions.

Stuttgart und Zürich,
February 2007

Annette Bussmann-Holder
Hugo Keller

List of Authors

A.S. Alexandrov
Department of Physics
Loughborough University
LE11 3TU Loughborough
UK
A.S.Alexandrov@lboro.ac.uk

Y. Ando
Institute of Scientific
and Industrial Research
Osaka University
567-0047 Osaka
Japan

Y.S. Barash
Institute of Solid State Physics,
Russian Academy of Sciences
Chernogolovka
142432 Moscow District
Russia

J.G. Bednorz
IBM Research
Zurich Research Laboratory
Säumerstrasse 4
CH-8803 Rüschlikon
Switzerland
bed@zurich.ibm.com

A. Bill
Department of Physics & Astronomy
California State University Long Beach
90840 Long Beach, CA
USA

J. Bok
Solid State Physics Laboratory
ESPCI
10, rue Vauquelin
75231 Paris Cedex 05
France
julien.bok@espci.fr

J. Bouvier
Solid State Physics Laboratory
ESPCI
10, rue Vauquelin
75231 Paris Cedex 05
France

I. Bozovic
Brookhaven National Laboratory
11973 Upton, NY
USA
bozovic@bnl.gov

A. Bussmann-Holder
Max-Planck-Institut
für Festkörperforschung
Heisenbergstrasse 1D
70569 Stuttgart
Germany
a.bussmann-holder@fkf.mpg.de

R.J. Cava
Department of Chemistry
Princeton University
08544 Princeton, NJ
USA

C.W. Chu
TCSUH, University of Houston
Hong Kong University
of Science and Technology
& Lawrence Berkeley National Laboratory
USA
cwchu@ust.hk

K. Conder
Laboratory for Developments and Methods
Paul Scherrer Institut
CH-5232 Villigen PSI
Switzerland
kazimierz.conder@psi.ch

T. Cuk
Department of Physics, Applied Physics,
and Stanford Synchrotron
Radiation Laboratory
Stanford University
94305 Stanford, CA
USA

N.S. Dalal
Department of Chemistry & Biochemistry
and NHMFL
Florida State University
32306-4390 Tallahassee, FL
USA
dalal@chemmail.chem.fsu.edu

S. Deng
Max-Planck-Institut
für Festkörperforschung
Heisenbergstr. 1
70569
Stuttgart
Germany

G. Deutscher
School of Physics and Astronomy
Tel Aviv University
Ramat Aviv
69978 Tel Aviv
Israel
guyde@post.tau.ac.il

T.P. Devereaux
Department of Physics
University of Waterloo, Waterloo
N2L 3G1 Ontario
Canada

T. Egami
University of Tennessee, Science Alliance,
Department of Materials Science
and Engineering
Department of Physics and Astronomy
and the Joint Institute for Neutron Sciences
37996-1508 Knoxville, TN
USA
egami@utk.edu

R.A. Fisher
Lawrence Berkeley National Laboratory and
Department of Chemistry
University of California
94720 Berkeley, CA
USA

M.-L. Foo
Department of Chemistry
Princeton University
08544 Princeton, NJ
USA

K. Fossheim
Department of Physics
Norwegian University
of Science and Technology
Trondheim
Norway
Kristian.Fossheim@phys.ntnu.no

A. Furrer
Laboratory for Neutron Scattering
ETH Zurich & PSI Villigen
CH-5232 Villigen PSI
Switzerland
albert.furrer@psi.ch

J.E. Gordon
Physics Department
Amherst College
01002 Amherst, MA
USA

V. Hizhnyakov
Institute of Physics
University of Tartu
EE-2400 Tartu
Estonia
hizh@eeter.fi.tartu.ee

Z. Hussain
Advanced Light Source
Lawrence Berkeley National Lab
94720 Berkeley, CA, USA

H. Kamimura
Department of Applied Physics
Tokyo University of Science
1-3 Kagurazaka, Shinjuku-ku
162-8601 Tokyo
Japan
kamimura@rs.kagu.tus.ac.jp

J. Karpinski
Laboratory for Solid State Physics
ETH 8093 Zürich
Switzerland
karpinski@solid.phys.ethz.ch

H. Keller
Physik-Institut der Universität Zürich
Winterthurerstrasse 190
CH-8057 Zürich
Switzerland
keller@physik.unizh.ch

R. Khasanov
Physik-Institut der Universität Zürich
Winterthurerstrasse 190
CH-8057 Zürich
Switzerland
khasanov@physik.unizh.ch

J.S. Kim
Max-Planck-Institut
für Festkörperforschung
Heisenbergstr. 1
70569 Stuttgart
Germany

B.I. Kochelaev
Department of Physics
Kazan State University
420008 Kazan
Russia
Boris.Kochelaev@ksu.ru

J. Köhler
Max-Planck-Institut
für Festkörperforschung
Heisenbergstr. 1
70569 Stuttgart
Germany
j.koehler@fkf.mpg.de

T. Kopp
Center for Electronic Correlations
and Magnetism, EP6
Universität Augsburg
86135 Augsburg
Germany

R.K. Kremer
Max-Planck-Institut
für Festkörperforschung
Heisenbergstr. 1
70569 Stuttgart
Germany
rekre@fkf.mpg.de

P. Kusar
Dept. of Complex Matter
Jozef Stefan Institute
SI-1000 Ljubljana
Slovenia

W.S. Lee
Department of Physics, Applied Physics, and Stanford Synchrotron Radiation Laboratory
Stanford University
94305 Stanford, CA
USA
leews@stanford.edu

C.T. Lin
Max-Planck Institute für Festkörperforschung
70569 Stuttgart
Germany

D.H. Lu
Department of Physics, Applied Physics, and Stanford Synchrotron Radiation Laboratory
Stanford University
94305 Stanford, CA
USA

J. Mannhart
Center for Electronic Correlations and Magnetism, EP6
Universität Augsburg
86135 Augsburg
Germany
jochen.mannhart@physik.uni-augsburg.de

S. Matsuno
General Education Program Center
Tokai University
Shimizu Campus 3-20-1, Shimizu-Orido
424-8610 Shizuoka
Japan

W. Meevasane
Department of Physics, Applied Physics, and Stanford Synchrotron Radiation Laboratory
Stanford University
94305 Stanford, CA
USA

D. Mihailovic
Dept. of Complex Matter
Jozef Stefan Institute
SI-1000 Ljubljana
Slovenia
dragan.mihailovic@ijs.si

N. Oeschler
Lawrence Berkeley National Laboratory and Department of Chemistry
University of California
94720 Berkeley, CA
USA

H. Oyanagi
Photonics Research Institute
AIST
1-1-1 Umezono, Tsukuba
305-8568 Ibaraki
Japan
h.oyanagi@aist.go.jp

N.E. Phillips
Lawrence Berkeley National Laboratory and Department of Chemistry
University of California
94720 Berkeley, CA
USA
nephill@cchem.berkeley.edu

T. Schneider
Physik-Institut der Universität Zürich
Winterthurerstrasse 190
CH-8057 Zürich
Switzerland
toni.schneider@swissonline.ch

J.F. Scott
Centre for Ferroics
Earth Sciences Department,
University of Cambridge
CB2 3EQ Cambridge
UK
jsco99@esc.cam.ac.uk

G. Seibold
Lehrstuhl für Theoretische Physik
BTU Cottbus
03013 Cottbus
Germany

K.M. Shen
Department of Physics, Applied Physics, and Stanford Synchrotron Radiation Laboratory
Stanford University
94305 Stanford, CA
USA

Z.-X. Shen
Department of Physics, Applied Physics, and Stanford Synchrotron Radiation Laboratory
Stanford University
94305 Stanford, CA
USA

A. Shengelaya
Physics Institute of Tbilisi State University
Chavchavadze 3
GE-0128 Tbilisi
Georgia
salex@physik.unizh.ch

J.-I. Shimoyama
Department of Applied Chemistry
University of Tokyo
113-8656 Tokyo
Japan

E. Sigmund
Lehrstuhl für Theoretische Physik
BTU Cottbus
03013 Cottbus
Germany

A. Simon
Max-Planck-Institut für Festkörperforschung
Heisenbergstr. 1
70569 Stuttgart
Germany

C.P. Slichter
Department of Physics
University of Illinois
Urbana
61801-3080 Illinois
USA
cslichte@uiuc.edu

M. Takashige
Department of Electronics
Iwaki Meisei University
Chuohdai Iino 5-5-1, Iwaki
970-8551 Fukushima
Japan
masa@iwakimu.ac.jp

H. Thomas
Universität Basel
Institut für Physik
Klingelbergstrasse 82
CH-4056 Basel
Switzerland
Harry.Thomas@unibas.ch

H. Ushio
Tokyo National College of Technology
1220-2 Kunugidai-chou
193-0997 Hachioji
Japan

F. Waldner
Physics Institute
University of Zürich
CH-8057 Zürich
Switzerland

W.L. Yang
Advanced Light Source
Lawrence Berkeley National Lab
94720 Berkeley, CA
USA

X.J. Zhou
Advanced Light Source
Lawrence Berkeley National Lab
94720 Berkeley, CA
USA

Contents

Alexandrov AS (2007) In: Bussmann-Holder A, Keller H (eds)
High T_c Superconductors and Related Transition Metal Oxides.
Springer-Verlag Berlin Heidelberg, 1–15
DOI 10.1007/978-3-540-71023-3

Superlight Small Bipolarons: A Route to Room Temperature Superconductivity

A. S. Alexandrov

Department of Physics, Loughborough University, Loughborough LE11 3TU, UK
A.S.Alexandrov@lboro.ac.uk

Abstract Extending the BCS theory towards the strong electron-phonon interaction (EPI), a charged Bose liquid of small bipolarons has been predicted by us with a further prediction that the highest superconducting critical temperature is found in the crossover region of the EPI strength from the BCS-like to bipolaronic superconductivity. Later on we have shown that the unscreened (infinite-range) Fröhlich EPI combined with the strong Coulomb repulsion create *superlight* small bipolarons, which are several orders of magnitude lighter than small bipolarons in the Holstein–Hubbard model (HHM) with a zero-range EPI. The analytical and numerical studies of this Coulomb–Fröhlich model (CFM) provide the following recipes for room-temperature superconductivity: (a) The parent compound should be an ionic insulator with light ions to form high-frequency optical phonons, (b) the structure should be quasi two-dimensional to ensure poor screening of high-frequency phonons polarized perpendicular to the conducting planes, (c) a triangular lattice is required in combination with strong, on-site Coulomb repulsion to form the small superlight bipolaron, (d) moderate carrier densities are required to keep the system of small bipolarons close to the Bose-Einstein condensation regime. Clearly most of these conditions are already met in the cuprates.

Introduction

The discovery of high-temperature superconductivity in cuprates [1] has widened significantly our horizons of the theoretical understanding of the physical phenomenon. A great number of observations point to the possibility that the cuprate superconductors may not be conventional Bardeen–Cooper–Schrieffer (BCS) superconductors [2], but rather derive from the Bose-Einstein condensation (BEC) of real-space small bipolarons [3–5]. Importantly a first proposal for high temperature superconductivity by Ogg in 1946 [6], already involved real-space pairing of individual electrons into bosonic molecules with zero total spin. This idea was further developed as a natural explanation of conventional superconductivity by Schafroth [7] and Butler and Blatt [8]. Unfortunately the Ogg–Schafroth picture was practically forgotten because it neither accounted quantitatively for the critical behavior of conventional superconductors, nor did it explain the microscopic nature of attractive forces which could overcome the Coulomb repulsion between two electrons constituting a pair. On the contrary highly successful for low-T_c metals and alloys the BCS theory, where two electrons were indeed correlated, but at a very large distance of about 10^3 times of the average inter-electron spacing, led many researchers to believe that any superconductor is a "BCS-like".

However it has been found – unexpectedly for many researchers – that the BCS theory and its extension [9–12] towards the intermediate coupling regime, $\lambda \lesssim 1$, break down already at $\lambda \gtrsim 1$ [13, 14]. It happens since the Migdal "noncrossing" approximation [15, 16] of the theory is not applied at $\lambda \gtrsim 1$. In fact, the small parameter of the theory, $\lambda\omega_0/E_F$, becomes large at $\lambda \gtrsim 1$ because the bandwidth is narrowed and the Fermi energy, E_F is renormalised down exponentially due to the small polaron formation [13, 14, 17] (here ω_0 is the characteristic phonon frequency, and we take $\hbar = c = k_B = 1$). Extending the BCS theory towards the strong interaction between electrons and ion vibrations, $\lambda \gg 1$, a charged Bose gas of tightly bound small bipolarons was predicted [18, 19] instead of Cooper pairs, with a further prediction that the highest superconducting transition temperature is attained in the crossover region of EPI strength, $\lambda \approx 1$, between the BCS and bipolaronic superconductivity [13, 14].

For a very strong EPI polarons become self-trapped on a single lattice site and bipolarons are on-site singlets. A finite on-site bipolaron mass appears only in the second order of polaron hopping [18, 19], so that on-site bipolarons might be very heavy in HHM, where EPI is short-ranged. Actually HHM led some authors to the conclusion that the formation of itinerant small polarons and bipolarons in real materials is unlikely [20], and high-temperature bipolaronic superconductivity is impossible [21]. Nevertheless treating the onsite repulsion (Hubbard U) and the short-range EPI on an equal footing led several authors to the opposite conclusion with respect to bipolaron mobility even in HHM, which is generally unfavorable for coherent tunnelling. Aubry [22] found along with the onsite bipolaron ($S0$) also an anisotropic pair of polarons lying on two neighboring sites (i.e. the intersite bipolaron, $S1$) with classical phonons in the extreme adiabatic limit. Such bipolarons were originally hypothesized in [23–25] to explain the anomalous nuclear magnetic relaxation (NMR) in cuprate superconductors. The intersite bipolaron could take a form of a "quadrisinglet" (QS) in 2D HHM, where the electron density at the central site is 1 and "1/4" on the four nearest neigbouring sites. In a certain region of U, where QS is the ground state, the double-well potential barrier which usually pins polarons and bipolarons to the lattice depresses to almost zero, so that adiabatic lattice bipolarons can be rather mobile.

Mobile $S1$ bipolarons were found in 1D HHM using variational methods also in the non and near-adiabatic regimes with dynamical quantum phonons [26, 27]. The intersite bipolaron with a relatively small effective mass is stable in a wide region of the parameters of HHM due to both exchange and nonadiabaticity effects [27]. Near the strong coupling limit the mobile $S1$ bipolaron has an effective mass of the order of a single Holstein polaron mass, so that one should not rule out the possibility of a superconducting state of $S1$ bipolarons with s or d-wave symmetry in HHM [26]. More recent diagrammatic Monte Carlo study [28] found $S1$ bipolarons for large U at intermediate and large EPI and established the phase diagram of 2D HHM, comprising unbound polarons, $S0$ and $S1$ domains [28] emphasised that the transition to the bound state and the properties of the bipolaron in HHM are very different from bound states in the attractive (negative U) Hubbard model without EPI [29].

In any case the Holstein model is an extreme polaron model, with typically highest possible values of the (bi)polaron mass in the strong coupling regime [30–33]. Many doped ionic lattices, including cuprates, are characterized by poor screening of high-frequency optical phonons and they are more appropriately described by the finite-range Fröhlich EPI. The unscreened Fröhlich EPI provides relatively light lattice polarons and combined with the Coulomb repulsion also "superlight" but yet small (intersite) bipolarons. In contrast with the crawler motion of on-site bipolarons, the intersite-bipolaron tunnelling is a crab-like, so that the effective mass scales linearly with the polaron mass. Such bipolarons are several orders of magnitude lighter than small bipolarons in HHM [30]. Here I review a few analytical [30, 34] and more recent Quantum Monte-Carlo (QMC) [35, 36] studies of CFM which have found superlight bipolarons in a wide parameter range with achievable phonon frequencies and couplings. They could have a superconducting transition in excess of room temperature.

Coulomb–Fröhlich Model

Any realistic theory of doped narrow-band ionic insulators should include both the finite-range Coulomb repulsion and the strong finite-range EPI. From a theoretical standpoint, the inclusion of the finite-range Coulomb repulsion is critical in ensuring that the carriers would not form clusters. The Coulomb repulsion, V_c, makes the clusters unstable and lattice bipolarons *more mobile*.

To illustrate the point let us consider a generic multi-polaron "Coulomb–Fröhlich" model (CFM) on a lattice, which explicitly includes the finite-range Coulomb repulsion, V_c, and the strong long-range EPI [30, 34]. The implicitly present (infinite) Hubbard U prohibits double occupancy and removes the need to distinguish the fermionic spin, if we are interested in the charge rather than spin excitations. Introducing spinless fermion operators c_n and phonon operators d_m, the Hamiltonian of CFM is written in the real-space representation as [34]

$$
\begin{aligned}
H = & \sum_{n \neq n'} T(\boldsymbol{n}-\boldsymbol{n}') c_{\boldsymbol{n}}^{\dagger} c_{\boldsymbol{n}'} + \frac{1}{2} \sum_{n \neq n'} V_{\mathrm{c}}(\boldsymbol{n}-\boldsymbol{n}') c_{\boldsymbol{n}}^{\dagger} c_{\boldsymbol{n}} c_{\boldsymbol{n}'}^{\dagger} c_{\boldsymbol{n}'} \\
& + \omega_0 \sum_{n \neq m} g(\boldsymbol{m}-\boldsymbol{n})(\boldsymbol{e} \cdot \boldsymbol{e}_{\boldsymbol{m}-\boldsymbol{n}}) c_{\boldsymbol{n}}^{\dagger} c_{\boldsymbol{n}} (d_{\boldsymbol{m}}^{\dagger} + d_{\boldsymbol{m}}) \\
& + \omega_0 \sum_{m} \left(d_{\boldsymbol{m}}^{\dagger} d_{\boldsymbol{m}} + \frac{1}{2} \right) ,
\end{aligned} \tag{1}
$$

where $T(\boldsymbol{n})$ is the bare hopping integral in a rigid lattice. In general, this many-body model is of considerable complexity. However, if we are interested in the non or near adiabatic limit and the strong EPI, the kinetic energy is a perturbation. Then the model can be grossly simplified using the Lang–Firsov canonical transformation [37, 38] in the Wannier representation for electrons and phonons,

$$
S = \sum_{m \neq n} g(\boldsymbol{m}-\boldsymbol{n})(\boldsymbol{e} \cdot \boldsymbol{e}_{\boldsymbol{m}-\boldsymbol{n}}) c_{\boldsymbol{n}}^{\dagger} c_{\boldsymbol{n}} (d_{\boldsymbol{m}}^{\dagger} - d_{\boldsymbol{m}}) . \tag{2}
$$

Here we consider a particular lattice structure, where intersite lattice bipolarons tunnel already in the first order in $T(\boldsymbol{n})$. That allows us to average the transformed Hamiltonian, $\tilde{H} = \exp(S) H \exp(-S)$ over phonons to obtain

$$
\tilde{H} = H_0 + H_{\mathrm{pert}} , \tag{3}
$$

where

$$
H_0 = -E_{\mathrm{p}} \sum_{n} c_{\boldsymbol{n}}^{\dagger} c_{\boldsymbol{n}} + \frac{1}{2} \sum_{n \neq n'} v(\boldsymbol{n}-\boldsymbol{n}') c_{\boldsymbol{n}}^{\dagger} c_{\boldsymbol{n}} c_{\boldsymbol{n}'}^{\dagger} c_{\boldsymbol{n}'} + \omega_0 \sum_{m} \left(d_{\boldsymbol{m}}^{\dagger} d_{\boldsymbol{m}} + \frac{1}{2} \right) , \tag{4}
$$

and

$$
H_{\mathrm{pert}} = \sum_{n \neq n'} t(\boldsymbol{n}-\boldsymbol{n}') c_{\boldsymbol{n}}^{\dagger} c_{\boldsymbol{n}'} , \tag{5}
$$

is a perturbation. E_{p} is the familiar polaron level shift,

$$
E_{\mathrm{p}} = \omega_0 \sum_{m\nu} g^2(\boldsymbol{m}-\boldsymbol{n})(\boldsymbol{e} \cdot \boldsymbol{e}_{\boldsymbol{m}-\boldsymbol{n}})^2 , \tag{6}
$$

which is independent of $\boldsymbol{n}$. The polaron-polaron interaction is

$$
v(\boldsymbol{n}-\boldsymbol{n}') = V_{\mathrm{c}}(\boldsymbol{n}-\boldsymbol{n}') - V_{\mathrm{ph}}(\boldsymbol{n}-\boldsymbol{n}') , \tag{7}
$$

where

$$
V_{\mathrm{ph}}(\boldsymbol{n}-\boldsymbol{n}') = 2\omega_0 \sum_{m} g(\boldsymbol{m}-\boldsymbol{n}) g(\boldsymbol{m}-\boldsymbol{n}')(\boldsymbol{e} \cdot \boldsymbol{e}_{\boldsymbol{m}-\boldsymbol{n}})(\boldsymbol{e} \cdot \boldsymbol{e}_{\boldsymbol{m}-\boldsymbol{n}'}) . \tag{8}
$$

The transformed hopping integral is $t(\boldsymbol{n}-\boldsymbol{n}') = T(\boldsymbol{n}-\boldsymbol{n}')\exp[-g^2(\boldsymbol{n}-\boldsymbol{n}')]$ with

$$g^2(\boldsymbol{n}-\boldsymbol{n}') = \sum_{\boldsymbol{m},\nu} g(\boldsymbol{m}-\boldsymbol{n})(\boldsymbol{e}\cdot\boldsymbol{e}_{\boldsymbol{m}-\boldsymbol{n}}) \times \left[g(\boldsymbol{m}-\boldsymbol{n})(\boldsymbol{e}\cdot\boldsymbol{e}_{\boldsymbol{m}-\boldsymbol{n}}) - g(\boldsymbol{m}-\boldsymbol{n}')(\boldsymbol{e}\cdot\boldsymbol{e}_{\boldsymbol{m}-\boldsymbol{n}'})\right] \quad (9)$$

at low temperatures. The mass renormalization exponent can be expressed via E_{p} and V_{ph} as

$$g^2(\boldsymbol{n}-\boldsymbol{n}') = \frac{1}{\omega_0}\left[E_{\mathrm{p}} - \frac{1}{2}V_{\mathrm{ph}}(\boldsymbol{n}-\boldsymbol{n}')\right]. \quad (10)$$

The Hamiltonian $\tilde{H}$, (Eq. 3), in zero order with respect to the hopping describes localised polarons and independent phonons, which are vibrations of ions relative to new equilibrium positions depending on the polaron occupation numbers. Importantly the phonon frequencies remain unchanged in this limit at any polaron density, n. At finite λ and n there is a softening of phonons $\delta\omega_0$ of the order of $\omega_0 n/\lambda^2$ [39]. Interestingly the optical phonon can be mixed with a low-frequency polaronic plasmon forming a new excitation, "plasphon", which was proposed in [39, 40] as an explanation of the anomalous phonon mode splitting observed in cuprates [41]. The middle of the electron band is shifted down by the polaron level-shift E_{p} due to the potential well created by lattice deformation.

When V_{ph} exceeds V_{c} the full interaction becomes negative and polarons form pairs. The real space representation allows us to elaborate more physics behind the lattice sums in V_{ph} [34]. When a carrier (electron or hole) acts on an ion with a force $\boldsymbol{f}$, it displaces the ion by some vector $\boldsymbol{x} = \boldsymbol{f}/k$. Here k is the ion's force constant. The total energy of the carrier-ion pair is $-f^2/(2k)$. This is precisely the summand in Eq. 6 expressed via dimensionless coupling constants. Now consider two carriers interacting with the *same* ion. The ion displacement is $\boldsymbol{x} = (\boldsymbol{f}_1 + \boldsymbol{f}_2)/k$ and the energy is $-f_1^2/(2k) - f_2^2/(2k) - (\boldsymbol{f}_1 \cdot \boldsymbol{f}_2)/k$. Here the last term should be interpreted as an ion-mediated interaction between the two carriers. It depends on the scalar product of $\boldsymbol{f}_1$ and $\boldsymbol{f}_2$ and consequently on the relative positions of the carriers with respect to the ion. If the ion is an isotropic harmonic oscillator, then the following simple rule applies. If the angle ϕ between $\boldsymbol{f}_1$ and $\boldsymbol{f}_2$ is less than $\pi/2$ the polaron-polaron interaction will be attractive, if otherwise it will be repulsive. In general, some ions will generate attraction, and some repulsion between polarons.

The overall sign and magnitude of the interaction is given by the lattice sum in Eq. 8. One should note that according to Eq. 10 an attractive EPI reduces the polaron mass (and consequently the bipolaron mass), while repulsive EPI enhances the mass. Thus, the long-range EPI serves a double purpose. Firstly, it generates an additional inter-polaron attraction because the distant ions have small angle ϕ. This additional attraction helps to overcome the direct Coulomb repulsion between polarons. And secondly, the Fröhlich EPI makes lattice (bi)polarons lighter. Here, following [30, 34–36], we consider a few examples of intersite superlight bipolarons.

Apex Bipolarons

High-T_c oxides are doped charged-transfer ionic insulators with narrow electron bands. Therefore, the interaction between holes can be analyzed using computer simulation techniques based on a minimization of the ground state energy of an ionic insulator with two holes, the lattice deformations and the Coulomb repulsion fully taken into account, but neglecting the kinetic energy terms. Using these techniques net inter-site interactions of the in-plane oxygen hole with the *apex* hole, Fig. 1, and of two in-plane oxygen holes, Fig. 2, were found to be attractive in La_2CuO_4 [42] with the binding energies

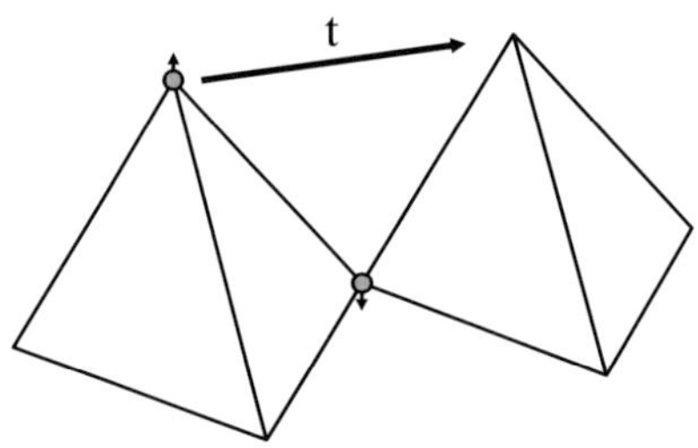

Fig. 1 Apex bipolaron tunnelling in perovskites (after [30])

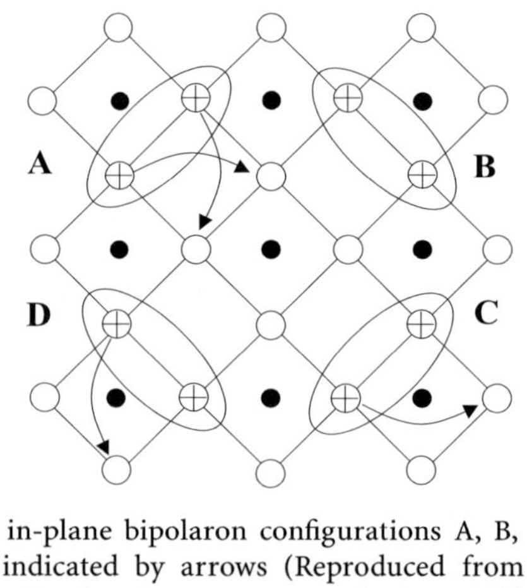

Fig. 2 Four degenerate in-plane bipolaron configurations A, B, C, and D. Some single-polaron hoppings are indicated by arrows (Reproduced from Alexandrov AS, Kornilovitch PE (2002) J Phys: Condens Matter 14:5337, © IOP Publishing Limited, 2007)

$\Delta = 119$ meV and $\Delta = 60$ meV, respectively. All other interactions were found to be repulsive.

Both apex and in-plane bipolarons can tunnel from one unit cell to another via the *single*-polaron tunnelling from one apex oxygen to its apex neighbor in case of the apex bipolaron [30], Fig. 1, or via the next-neighbor hopping in case of the in-plane bipolaron [34], Fig. 2.

The Bloch bands of these bipolarons are obtained using the canonical transformation, described above, projecting the transformed Hamiltonian, (Eq. 3), onto a reduced Hilbert space containing only empty or doubly occupied elementary cells [4]. The wave function of the apex bipolaron localized, say in cell $\boldsymbol{m}$ is written as

$$|\boldsymbol{m}\rangle = \sum_{i=1}^{4} A_i c_i^\dagger c_{\text{apex}}^\dagger |0\rangle \,, \tag{11}$$

where i denotes the $p_{x,y}$ orbitals and spins of the four plane oxygen ions in the cell, Fig. 1, and $c_{\text{apex}}^\dagger$ is the creation operator for the hole in one of the three apex oxygen orbitals with the spin, which is the same or opposite to the spin of the in-plane hole depending on the total spin of the bipolaron. The probability amplitudes A_i are normalized by the condition $|A_i| = 1/2$, if four plane orbitals p_{x1}, p_{y2}, p_{x3} and p_{y4} are involved, or by $|A_i| = 1/\sqrt{2}$ if only two of them are relevant. Then a matrix element of the Hamiltonian Eq. 3 describing the *bipolaron* tunnelling to the nearest neighbor cell $\boldsymbol{m} + \boldsymbol{a}$ is found as

$$t_{\text{b}} = \langle \boldsymbol{m}|\tilde{H}|\boldsymbol{m} + \boldsymbol{a}\rangle = |A_i|^2 T_{\text{pp}'}^{\text{apex}} e^{-g^2} \,, \tag{12}$$

where $T_{\text{pp}'}^{\text{apex}} e^{-g^2}$ is a *single polaron* hopping integral between two apex ions. The inter-site bipolaron tunnelling appears already in *the first order* with respect to the single-hole transfer $T_{\text{pp}'}^{\text{apex}}$, and the bipolaron energy spectrum consists of two subbands $E^{x,y}(\boldsymbol{K})$, formed by the overlap of p_x and p_y *apex* oxygen orbitals, respectively (here we take the lattice constant $a = 1$):

$$\begin{aligned} E^x(\boldsymbol{K}) &= t\cos(K_x) - t'\cos(K_y), \\ E^y(\boldsymbol{K}) &=- t'\cos(K_x) + t\cos(K_y) \,. \end{aligned} \tag{13}$$

They transform into one another under $\pi/2$ rotation. If $t, t' > 0$, "x" bipolaron band has its minima at $\boldsymbol{K} = (\pm\pi, 0)$ and y-band at $\boldsymbol{K} = (0, \pm\pi)$. In these equations t is the renormalized hopping integral between p orbitals of the same symmetry elongated in the direction of the hopping ($pp\sigma$) and t' is the renormalized hopping integral in the perpendicular direction ($pp\pi$). Their ratio $t/t' = T_{\text{pp}'}^{\text{apex}}/T_{pp'}^{\prime\text{apex}} = 4$ as follows from the tables of hopping integrals in solids. Two different bands are not mixed because $T_{p_x,p_y'}^{\text{apex}} = 0$ for the nearest neighbors. A random potential does not mix them either, if it varies smoothly on the lattice scale. Hence, we can distinguish "x" and "y" bipolarons with a lighter effective mass in x or y direction, respectively. The apex z bipolaron, if formed, is ca. four times less mobile than x and y bipolarons. The bipolaron bandwidth is *of the same order* as the polaron one, which is a specific feature of the inter-site bipolaron. For a large part of the Brillouin zone near $(0, \pi)$ for "x" and $(\pi, 0)$ for "y" bipolarons, one can adopt the effective mass approximation

$$E^{x,y}(\boldsymbol{K}) = \frac{K_x^2}{2m_{x,y}^{**}} + \frac{K_y^2}{2m_{y,x}^{**}} \tag{14}$$

with $K_{x,y}$ taken relative to the band bottom positions and $m_x^{**} = 1/t$, $m_y^{**} = 4m_x^{**}$.

X and y bipolarons bose-condense at the boundaries of the center-of-mass Brillouin zone with $K = (\pm\pi, 0)$ and $K = (0, \pm\pi)$, respectively, which explains the d-wave symmetry and the checkerboard modulations of the order parameter in cuprates [43].

In-Plane Bipolarons

Now let us consider in-plane bipolarons in a two-dimensional lattice of ideal octahedra that can be regarded as a simplified model of the copper-oxygen perovskite layer, Fig. 3 [34]. The lattice period is $a = 1$ and the distance between the apical sites and the central plane is $h = a/2 = 0.5$. For mathematical transparency we assume that all in-plane atoms, both copper and oxygen, are static but apex oxygens are independent three-dimensional isotropic harmonic oscillators.

Due to poor screening, the hole-apex interaction is purely coulombic,

$$g_\alpha(\boldsymbol{m} - \boldsymbol{n}) = \frac{\kappa_\alpha}{|\boldsymbol{m} - \boldsymbol{n}|^2},$$

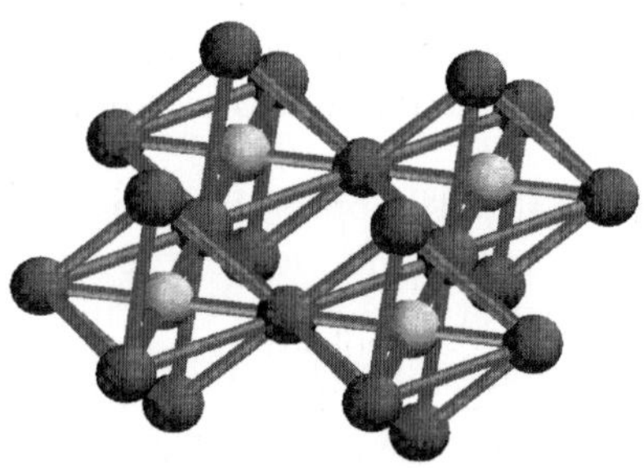

Fig. 3 Simplified model of the copper-oxygen perovskite layer. (Reproduced from Alexandrov AS, Kornilovitch PE (2002) J Phys: Condens Matter 14:5337, © IOP Publishing Limited, 2007)

where $\alpha = x, y, z$. To account for the fact that c axis-polarized phonons couple to the holes stronger than others due to a poor screening [30, 31, 34], we choose $\kappa_x = \kappa_y = \kappa_z/\sqrt{2}$. The direct hole-hole repulsion is

$$V_c(\boldsymbol{n} - \boldsymbol{n}') = \frac{V_c}{\sqrt{2}|\boldsymbol{n} - \boldsymbol{n}'|}$$

so that the repulsion between two holes in the nearest neighbor (NN) configuration is V_c. We also include the bare NN hopping T_{NN}, the next nearest neighbor (NNN) hopping across copper T_{NNN} and the NNN hopping between the pyramids T'_{NNN}.

The polaron shift is given by the lattice sum Eq. 6, which after summation over polarizations yields

$$E_p = 2\kappa_x^2 \omega_0 \sum_m \left(\frac{1}{|\boldsymbol{m} - \boldsymbol{n}|^4} + \frac{h^2}{|\boldsymbol{m} - \boldsymbol{n}|^6} \right) = 31.15\kappa_x^2 \omega_0 \,, \tag{15}$$

where the factor 2 accounts for two layers of apical sites. For reference, the Cartesian coordinates are $\boldsymbol{n} = (n_x + 1/2, n_y + 1/2, 0)$, $\boldsymbol{m} = (m_x, m_y, h)$, and n_x, n_y, m_x, m_y are integers. The polaron-polaron attraction is

$$V_{ph}(\boldsymbol{n} - \boldsymbol{n}') = 4\omega\kappa_x^2 \sum_m \frac{h^2 + (\boldsymbol{m} - \boldsymbol{n}') \cdot (\boldsymbol{m} - \boldsymbol{n})}{|\boldsymbol{m} - \boldsymbol{n}'|^3 |\boldsymbol{m} - \boldsymbol{n}|^3} \,. \tag{16}$$

Performing the lattice summations for the NN, NNN, and NNN' configurations one finds $V_{ph} = 1.23E_p$, $0.80E_p$, and $0.82E_p$, respectively. As a result, we obtain a net inter-polaron interaction as $v_{NN} = V_c - 1.23E_p$, $v_{NNN} = \frac{V_c}{\sqrt{2}} - 0.80E_p$, $v'_{NNN} = \frac{V_c}{\sqrt{2}} - 0.82E_p$, and the mass renormalization exponents as $g^2_{NN} = 0.38(E_p/\omega)$, $g^2_{NNN} = 0.60(E_p/\omega)$ and $(g'_{NNN})^2 = 0.59(E_p/\omega)$.

Let us now discuss different regimes of the model. At $V_c > 1.23E_p$, no bipolarons are formed and the system is a polaronic Fermi liquid. Polarons tunnel in the *square* lattice with $t = T_{NN} \exp(-0.38E_p/\omega)$ and $t' = T_{NNN} \exp(-0.60E_p/\omega)$ for NN and NNN hoppings, respectively. Since $g^2_{NNN} \approx (g'_{NNN})^2$ one can neglect the difference between NNN hoppings within and between the octahedra. A single polaron spectrum is therefore

$$E_1(\boldsymbol{k}) = -E_p - 2t'[\cos k_x + \cos k_y] - 4t\cos(k_x/2)\cos(k_y/2) \,. \tag{17}$$

The polaron mass is $m^* = 1/(t + 2t')$. Since in general $t > t'$, the mass is mostly determined by the NN hopping amplitude t.

If $V_c < 1.23E_p$ then intersite NN bipolarons form. The bipolarons tunnel in the plane via four resonating (degenerate) configurations A, B, C, and D, as shown in Fig. 2. In the first order of the renormalized hopping integral, one should retain only these lowest energy configurations and discard all the processes that involve configurations with higher energies. The result of such a projection is the bipolaron Hamiltonian,

$$H_b = (V_c - 3.23E_p)\sum_l \left[A_l^\dagger A_l + B_l^\dagger B_l + C_l^\dagger C_l + D_l^\dagger D_l\right] \\ - t'\sum_l \left[A_l^\dagger B_l + B_l^\dagger C_l + C_l^\dagger D_l + D_l^\dagger A_l + H.c.\right] \\ - t'\sum_n \left[A_{l-x}^\dagger B_l + B_{l+y}^\dagger C_l + C_{l+x}^\dagger D_l + D_{l-y}^\dagger A_l + H.c.\right] , \tag{18}$$

where $\boldsymbol{l}$ numbers octahedra rather than individual sites, $\boldsymbol{x} = (1,0)$, and $\boldsymbol{y} = (0,1)$. A Fourier transformation and diagonalization of a 4×4 matrix yields the bipolaron spectrum:

$$E_2(\boldsymbol{K}) = V_c - 3.23E_p \pm 2t'\left[\cos(K_x/2) \pm \cos(K_y/2)\right] . \tag{19}$$

There are four bipolaronic subbands combined in the band of the width $8t'$. The effective mass of the lowest band is $m^{**} = 2/t'$. The bipolaron binding energy is $\Delta \approx 1.23E_p - V_c$. Inter-site bipolarons already move in the *first* order of the single polaron hopping. This remarkable property is entirely due to the strong on-site repulsion and long-range electron-phonon interactions that leads to a non-trivial connectivity of the lattice. This fact combines with a weak renormalization of t' yielding a *superlight* bipolaron with the mass $m^{**} \propto \exp(0.60E_p/\omega)$. We recall that in the Holstein model $m^{**} \propto \exp(2E_p/\omega)$ [18, 19]. Thus the mass of the Fröhlich bipolaron in the perovskite layer scales approximately as a *cubic root* of that of the Holstein bipolaron.

At even stronger EPI, $V_c < 1.16E_p$, *NNN* bipolarons become stable. More importantly, holes can now form 3- and 4-particle clusters. The dominance of the potential energy over kinetic in the transformed Hamiltonian enables us to readily investigate these many-polaron cases. Three holes placed within one oxygen square have four degenerate states with the energy $2(V_c - 1.23E_p) + V_c/\sqrt{2} - 0.80E_p$. The first-order polaron hopping processes mix the states resulting in a ground state linear combination with the energy $E_3 = 2.71V_c - 3.26E_p - \sqrt{4t^2 + t'^2}$. It is essential that between the squares such triads could move only in higher orders of polaron hopping. In the first order, they are immobile. A cluster of four holes has only one state within a square of oxygen atoms. Its energy is $E_4 = 4(V_c - 1.23E_p) + 2(V_c/\sqrt{2} - 0.80E_p) = 5.41V_c - 6.52E_p$. This cluster, as well as all bigger ones, are also immobile in the first order of polaron hopping. We would like to stress that at distances much larger than the lattice constant the polaron-polaron interaction is always repulsive, and the formation of infinite clusters, stripes or strings is prohibited. We conclude that at $V_c < 1.16E_p$ the system quickly becomes a charge segregated insulator.

The fact that within the window, $1.16E_p < V_c < 1.23E_p$, there are no three or more polaron bound states, indicates that bipolarons repel each other. The system is effectively a charged Bose-gas, which is a superconductor [6, 7]. This superconducting state requires a rather fine balance between electronic and ionic interactions in cuprates.

All-Coupling Lattice Bipolarons

The multi-polaron CFM model discussed above is analytically solvable in the strong-coupling nonadibatic ($\omega_0 \gtrsim T(a)$) limit using the Lang-Firsov transformation of the Hamiltonian, (Eq. 1), and projecting it on the inter-site pair Hilbert space [30, 34]. To extend the theory for the whole parameter space an advanced continuous time QMC technique (CTQMC) has been recently developed for bipolarons [35, 36]. Using CTQMC [35, 36] simulated the CFM Hamiltonian on a staggered triangular ladder (1D), triangular (2D) and strongly anisotropic hexagonal (3D) lattices including triplet pairing [36]. On such lattices, bipolarons are found to move with a crab like motion (Fig. 1), which is distinct from the crawler motion found on cubic lattices [18, 19]. Such bipolarons are small but very light for a wide range of electron-phonon couplings and phonon frequencies. EPI has been modeled using the force function in the site-representation as

$$H_{\text{e-ph}} = -\sum_{nm\sigma} f_m(\boldsymbol{n}) c^\dagger_{\boldsymbol{n}\sigma} c_{\boldsymbol{n}\sigma} \xi_{\boldsymbol{m}} \,. \tag{20}$$

Each vibrating ion has one phonon degree of freedom $\xi_{\boldsymbol{m}}$ associated with a single atom. The sites are numbered by the indices $\boldsymbol{n}$ or $\boldsymbol{m}$ for electrons and ions respectively. Operators $c^\dagger_{\boldsymbol{n}\sigma}$ creates an electron on site $\boldsymbol{n}$ with spin σ. Coulomb repulsion $V(\boldsymbol{n} - \boldsymbol{n}')$ has been screened up to the first nearest neighbors, with on site repulsion U and nearest-neighbor repulsion V_c. In contrast, the Fröhlich interaction is assumed to be long-range, due to unscreened interaction with c-axis high-frequency phonons [30]. The form of the interaction with c-axis polarized phonons has been specified via the force function [31], $f_{\boldsymbol{m}}(\boldsymbol{n}) = \kappa\left[(\boldsymbol{m} - \boldsymbol{n})^2 + 1\right]^{-3/2}$, where κ is a constant. The dimensionless electron-phonon coupling constant λ is defined as $\lambda = \sum_{\boldsymbol{m}} f^2_{\boldsymbol{m}}(0)/2M\omega^2 zT(a)$ which is the ratio of the polaron binding energy to the kinetic energy of the free electron $zT(a)$, and the lattice constant is taken as $a = 1$.

In the limit of high phonon frequency $\omega \gg T(a)$ and large on-site Coulomb repulsion (Hubbard U), the model is reduced to an extended Hubbard model with intersite attraction and suppressed double-occupancy [34] by applying the Lang-Firsov canonical transformation (Section 2). Then the Hamiltonian can be projected onto the subspace of nearest neighbor intersite *crab* bipolarons (Section 4). In contrast with the crawler bipolaron, the crab bipolaron's mass scales linearly with the polaron mass ($m^{**} = 4m^*$ on the staggered chain and $m^{**} = 6m^*$ on the triangular lattice).

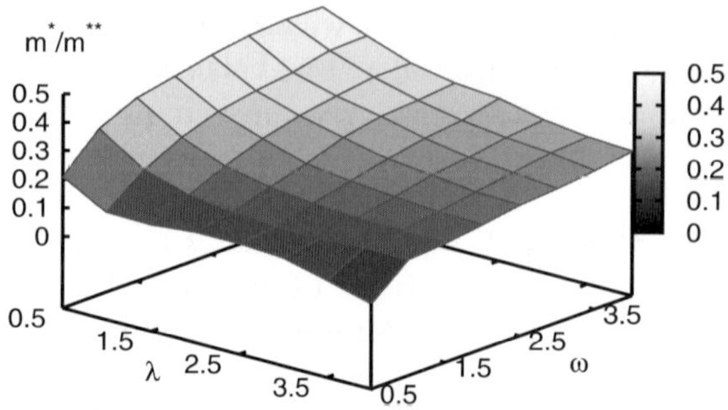

Fig. 4 Polaron to bipolaron mass ratio for a range of $\bar{\omega} = \omega_0/T(a)$ and λ on the staggered ladder. Mobile small bipolarons are seen even in the adiabatic regime $\bar{\omega} = 0.5$ for couplings λ up to 2.5 (Reproduced from Hague JP et al. (2007) Phys Rev Lett 98:037002, © American Physical Society, 2007)

Extending the CTQMC algorithm to systems of two particles with strong EPI and Coulomb repulsion solved the bipolaron problem on a staggered ladder, triangular and anisotropic hexagonal lattices from weak to strong coupling in a realistic parameter range where usual strong and weak-coupling limiting approximations fail. Importantly small but light bipolarons have been found for more realistic intermediate values of EPI, $\lambda \lesssim 1$ and phonon frequency, $\omega \lesssim T(a)$ [35, 36].

Figure 4 shows the ratio of the polaron to bipolaron masses on the staggered ladder as a function of effective coupling and phonon frequency for $V_c = 0$. The bipolaron to polaron mass ratio is about 2 in the weak coupling regime ($\lambda \ll 1$) as it should be for a large bipolaron [44, 45]. In the strong-coupling, large phonon frequency limit the mass ratio approaches 4, in agreement with strong-coupling arguments given above. In a wide region of parameter space, we find a bipolaron/polaron mass ratio of between 2 and 4

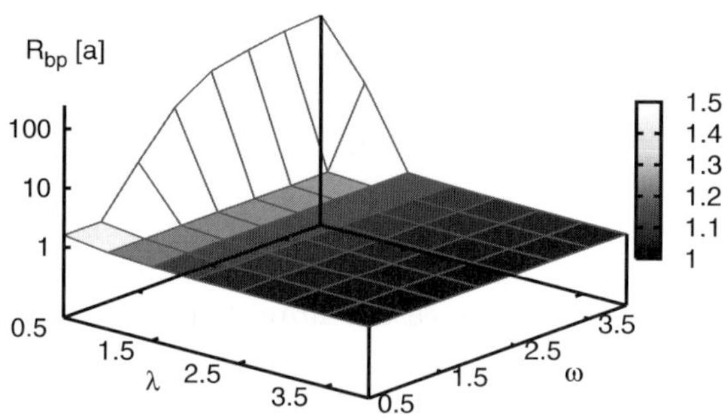

Fig. 5 Bipolaron radius (in units of a) for a range of $\bar{\omega}$ and λ on the staggered ladder (Reproduced from Hague JP et al. (2007) Phys Rev Lett 98:037002, © American Physical Society, 2007)

and a bipolaron radius similar to the lattice spacing, see Figs. 4 and 5. Thus the bipolaron is small and light at the same time. Taking into account additional intersite Coulomb repulsion V_c does not change this conclusion. The bipolaron is stable for $V_c < 4T(a)$. As V_c increases the bipolaron mass decreases but the radius remains small, at about 2 lattice spacings. Importantly, the absolute value of the small bipolaron mass is only about 4 times of the bare electron mass m_0, for $\lambda = \omega/T(a) = 1$ (Fig. 4).

Simulations of the bipolaron on an infinite triangular lattice including exchanges and large on-site Hubbard repulsion $U = 20T(a)$ also lead to the bipolaron mass of about $6m_{0xy}$ and the bipolaron radius $R_{bp} \approx 2a$ for a moderate coupling $\lambda = 0.5$ and a large phonon frequency $\omega = T(a)$ (for the triangular lattice, $m_{0xy} = 1/3a^2T(a)$). Finally, the bipolaron in a hexagonal lattice with out-of-plane hopping $T' = T(a)/3$ has also a light in-plane inverse mass, $m_{xy}^{**} \approx 4.5m_{0xy}$ but a small size, $R_{bp} \approx 2.6a$ for experimentally achievable values of the phonon frequency $\omega = T(a) = 200$ meV and EPI, $\lambda = 0.36$. Out-of-plane $m_z^{**} \approx 70m_{0z}$ is Holstein like, where $m_{0z} = 1/2d^2T'$, (d is the inter-plane spacing). When bipolarons are small and pairs do not overlap, the pairs can form a BEC at $T_{BEC} = 3.31(2n_B/a^2\sqrt{3}d)^{2/3}/(m_{xy}^{2/3}m_z^{1/3})$. If we choose realistic values for the lattice constants of 0.4 nm in the plane and 0.8 nm out of the plane, and allow the density of bosons to be $n_B = 0.12$ per lattice site, which easily avoids overlap of pairs, then $T_{BEC} \approx 300$ K.

Summary

For a very strong electron-phonon coupling in the Holstein model with the zero-range EPI, polarons become self-trapped on a single lattice site and bipolarons are on-site singlets. The on-site bipolaron mass appears only in the second order of polaron hopping [18, 19], so that on-site bipolarons are very heavy. This estimate led some authors to the conclusion that *high-temperature* bipolaronic superconductivity is impossible.

However we have found that small but relatively light bipolarons could exist within the realistic range of the *finite-range* EPI with high-frequency optical phonons. The effect appears since the finite-range Fröhlich interaction combined with the long-range Coulomb repulsion provides an effective interaction with a deep attraction minimum for two holes on the neighbouring sites, and repulsive for other hole configurations. Bipolarons which are both light and small give rise to Ogg–Schafroth's bose-condensed state of charged bosons at high-temperatures, since the Bose–Einstein condensate has transition temperature that is inversely proportional to mass. Our conclusion is backed up by analytical [30, 34] and CTQMC studies [35, 36]. These studies let us believe that the following recipes is worth investigating to look for room-temperature superconductivity [30, 31, 35]: (a) The parent compound should be an ionic insulator with light ions to form high-frequency optical phonons;

(b) The structure should be quasi two-dimensional to ensure poor screening of high-frequency phonons polarized perpendicular to the conducting planes; (c) A triangular lattice is required in combination with strong, on-site Coulomb repulsion to form the small superlight crab bipolaron; (d) Moderate carrier densities are required to keep the system of small bipolarons close to the dilute regime. Clearly most of these conditions are already met in the cuprate superconductors.

Acknowledgements I would like to thank Jim Hague, Pavel Kornilovitch, and John Samson for collaboration and helpful discussions, and to acknowledge support of EPSRC (UK) (grant numbers EP/C518365/1 and EP/D07777X/1).

References

1. Bednorz JG, Müller KA (1986) Z Phys B 64:189
2. Bardeen J, Cooper LN, Schrieffer JR (1957) Phys Rev 108:1175
3. Alexandrov AS, Mott NF (1994) Rep Prog Phys 57:1197
4. Alexandrov AS (2003) Theory of Superconductivity: From Weak to Strong Coupling. IoP Publishing, Bristol
5. Edwards PP, Rao CNR, Kumar N, Alexandrov AS (2006) Chem Phys Chem 7:2015
6. Ogg RA Jr (1946) Phys Rev 69:243
7. Schafroth MR (1955) Phys Rev 100:463
8. Blatt JM, Butler ST (1955) Phys Rev 100:476
9. Eliashberg GM (1960) Zh Eksp Teor Fiz 38:966
10. Eliashberg GM (1960) Zh Eksp Teor Fiz 39:1437
11. Eliashberg GM (1960) Sov Phys JETP 11:696
12. Eliashberg GM (1960) Sov Phys JETP 12:1000
13. Alexandrov AS (1983) Zh Fiz Khim 57:273
14. Alexandrov AS (1983) Russ J Phys Chem 57:167
15. Migdal AB (1958) Zh Eksp Teor Fiz 34:1438
16. Migdal AB (1958) Sov Phys JETP 7:996
17. Alexandrov AS (2001) Europhys Lett 56:92
18. Alexandrov AS, Ranninger J (1981) Phys Rev B 23:1796
19. Alexandrov AS, Ranninger J (1981) Phys Rev B 24:1164
20. de Mello EVL, Ranninger J (1998) Phys Rev B 58:9098
21. Anderson PW (1997) The Theory of Superconductivity in the Cuprates. Princeton Univ. Press, Princeton NY
22. Aubry S (1995) In: Salje EKH, Alexandrov AS, Liang WY (eds) Polarons and Bipolarons in High Tc Superconductors and related materials. Cambridge University Press, Cambridge, p 271
23. Alexandrov AS (1991) Physica C (Amsterdam) 182:327
24. Alexandrov AS (1992) J Low Temp Phys 87:721
25. Alexandrov AS, Mott NF (1994) J Superconductivity: Incorporating Novel Magnetism 7:599
26. Bonča J, Katrasnic T, Trugman SA (2000) Phys Rev Lett 84:3153
27. La Magna A, Pucci R (1997) Phys Rev B 55:14886
28. Macridin A, Sawatzky GA, Jarrell M (2004) Phys Rev B 69:245111
29. Robaszkiewicz S, Micnas R, Chao KA (1981) Phys Rev B 23:1447

30. Alexandrov AS (1996) Phys Rev B 53:2863
31. Alexandrov AS, Kornilovitch PE (1999) Phys Rev Lett 82:807
32. Spencer PE, Samson JH, Kornilovitch PE, Alexandrov AS (2005) Phys Rev B 71:184319
33. Hague JP, Kornilovitch PE, Alexandrov AS, Samson JH (2006) Phys Rev B 73:054303
34. Alexandrov AS, Kornilovitch PE (2002) J Phys Condens Matter 14:5337
35. Hague JP, Kornilovitch PE, Samson JH, Alexandrov AS (2007) Phys Rev Lett 98:037002
36. Hague JP, Kornilovitch PE, Samson JH, Alexandrov AS (2007) submitted to the special Mott's issue of J Phys Condens Matter
37. Lang IG, Firsov YA (1962) Zh Eksp Teor Fiz 43:1843
38. Lang IG, Firsov YA (1962) Sov Phys JETP 16:1301
39. Alexandrov AS (1992) Phys Rev B 46:2838
40. Alexandrov AS (1992) Sol St Commun 81:965
41. Rietschel H, Pintschovius L, Reichardt W (1989) Physica C (Amsterdam) 162:1705
42. Catlow CRA, Islam MS, Zhang X (1998) J Phys Condens Matter 10:L49
43. Alexandrov AS (2004) J Superconductivity: Incorporating Novel Magnetism 17:53
44. Verbist G, Peeters FM, Devreese JT (1991) Phys Rev B 43:2712
45. Verbist G, Peeters FM, Devreese JT (1990) Solid State Commun 76:1005

Ando Y (2007) In: Bussmann-Holder A, Keller H (eds)
High T_c Superconductors and Related Transition Metal Oxides.
Springer-Verlag Berlin Heidelberg, 17–28
DOI 10.1007/978-3-540-71023-3

Mott Limit in High-T_c Cuprates

Y. Ando

Institute of Scientific and Industrial Research, Osaka University, 567-0047 Osaka, Japan

Abstract In high-T_c cuprates, the conventional criteria for the metallic transport, the Mott limit, is apparently violated. In this article, the basic idea of the Mott limit is reviewed and its applicability to the cuprates is carefully examined in the light of the recent knowledge about the electronic structure that involves the unusual "Fermi arcs". Also, a possible way to apply the Boltzmann transport theory to the Fermi arcs is proposed, and it is discussed that the apparent Mott-limit violation can in principle be understood in this conventional framework. Lastly, it is shown that the peculiarities of the charge transport in cuprates is strikingly manifested in the phase diagram that features an unusual insulating phase at low temperature throughout the doping range where the Fermi arcs govern the low-energy physics.

Introduction

Intriguingly, the applicability of such simple notions as metal and insulator is rather ambiguous in high-T_c cuprates: On one hand, cuprates show "metallic" behavior with unusually large resistivity at high temperatures (so-called bad metal behavior) [1], while an "insulating" behavior has been found at low temperature in samples with low resistivity that would normally correspond to a metal, when the superconductivity is suppressed by high enough magnetic fields [2, 3]. In this context, an important principle for discussing whether a system should be a metal or an insulator is the *Mott limit* (or Mott–Ioffe–Regel limit) [4], which is based on very simple physics. To understand the rather confusing transport phenomena in cuprates regarding the metallic/insulating behavior, it is prudent to carefully examine the applicability of the Mott limit to cuprates and see whether/how this basic criteria is violated. Because the understanding of the emergence of metallic transport in cuprates upon doping charge carriers to the parent Mott insulator obviously holds a key to elucidating the superconductivity mechanism, in this article dedicated to Prof. K. Alex Müller to celebrate his 80th birthday, I present the forefront of our understanding of the very basics of the high-T_c cuprates, that is, their metallic transport.

What is Mott Limit?

An essential characteristics of a metallic state in crystalline systems is that electrons are expressed in terms of extended Bloch waves and, hence, the wave vector $\boldsymbol{k}$ (rather than the position $\boldsymbol{r}$) is a good quantum number. In real materials, electrons are always scattered by disorder or some quasiparticles, which sets a finite mean free path ℓ for the electrons and restricts the coherent extent of their Bloch waves. Since the electrons that are responsible for the charge transport are those with energies close to the Fermi energy ε_F, the Fermi wave vector k_F gives the relevant wave number for the charge transport by the Bloch waves.

Naturally, for the metallic transport to be realized in a material, the Bloch waves of the electrons near ε_F must be well-defined, which dictates that their mean free path ℓ must be longer than the wave length λ_F ($= 2\pi/k_F$). This condition provides the well-known Mott limit that is expressed as $\ell \gtrsim \lambda_F$, which is equivalent to

$$k_F \ell \gtrsim 2\pi \,. \tag{1}$$

Since this argument is quite crude, the numerical factor 2π is usually not taken so seriously, and it is customary to relax the condition a little and consider the Mott limit to be given by

$$k_F \ell \gtrsim 1 \,. \tag{2}$$

Hence, it should be kept in mind that the discussions involving the Mott limit are always ambiguous up to the factor of 2π.

Mott Limit in 2D Metals

To relate the resistivity value to the Mott limit, let us first consider a simple example, the two-dimensional (2D) free electron system. In this case, k_F is simply determined by the 2D density of electrons n_{2D} as

$$k_F^2 = 2\pi n_{2D} \,. \tag{3}$$

We further employ the Drude model

$$\rho = \frac{c_0 m^*}{e^2 n_{2D} \tau} \,, \tag{4}$$

where m^* is the electron effective mass and τ is the relaxation time. Note that we here consider a layered quasi-2D system with the layer distance c_0, and the resistivity ρ is measured in the three-dimensional (3D) unit Ω cm. Since ℓ is expressed as $v_F\tau$ by using the Fermi velocity v_F, one may write

$$k_F\ell = k_F(v_F\tau) = \frac{\hbar k_F^2 \tau}{m^*} \,. \tag{5}$$

By combining Eqs. 3–5, one obtains

$$k_F\ell = \frac{2\pi\hbar n_{2D}\tau}{m^*} = \frac{hc_0}{e^2\rho} \,. \tag{6}$$

This gives a simple means [2] to calculate $k_F\ell$ from ρ.

Although Eq. 6 is obtained for the free electron system, the Luttinger theorem [5] assures that the volume of the occupied states in the Brillouin zone is unchanged in the presence of interactions, and therefore one may extend the above result to any 2D Fermi liquid system which has a single band centered around the Γ point (center of the Brillouin zone) by considering k_F to be an average over the Fermi surface. Hence, rather generally, the $k_F\ell$ value of a 2D system can be estimated simply from the resistivity.

If Eq. 6 is used for the examination of the Mott limit Eq. 2, one may conclude that for a quasi-2D system to be metallic, the resistivity should satisfy the condition

$$\rho \lesssim \frac{hc_0}{e^2} \,. \tag{7}$$

It is useful to note that Eq. 7 would suggest a condition $\rho \lesssim 1.7\,\text{m}\Omega\,\text{cm}$ for metallic transport in the prototypical cuprate $La_{2-x}Sr_xCuO_4$ (LSCO), where the distance between the CuO_2 planes (c_0) is 0.66 nm [6].

Metallic Transport in LSCO

Figure 1 shows the temperature dependences of the in-plane resistivity ρ_{ab} of LSCO for some representative doping levels. One may immediately notice that, while at $x = 0$ the system is indeed an insulator [7], a metallic behavior ($d\rho_{ab}/dT > 0$, i.e., the resistivity decreases with decreasing temperature as the electrons become less scattered) is already observed for $x = 0.01$ at moderate temperature [8]. However, the value of ρ_{ab} at 400 K is as large as 22 mΩ cm for $x = 0.01$, which means that this metallic transport is apparently violating the Mott limit discussed above. Furthermore, since the Néel temperature T_N at $x = 0.01$ is 240 K [8] and there is no anomaly in $\rho_{ab}(T)$ at T_N, this metallic transport is apparently insensitive to the establishment of the long-range antiferromagnetic (AF) order, which is unusual because in LSCO the same Cu $3d^9$ state is responsible for both the magnetic order and the charge transport [9]. Therefore, the metallic transport observed in LSCO upon doping only 1% of holes is quite unusual in that it appears to violate the Mott limit and that it coexist with the long-range AF order.

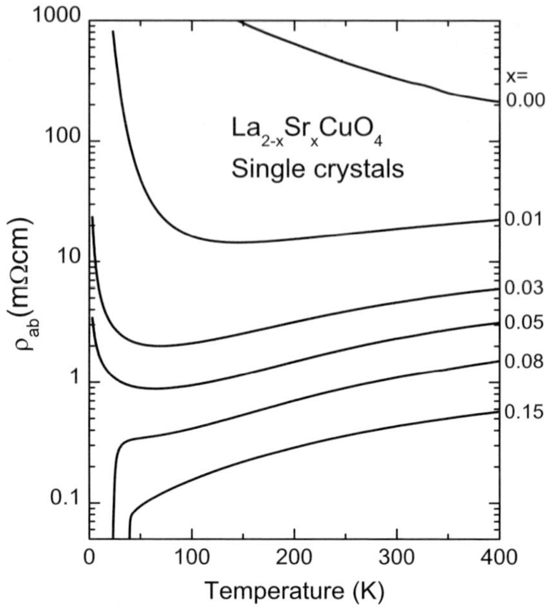

Fig. 1 Temperature dependences of ρ_{ab} of a series of high-quality LSCO single crystals measured up to 400 K. The small wiggle at 325 K in the $x = 0.00$ data is due to the Néel transition in the parent system; intriguingly, such an anomaly is absent for $x = 0.01$, where the Néel transition still exists at 240 K [8]

k_F in Low-Carrier Systems

In order to understand the apparent Mott limit violation in LSCO, one should remember that Eq. 6 is valid only when the relevant Fermi surface (FS) is centered around the Γ point, in which case k_F is straightforwardly dictated by the carrier density. However, if the FS is centered elsewhere, the relation $k_F^2 = 2\pi n_{2D}$ no longer holds, and k_F can take a relatively large value even for low-carrier systems. In Fig. 2, the difference between the free electron system and a low-carrier system with off-centered FS is depicted. Note that in

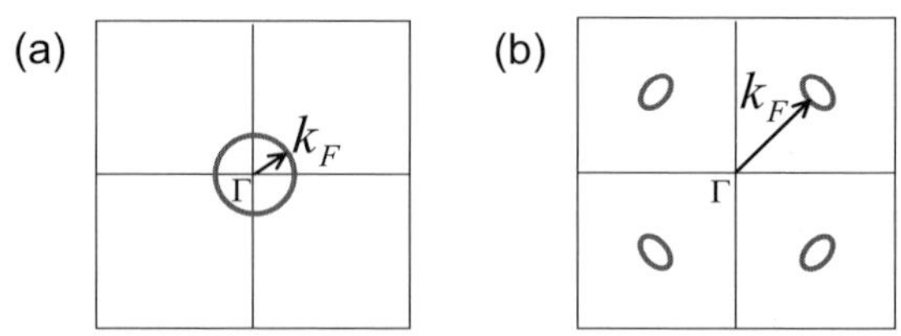

Fig. 2 Schematic pictures to show k_F for the free electron system (**a**) and a low-carrier system with off-centered Fermi surfaces (**b**)

the discussion of the Mott limit, the important physical quantity is the Bloch-wave length, and therefore the actual wave number that is measured from the center of the Brillouin zone is to be considered. In the literature, depending on the purposes of the discussions, sometimes k_F is measured from the bottom of the band, but one should not be confused with such convention in the present context.

Fermi Arc

In lightly-doped LSCO, the situation in the Brillouin zone has turned out to be somewhat similar to the case shown in Fig. 2b; namely, as depicted in Fig. 3, near ε_F some electronic states with well-defined $\boldsymbol{k}$ are observed in the middle of the Brillouin zone diagonals, forming four "Fermi arcs" at ε_F [10, 11]. Intriguingly, those arcs are portions of the non-interacting FS (i.e., the large FS that would be expected in the absence of the correlation gap) and do not appear to encircle small hole pockets [12]. Obviously, the conventional wisdom for the Fermi liquid, such as the Luttinger theorem, is not applicable to the Fermi-arc system and in this sense the lightly-doped LSCO is a non-Fermi-liquid system. Although the actual situation depicted in Fig. 3 is not exactly the same as that shown in Fig. 2b, the location of the Fermi arcs in the Brillouin zone defines the wave vector of the electronic states responsible for the charge transport and, hence, k_F of the Fermi arcs is relatively large even at low doping. Recent angle-resolved photoemission spectroscopy (ARPES) experiments [10] found that in lightly-doped LSCO the Fermi arcs have $k_F \approx 4\,\mathrm{nm}^{-1}$.

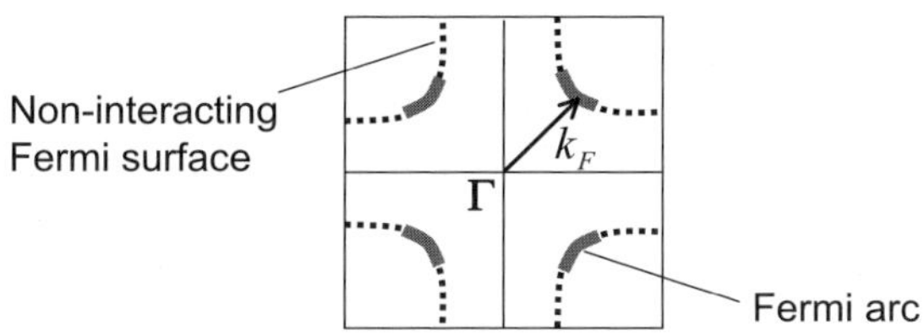

Fig. 3 Schematic picture of the Fermi arcs observed in lightly-doped LSCO

Examination of the Mott Limit in LSCO

With the knowledge about the electronic structure obtained from ARPES on lightly-doped LSCO, we can now rigorously examine the Mott limit in this system. To do that, it is better to know the exact number of carriers that are participating in the charge transport. In this regard, the Hall coefficient R_H at low doping is quite convenient, because it behaves like a *Hall*

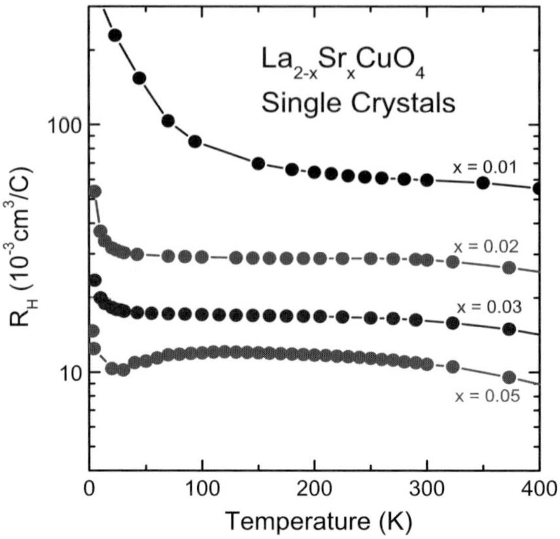

Fig. 4 Temperature dependences of the Hall coefficient R_H of high-quality LSCO single crystals in the lightly-doped regime ($x = 0.01 - 0.05$). Note that R_H is essentially T-independent in the temperature region where the resistivity shows the metallic behavior, and the Hall number n_{Hall} estimated from the nearly-T-independent R_H values via $n_{Hall} = 1/eR_H$ is exactly equal to $x/(a^2c_0)$ in this lightly-doped regime ([13])

constant of a conventional metal; namely, as is shown in Fig. 4, R_H is essentially T-independent in the temperature region where the resistivity shows the metallic behavior, and the value of the nearly-T-independent R_H indicates that the carrier number is almost exactly equal to x [13]. Therefore, by using the Drude formula Eq. 4 with $n_{2D} = x/a^2$ ($a = 0.38$ nm is the in-plane lattice constant) and $m^* = 3m_e$ (m_e is the free electron mass and this m^* has been obtained from optical conductivity studies [14]), one obtains $\tau \approx 5$ fs for $x = 0.01$ at 400 K where $\rho_{ab} = 22$ mΩ cm. Since the ARPES experiments found $v_F \approx 0.15$ eV nm ($= 2.3 \times 10^5$ m/s) for the Fermi arcs [10], ℓ can be estimated to be about 1 nm, which leads to $k_F\ell \approx 4$ and hence the Mott limit Eq. 2 is *not* violated.

Physically, in the lightly-doped LSCO the peculiar electronic structure dictates that electrons with a large wave number and a large velocity become responsible for the charge transport even when the number of mobile carriers are small, and this causes the mean free path of the carriers to be reasonably long ($\ell \approx 1$ nm $\approx 3a$) for a rather short relaxation time of about 5 fs.

In the real space picture, the fact that the Fermi arcs are responsible for the charge transport means that only those Bloch waves that travel with the wave number $\sim 4\ \mathrm{nm}^{-1}$ along the four directions diagonal to the Cu – O – Cu bond directions can carry current. Interestingly, this situation offers an intuitive understanding about why the metallic transport can coexist with the

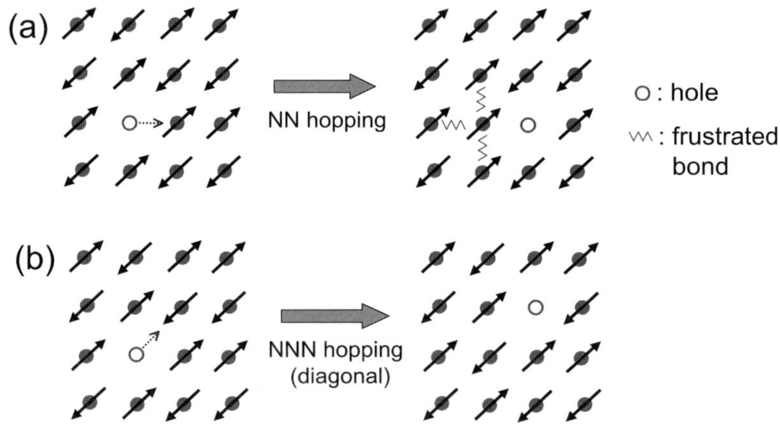

Fig. 5 Schematic picture to show why the nearest-neighbour (NN) hopping frustrates the AF order (**a**), while the next-nearest-neighbour (NNN) hopping does not (**b**)

long-range AF order at $x = 0.01$: When there is a hole in the antiferromagnetically ordered state, moving the hole into a nearest-neighbour site necessarily frustrates the antiferromagnetic order (Fig. 5a); however, moving a hole to a next-nearest-neighbour site (which corresponds to moving a charge along one of the four diagonal directions) does not frustrate the AF order (Fig. 5b), and hence the charge transport on the Fermi arc is naturally harmonious with the AF order.

Boltzmann Theory and Fermi Arcs

Let us now discuss whether the Boltzmann transport theory [15] can still be applied to the system that has Fermi arcs instead of an ordinary Fermi surface. In the following, I will show that with a suitable modeling of the Fermi arcs in the context of Fermi liquids, the Boltzmann theory could be extended to treat the transport on the Fermi arcs in a rather simple way.

To begin with, one should remember that the charge current J is calculated in the Boltzmann theory via

$$J = 2 \int_{\text{all } k \text{ states}} e v_k g_k d^3 \boldsymbol{k} \,, \tag{8}$$

where $g_k \equiv f_k - f_k^0$ is the deviation of the distribution function f_k from its equilibrium f_k^0. In the relaxation time approximation, when g_k is caused by the action of an electric field $\boldsymbol{E}$, $\boldsymbol{J}$ can be written as

$$\boldsymbol{J} = \frac{1}{4\pi^3} \int e^2 \tau_k \boldsymbol{v}_k (\boldsymbol{v}_k \cdot \boldsymbol{E}) \left(- \frac{\partial f_k^0}{\partial \varepsilon} \right) \frac{dS}{\hbar v_k} \, \mathrm{d}\varepsilon \tag{9}$$

$$= \frac{e^2}{4\pi^3 \hbar} \int_{\mathrm{FS}} \tau_k(\varepsilon_\mathrm{F}) \boldsymbol{v}_k \left(\frac{\boldsymbol{v}_k}{v_k} \cdot \boldsymbol{E} \right) \mathrm{d}S_\mathrm{F} \, . \tag{10}$$

Note here that $\partial f_k^0 / \partial \varepsilon$ is non-zero only near the Fermi surface, which is the reason why the volume integral over the Brillouin zone in Eq. 9 can be reduced to the surface integral on the Fermi surface in Eq. 10. Note also that in the present calculation we allow the relaxation time τ_k to vary over the Fermi surface.

In the 2D case, when $\boldsymbol{E}$ is along the x axis,

$$J_x = \frac{e^2}{4\pi^3 \hbar} \int_{\mathrm{FS}} \tau_k(\varepsilon_\mathrm{F}) \left(v_x^2 E_x / v \right) \mathrm{d}L_\mathrm{F} \tag{11}$$

$$= \frac{e^2}{8\pi^3 \hbar} v_\mathrm{F} E_x \int_{\mathrm{FS}} \tau_k(\varepsilon_\mathrm{F}) \, \mathrm{d}L_\mathrm{F} \, , \tag{12}$$

where $v = \sqrt{{v_x}^2 + {v_y}^2}$ and $\mathrm{d}L_\mathrm{F}$ is the line element of the Fermi "surface" in the 2D Brillouin zone.

When one tries to apply this formalism to the Fermi-arc system, a possible way to model the Fermi arcs is to suppose that there is an underlying Fermi surface across which the occupation number of the electronic states changes from one to zero, but only on the "arc" portion of this underlying Fermi surface τ_k is long enough ($\tau_k = \tau_\mathrm{arc}$) to allow well-defined quasiparticles; elsewhere on the underlying FS, one could suppose $\tau_k = 0$ which means that the states are incoherent. Note that upon deriving Eq. 10 from Eq. 9, we relied on the fact that $\partial f_k^0 / \partial \varepsilon$ is non-zero only near the Fermi surface, and the property $\partial f_k^0 / \partial \varepsilon \neq 0$ still holds at the underlying FS even when there are no well-defined quasiparticles on it; such a surface is called "Luttinger surface" in the recent literature [16, 17].

With this modeling of the Fermi arcs, one could obtain

$$\sigma_x = \frac{e^2}{8\pi^3 \hbar} v_\mathrm{F} \int_{\mathrm{FS}} \tau_k(\varepsilon_\mathrm{F}) \, \mathrm{d}L_\mathrm{F} \tag{13}$$

$$= \frac{e^2}{8\pi^3 \hbar} v_\mathrm{F} \tau_\mathrm{arc} \cdot L_\mathrm{arc} \, , \tag{14}$$

where L_arc is the total length of the Fermi arcs. Therefore, within this simple model, the large ρ_ab (i.e., small σ_x) for the metallic transport on the Fermi arcs can be understood to be essentially a result of the small L_arc compared to

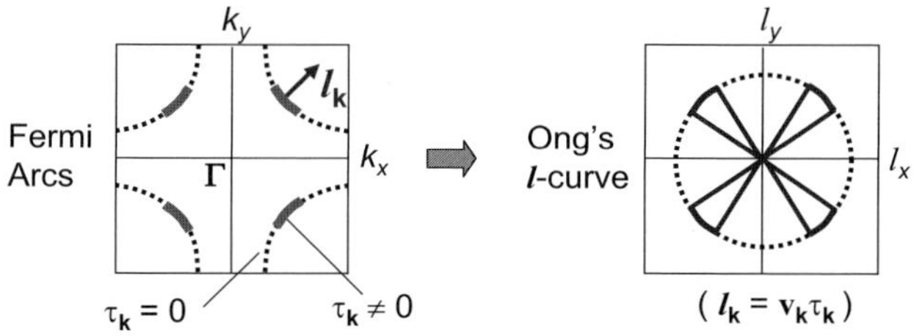

Fig. 6 Schematic diagram to show how to consider Ong's "l-curve" ([18]) for the Fermi arcs to calculate the Hall conductivity in the framework of the Boltzmann theory

the total length of the underlying FS, L_{full}. Quantitatively, however, the total length of the actual Fermi arcs observed in the lightly-doped LSCO is about a factor of three too long [11] to explain the measured ρ_{ab} with the above simple formalism. Obviously, more quantitative explanations of ρ_{ab} require a more elaborate modeling of the Fermi arcs; nevertheless, the present model allows an intuitive understanding of the transport involving the Fermi arcs in the framework of the conventional Boltzmann theory.

In passing, it is useful to note that one could employ the same treatment of the Fermi arcs (i.e., considering the Luttinger surface and supposing $\tau_k \neq 0$ only on the Fermi arcs) upon calculating the Hall conductivity σ_{xy}. For 2D metals, σ_{xy} is most conveniently calculated by using the geometrical formalism where σ_{xy} is given by the "Stokes area" traced out by the $\boldsymbol{l}_k$ vector which defines the Ong's "l-curve" [18]. Since $\boldsymbol{l}_k = \boldsymbol{v}_k\tau_k$, the $\boldsymbol{l}_k$ vector is non-zero only for those $\boldsymbol{k}$'s on the Fermi arcs (Fig. 6), and one obtains $\sigma_{xy}^{\text{arc}} = r\sigma_{xy}^{\text{full}}$ (where $r = L_{\text{arc}}/L_{\text{full}}$) in the simplest case of a circular underlying FS and constant τ_k on the Fermi arcs; in this case, the Hall coefficient can be written as

$$R_{\text{H}}^{\text{arc}} = \frac{\sigma_{xy}^{\text{arc}}}{\left(\sigma_{xx}^{\text{arc}}\right)^2} = R_{\text{H}}^{\text{full}}/r\,, \tag{15}$$

where $\sigma_{xx}^{\text{arc}} = r\sigma_{xx}^{\text{full}}$ from Eq. 14. This formula qualitatively explains why the Fermi arcs can give rise to a Hall constant that corresponds to a low carrier density when the underlying FS is what is expected for a high carrier density. But again, for a really quantitative understanding, one needs a more elaborate modeling of the Fermi arcs.

Unusual Localization

As is described above, the large k_{F} associated with the Fermi arcs makes it possible for the metallic transport to be realized at low doping levels, whatever the origin of the Fermi arc is. This in turn casts a new light on the strong localization behavior observed at low temperature for $x < 0.06$; namely, while it seems natural for the strong localized to occur when there are only a small number of carriers, the $k_F\ell$ value at the onset of the localization is unusually large. For example, at $x = 0.05$, $\rho_{ab}(T)$ shows strong localization [8] [characterized by a variable-range hopping (VRH) behavior [4]] below 50 K, where $\rho_{ab} = 0.9\,\mathrm{m\Omega\,cm}$ (Fig. 1) and the corresponding $k_F\ell$ value is as large as 13. If one remembers that in ordinary materials the VRH behavior is observed only when $k_F\ell < 1$ [4], the strong localization at $x = 0.05$ with $k_F\ell = 13$ is very unusual.

This consideration naturally reminds us of the $\log(1/T)$ insulating behavior that shows up when the superconductivity is suppressed by high magnetic fields in underdoped LSCO [2, 3]. It was already pointed out [3] that the $k_F\ell$ value estimated by using Eq. 6 at the onset of the insulating behavior was as large as 13 for $x = 0.15$ and hence the $\log(1/T)$ behavior in LSCO must be of unusual origin. In this regard, it is intriguing to note that the $\log(1/T)$ insulating behavior in LSCO has been observed for $x \leq 0.15$ [3], and the Fermi arcs have been found in exactly the same doping range [11]; in other words, at $x \geq 0.16$ where the transport is metallic down to low temperature, there are no Fermi arcs any more, but a large complete Fermi surface is observed by ARPES [11].

Phase Diagram

Based on the accumulated knowledge on LSCO, one can draw an electronic phase diagram as shown in Fig. 7; here, the solid lines delineate phase transitions between different phases in zero field [6], while the dotted line separates the "metallic" regime from the "insulating" regime that is observed *in the absence of superconductivity* either at low doping [8] or in high magnetic fields [2, 3] [the boundary is defined by the minimum in $\rho_{ab}(T)$]. This phase diagram depicts two prominent peculiarities in the charge transport in cuprates: (1) The Fermi arcs allow a metallic transport to be realized down to very low doping at moderate temperature, neglecting the onset of the Néel order; (2) but at the same time, unusual insulating behavior with large $k_F\ell$ values sets in at low temperature whenever the Fermi arcs govern the transport.

Summary

One should not apply the naive Mott-limit condition Eq. 7 to LSCO, because the Fermi arcs cause a large k_F to be associated with the electrons near ε_F even in the lightly-doped region. Because of this situation dictated by the peculiar electronic structure, the Mott limit is violated in cuprates *not* because a metallic behavior is observed for $k_F\ell \ll 1$, but because unusual localization occurs for $k_F\ell \approx 10$. Furthermore, since this unusual localization is intimately

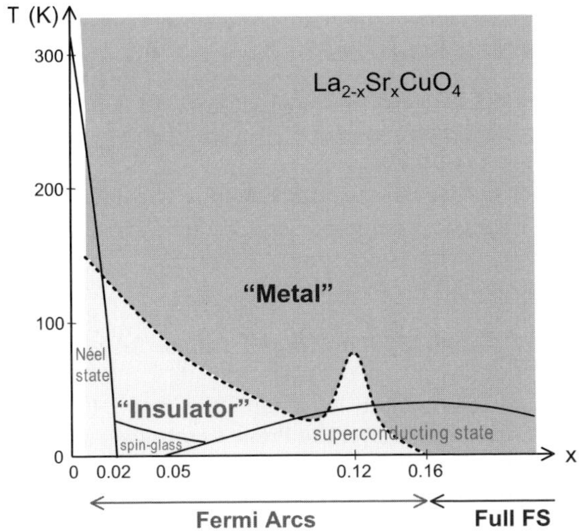

Fig. 7 Electronic phase diagram of LSCO. The solid lines delineate phase transitions between different phases in zero field, while the dotted line separates the "insulating" regime from the "metallic" regime. Since the Fermi arcs have been observed by ARPES in LSCO for $x \leq 0.15$ ([11]), which exactly matches the doping range where the insulating phase is found at low temperature, one may conclude that the Fermi arcs are inherently insulating at $T = 0$ in the absence of superconductivity

tied to the existence of the Fermi arcs, one may conclude that the Fermi arcs are inherently insulating at $T = 0$ in the absence of superconductivity. If the hole doping to a Mott-insulating cuprate initially leads to the formation of charge stripes [19–21] as is suggested by a number of experiments [22–30], both the appearance of the Fermi arcs and the Mott-limit-violating localization behavior should be understood to be a consequence of the charge stripe formation.

Acknowledgements I am deeply indebted to the collaborations with my colleagues at CRIEPI in the past 10 years; many thanks to K. Segawa, S. Komiya, S. Ono, X.F. Sun, A.N. Lavrov, J. Takeya, I. Tsukada, and A.A. Taskin. Also, I acknowledge D.N. Basov, G.S. Boebinger, A. Fujimori, T.H. Geballe, S.A. Kivelson, K.A. Müller, T. Tohyama, S. Uchida, and T. Yoshida for helpful discussions. This work was supported by a Grant-in-Aid for Science provided by the Japan Society for the Promotion of Science.

References

1. Emery VJ, Kivelson SA (1995) Phys Rev Lett 74:3253
2. Ando Y, Boebinger GS, Passner A, Kimura T, Kishio K (1995) Phys Rev Lett 75:4662
3. Boebinger GS, Ando Y, Passner A, Kimura T, Okuya M, Shimoyama J, Kishio K, Tamasaku K, Ichikawa N, Uchida S (1996) Phys Rev Lett 77:5417

4. Mott NF (1990) Metal-Insulator Transitions. Taylor & Francis, London, 2nd ed
5. Luttinger JM (1960) Phys Rev 119:1153
6. Kastner MA, Birgeneau RJ, Shirane G, Endoh Y (1998) Rev Mod Phys 70:897
7. Ono S, Komiya S, Ando Y (2007) Phys Rev B 75:024515
8. Ando Y, Lavrov AN, Komiya S, Segawa K, Sun XF (2001) Phys Rev Lett 87:017001
9. Dagotto E (1994) Rev Mod Phys 66:763
10. Yoshida T et al. (2003) Phys Rev Lett 91:027001
11. Yoshida T, Zhou XJ, Lu DH, Komiya S, Ando Y, Eisaki H, Kakeshita T, Uchida S, Hussain Z, Shen Z-X, Fujimori A (2006) cond-mat/0610759
12. Damascelli A, Hussain Z, Shen Z-X (2003) Rev Mod Phys 75:473
13. Ando Y, Kurita Y, Komiya S, Ono S, Segawa K (2004) Phys Rev Lett 92:197001
14. Padilla WJ, Lee YS, Dumm M, Blumberg G, Ono S, Segawa K, Komiya S, Ando Y, Basov DN (2005) Phys Rev B 72:060511(R)
15. A comprehensive description can be found in: Ziman JM (1972) Principles of the Theory of Solids. Cambridge University Press, Cambridge, 2nd ed
16. Dzyaloshinskii I (2003) Phys Rev B 68:085113
17. Stanescu TD, Phillips PW, Choy T-P (2006) cond-mat/0602280
18. Ong NP (1991) Phys Rev B 43:193
19. Zaanen J, Gunnarsson O (1989) Phys Rev B 40:7391
20. Machida K (1989) Physica C 158:192
21. Emery V, Kivelson SA (1993) Physica C 209:597
22. Cho JH, Chou FC, Johnston DC (1993) Phys Rev Lett 70:222
23. Borsa F et al. (1995) Phys Rev B 52:7334
24. Ando Y, Lavrov AN, Segawa K (1999) Phys Rev Lett 83:2813
25. Müller KA (2000) Physica C 11:341–348
26. Ando Y, Segawa K, Komiya S, Lavrov AN (2002) Phys Rev Lett 88:137005
27. Ando Y, Lavrov AN, Komiya S (2003) Phys Rev Lett 90:247003
28. Shengelaya A, Bruun M, Kochelaev BI, Safina A, Conder K, Müller KA (2004) Phys Rev Lett 93:017001
29. Padilla WJ, Dumm M, Komiya S, Ando Y, Basov DN (2005) Phys Rev B 72:205101
30. Kivelson SA, Bindloss IP, Fradkin E, Oganesyan V, Tranquada JM, Kapitulnik A, Howald C (2003) Rev Mod Phys 75:1201

Bednorz JG (2007) In: Bussmann-Holder A, Keller H (eds)
High T_c Superconductors and Related Transition Metal Oxides.
Springer-Verlag Berlin Heidelberg, 29–34
DOI 10.1007/978-3-540-71023-3

Perovskites and Their New Role in Oxide Electronics

J. G. Bednorz

IBM Research, Zurich Research Laboratory, Säumerstrasse 4, CH-8803 Rüschlikon, Switzerland
bed@zurich.ibm.com

Ever since the discovery of ferroelectricity in $BaTiO_3$, oxides with perovskite structure have been a fascinating research object in solid-state physics. Generations of scientists were attracted by the intriguing properties stemming from their enormous (large) chemical diversity. It was as early as 1963 that research on perovskites and related compounds was established at the IBM Zurich Research Laboratory (ZRL) through Alex Müller. In more than two decades, the study of charge-transfer processes and local properties of transition-metal dopants, of ferroelectric and soft-mode properties, and of the critical behavior of structural phase transitions created substantial insight into how to manipulate and tune specific properties. One compound, namely $SrTiO_3$, predominated in many of these studies, being the workhorse as a model system. Therefore it is no surprise that this crystal played a remarkable role in Alex's life, and he understood it to infuse others with his fascination. Triggered by the work on superconductivity in niobium-doped $SrTiO_3$ and by reading the signs set by two other superconducting oxides, $Li_{1+x}Ti_{2-x}O_4$ and $BaPb_{1-x}Bi_xO_3$, the focus of the perovskite effort at ZRL was shifted to metallic compounds. The concept for the search of new superconducting oxides was based on the Jahn–Teller effect, which had been intensively studied in insulating perovskites, and successfully led to the cuprate superconductors.

With the discovery of high-temperature superconductivity [1] it became apparent that the potential for further discoveries in perovskites and related materials was by far not exhausted. In addition to the rapidly growing research activities on superconducting oxides, the classical research fields of ferroelectric, magnetic and metallic oxides experienced a revival. The reason for this was that the development of thin-film deposition techniques for the high-T_c superconductors and the success in their epitaxial growth had opened up the perspective of much broader applications for the entire field of perovskites and related compounds. The application of semiconductor technology concepts to thin films and heterostructures, i.e., changing the state of a material by external stimuli such as electric or magnetic fields or optical excitation, is leading to a variety of new actuator and sensor devices having the potential for integration in microelectronic circuits. The manganites, which had already been investigated in the 50s, have become attractive again through the discovery of the colossal magnetoresistance in thin films [2, 3] and the occurrence of new magnetic and nonmagnetic ordering phenomena in both three-dimensional and layered compounds. Nonvolatile resistance changes, which are of particular interest for memory applications, are obtained in epitaxial heterostructures by electrostatic modulation of the carrier concentration in a metallic oxide by switching the polarization of an adjacent ferroelectric layer [4].

Recently perovskite oxides in epitaxial metal-insulator-metal (MIM) heterostructures have been shown to allow reversible resistive switching between different states, demonstrating another possible application as nonvolatile memory [5, 6]. The capacitor-like structures consist of a high-dielectric-constant oxide film (typical thickness: 100–300 nm) such as transition-metal (TM) doped (Ba,Sr)TiO_3 or $SrZrO_3$ sandwiched between a conducting $SrRuO_3$ base electrode grown on a $SrTiO_3$ substrate and a Au or Pt top electrode. An electric field applied across the insulator leads to a continuous change from its original high-resistance (MΩ) to a conducting state (kΩ) with a pronounced hysteretic current–voltage characteristic (IVC) as shown in Fig. 1. Short voltage pulses (of micro- to nanosecond duration) of opposite polarity can reversibly switch between IVCs with distinctly different slopes, representing the "high" and "low" memory states [6]. These states, which can have retention times of more than four years, can be read out resistively at small voltages of typically 0.2 V (Fig. 2). Attractive for large-scale applications is that for simple submicron-sized crosspoint cells obtained with Si technology (CMOS) processes the operating parameters will differ only slightly. While significant progress has been made with respect to the creation of operable devices, the conduction mechanism and the origin of the bistable resistance change are still under debate. A phenomenological model involves a nonpercolating domain structure and tunneling between different domains [7]. However, also extended defects [8] or changes in the

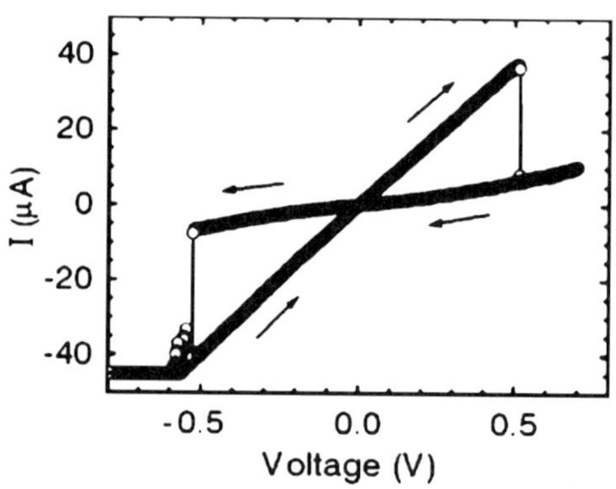

Fig. 1 I-V characteristic with the low- and high-resistance state

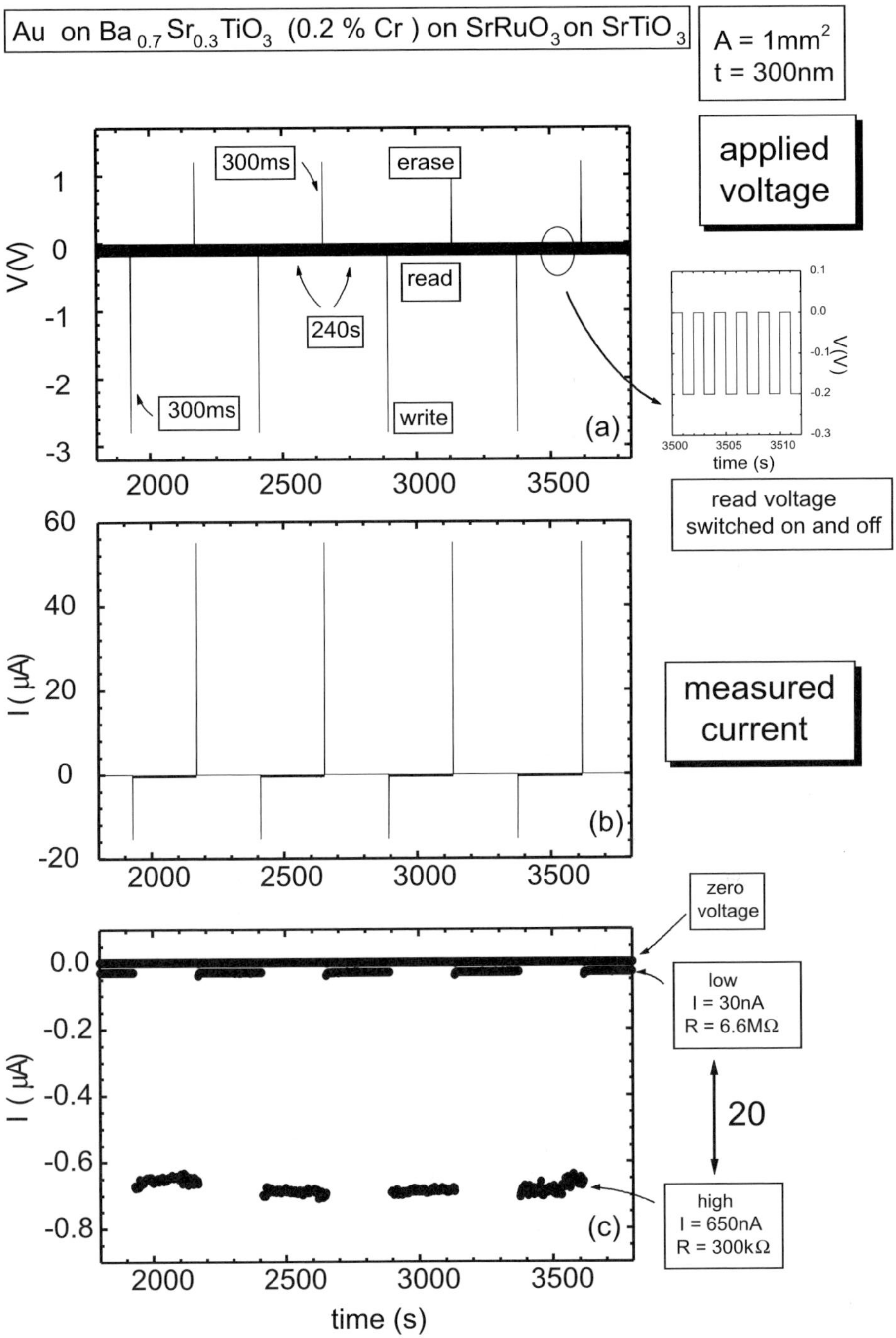

Fig. 2 Operation of a thin-film memory cell (Ba/Sr)TiO_3:0.2% Cr. **a** Voltage pulses for write-read-erase, **b** corresponding currents, and **c** read-out current levels at 0.2 V

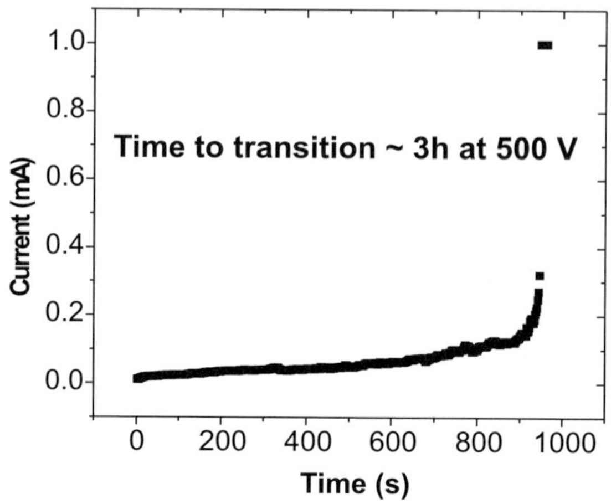

Fig. 3 Final period of a three-hour electrical stress procedure on a $SrTiO_3:Cr$ crystal, reaching a current compliance set at 1 mA

oxygen-vacancy concentration [9, 10] are considered to lead to a conducting state. On the other hand, modifications of the interface [11, 12], local inhomogeneities [13] or local reduction/oxidation processes [14] are being discussed as the origin of the bistable resistance state.

It is not too surprising that once again $SrTiO_3$ shows up in the ZRL memory research project and serves as a model system to study the contribution of possible volume and interface effects to the conduction and the switching phenomenon. TM-doped $SrTiO_3$ single crystals require the application of an electrical stress (DC or pulses) to increase the conductivity (Fig. 3) and induce an insulator–metal transition – the requirement for memory switching. The use of bulk single-crystal capacitors allows us to access the volume between the electrodes and to follow possible changes during the stressing phase, for example by simple optical microscopy. Field-induced changes are visualized in polarized light by birefringence pattern which are suggesting that lattice polarization effects occur in part of the volume. This volume grows with time as the current increases, which can be taken as an indication that the transition occurs in the bulk. It is not excluded, however, that because of imperfections the current can occasionally flow in confined regions, as is revealed by infrared microscopy. Although, as detected by X-ray absorption spectroscopy, valence changes of the TM dopant (Cr) from 3+ to 4+ occur at the anode at the beginning of the stressing process, the role that dopants may play in an electronic transition remains unclear. In a sense, this remarkably closes the circle: decades after the original work at ZRL by Alex Müller [15], the topic of charge-transfer transitions and local properties of dopants probed by Electron Spin Resonance has again gained high relevance – this time with a different formulation of the scientific problem.

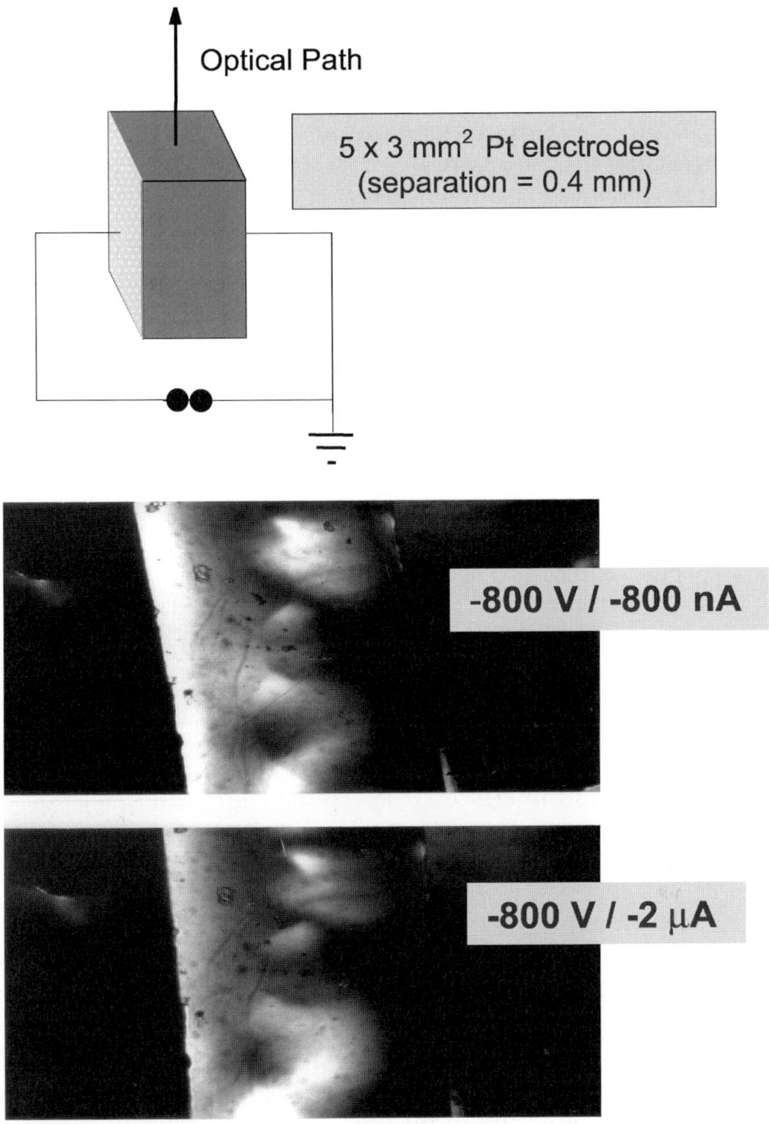

Fig. 4 Birefringence pattern indicating changes in the bulk of a crystal during transformation to the conducting state

References

1. Bednorz JG, Mueller KA (1986) Z Phys B 64:189
2. von Helmholt R et al. (1993) Phys Rev Lett 71:2331
3. Jin S et al. (1994) Science 264:413
4. Ahn CH et al. (1995) Science 269:373
5. Liu SQ et al. (2000) Appl Phys Lett 76:2749
6. Beck A et al. (2000) Appl Phys Lett 77:139
7. Rozenberg MJ et al. (2004) Phys Rev Lett 92:178302
8. Szot K et al.(2002) Phys Rev Lett 88:75508
9. Meijer GI et al. (2005) PRB 72:155102
10. Karg S et al. (2006) Appl Phys Lett 89:072106
11. Fujii T et al. (2005) Appl Phys Lett 86:012107
12. Baikalovet A et al. (2003) Appl Phys Lett 83:957
13. Rossel C et al. (2001) J Appl Phys 90:2892
14. Szot K et al. (2006) Nature Mater 5:312
15. Müller KA (1963) In: Proc 1st Int'l Conf. on Paramagnetic Resonance. Academic Press Inc, New York, Vol. 1, pp 17-43

Bok J, Bouvier J (2007) In: Bussmann-Holder A, Keller H (eds)
High T_c Superconductors and Related Transition Metal Oxides.
Springer-Verlag Berlin Heidelberg, 35–41
DOI 10.1007/978-3-540-71023-3

Van Hove Scenario for High T_c Superconductors

J. Bok (✉) · J. Bouvier

Solid State Physics Laboratory, ESPCI, 10, rue Vauquelin, 75231 Paris Cedex 05, France
julien.bok@espci.fr

Abstract All the high T_c cuprates (HTSC) have a lamellar structure and hence almost two dimensional properties. This 2D character leads to the existence of Van Hove singularities (VHs) or saddle points in their electronic band structure. These VHs have been observed experimentally in all the HTSC and we show that they explain many physical properties of these compounds, both in the normal and superconducting states. This feature induces a topological transition for a hole doping of $p_o \simeq 0.20$ hole/copper atom. The constant energy curves going from hole-like to electron-like.

The discovery of high temperature superconductivity in cuprates compounds in 1986 [1] has been a great sensation in the physics community and has raised great expectations, which, we hope, will one day be fulfilled. Twenty years after this discovery, the exact mechanism of HTSC is still not yet understood and remains a great challenge for solid-state physicists ... All these compounds are strongly anisotropic and almost two dimensional, due to their CuO_2 planes, where superconductivity mainly occurs. It is well known that in 2 dimensions, electrons in a periodic potential show a logarithmic density of states (DOS), named Van Hove singularity (VHs) (Van Hove (1953) [2]). The Van Hove scenario is based on the assumption that, in high critical temperature superconductors (HTSC), the Fermi Level (FL) lies close to such a singularity (Labbé-Bok 1987) [3]. The constant energy curves are hole-like for a doping less than 0.20 and become electron-like when the doping is increased. This topological transition occurs for a hole doping around 0.20, 0.21 hole/copper atom (Fig. 1). This hypothesis has been confirmed by many experiments, in particular by Angular Resolved Photoemission Spectroscopy in different compounds [4].

We want to stress that the model of 2D itinerant electrons in presence of VHs in the band structure has already explained a great number of experimental facts. Detailed calculations are presented in a review paper [5]:

- **critical temperature and gap anisotropy**
  We have computed the critical temperature T_c and its variation with doping. The optimum T_c is obtained for a hole doping of $p = 0.16$ and not $p = 0.20$ (when $E_{Fs} = Es$) as predicted in our first simple model. This is because we did not take into account the screening in the beginning. By

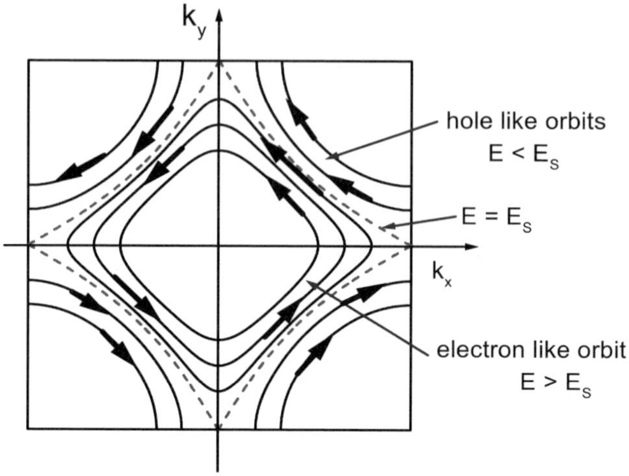

Fig. 1 Constant energy curves in the first Brillouin zone, in CuO_2 planes, representation of hole and electron like orbits

taking into account this screening and its variation with doping, our new results agree completely with the experimental results (Figs. 2, 3).
The superconducting gap is strongly anisotropic [6, 7]. It takes a high value in the directions of the saddle points (0,1) or (1,0), where the DOS is high and a low value in the (1,1) direction, where the DOS is low. It becomes negative if a repulsive interaction between electrons is included.
It has also been shown that the singularity is in the middle of a large band and that, in these circumstances, the Coulomb repulsion μ is renormalized and replaced by a smaller value [8]. This is important, because it shows that in the case of a VHs, Coulomb repulsion does not lower T_c as it does in a narrow band.

- **anomalous isotope effect and coherence length**
  Due to the peak in the DOS at the VHs, we have shown that the isotope effect should be very small at optimum doping, even if the electron-phonon interaction is responsible for the pairing of electrons [3, 5].This results from the fact that the cut-off in the BCS formula is mainly the width of the singularity and not the phonon frequency. The isotope effect reappears at low doping; this was observed experimentally [3]. We also were able to explain the very small values of the coherence length ξ [8, 9]. This is due to the fact that ξ is proportional to the Fermi velocity and that this velocity is zero at a saddle point.
- **tunnelling conductance**
  We have computed the conductance of tunnelling junctions with our calculated DOS and compared the results to the experimental curves obtained with various types of superconducting junctions. The results are satisfactory [10].

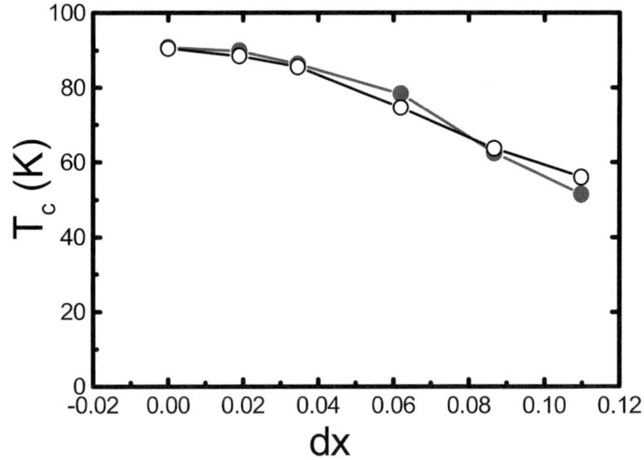

Fig. 2 Comparison of the variation of Tc versus the variation of doping dx, from the optimal doping; calculated in our model (*red filled circles*) and the experimental results of Koïke et al. (*black open circles*) [16]

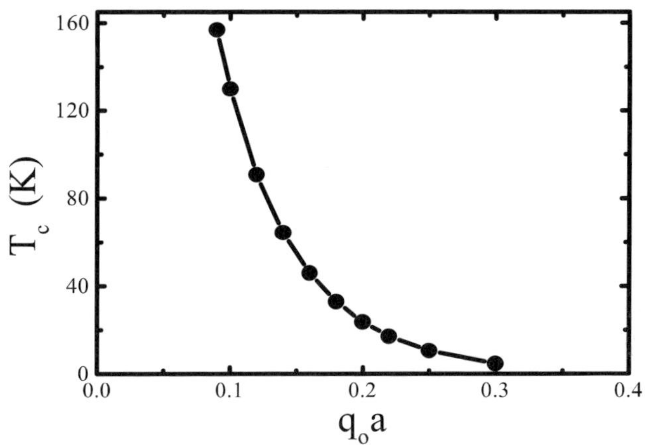

Fig. 3 Tc versus the screening parameter $q_0 a$ (cf. [11])

- **the specific heat** [7] **and the magnetic susceptibility** [11].
  These two thermodynamical quantities go through a maximum as the temperature varies. The variation of all these properties with hole doping (from underdoped to overdoped samples) and temperature are obtained and compared with the experiments. The agreement is very satisfactory. The variation with the doping is linked to $(E_F - E_S)$, so does the variation with the temperature due to the Fermi-Dirac distribution. When kT becomes of the order of $E_F - E_S$, the high density of states in the singularity is populated and contributes to this maximum. The good fit between theory and experiment is shown in Fig. 4.

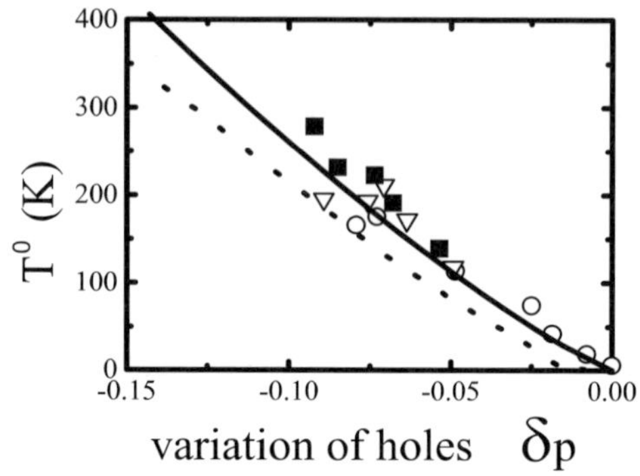

Fig. 4 The temperature, T° (directly linked to VHs), where the calculated χ_p (*dashed line*) and the specific heat (*solid line*) go through a maximum, versus δp. For comparison we show results presented in Fig. 27 of [17], the symbols are the same. (*solid squares*: from thermoelectric power, *circles*: from specific heat, *triangles*: from NMR Knight shift data)

- **transport properties**
  Transport properties in the normal state are described. We show that $E_\mathrm{F} - E_\mathrm{S}$ is critical for these properties, leading to Fermi liquid or marginal Fermi liquid behaviour [12]. We compute the Hall coefficient and its variation with doping and temperature [13]. We show that the experimental results may be explained by the topology of the Fermi surface (FS) which goes from hole-like to electron-like as the hole doping is increased. When the temperature is increased, the electron-like orbits start being populated. They give a negative contribution to the Hall coefficient which

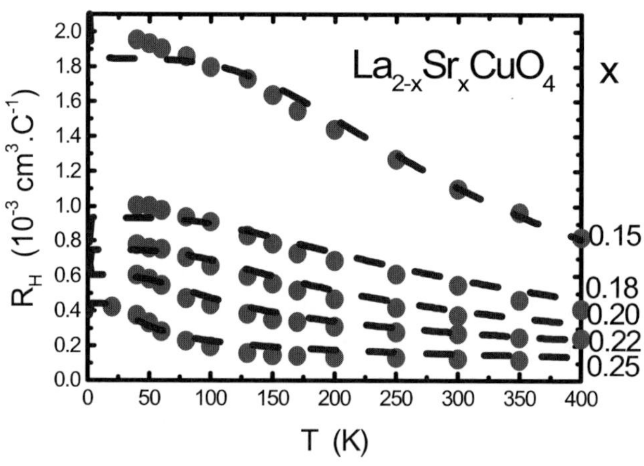

Fig. 5 *Filled circles*: experimental $R_\mathrm{H}(T)$ given by [18]. In polycrystalline $La_{2-x}Sr_xCuO_4$, for $x = 0.15, 0.18, 0.20, 0.22, 0.25$. *Dashed lines*: theoretical fits, the theoretical hole levels as the same as the experimental. The calculations are made with: $t = 0.23$ eV, $t' = 0.06$ eV, $2t'/t = 0.52$, $\Gamma(E_\mathrm{S}) = 0.1$ (cf. [13])

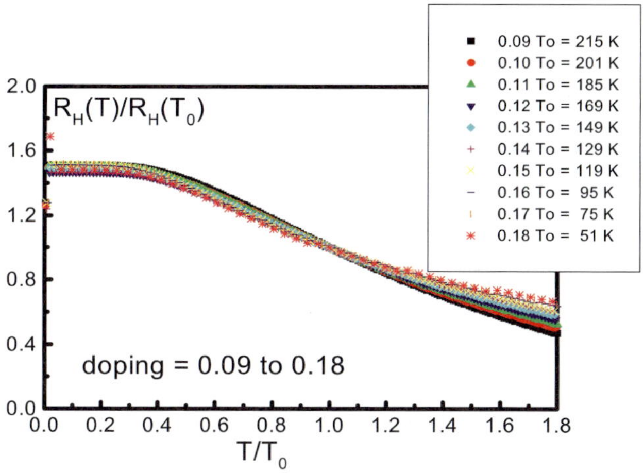

Fig. 6 Universal law $R_H(T)/R_H(T_0)$ versus T/T_0 for various hole doping levels, from 0.09 to 0.18 (cf. [13])

decreases as temperature increases as observed in all experiments (Figs. 5, 6). The critical doping, for which a topological transition is observed and calculated is $p = 0.21$ hole per CuO_2 plane.

- **effect of disorder**
  In the normal states the cuprates, especially the underdoped ones, may be considered as bad metals in the sense of Altshuler and Aronov [14], because the diffusion coefficient is very low. The diffusion coefficient is proportional to the Fermi velocity which is zero at the saddle points. The

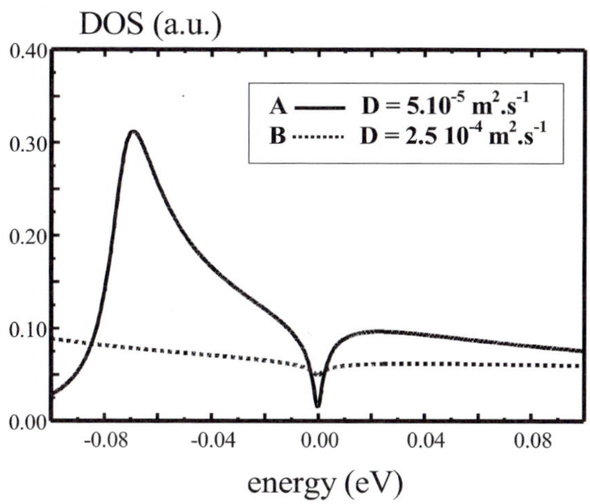

Fig. 7 Calculated DOS with Coulomb interaction with different sets of values of D: in the (1,0) direction, and equivalent directions – B: in the (1,1) direction

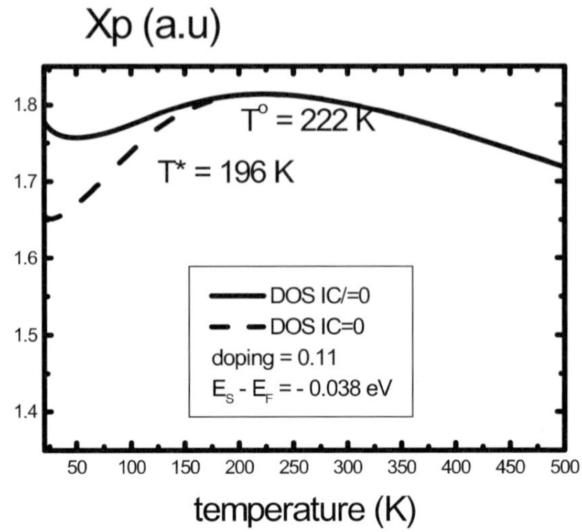

Fig. 8 Calculated Pauli Susceptibilities. *Full line* = without disorder effect – *Dashed line* = with disorder effect, for 0.11 hole doping

Coulomb interaction between electrons must then be taken into account The main effect is to open a dip in the DOS at the Fermi level [14, 15], see Fig. 7. We show that this may explain many observed features of the "pseudo-gap": its value, anisotropy and variation with doping [15]. The anisotropy is directly related to the anisotropy of the Fermi velocity. The loss of states at the Fermi level explains the decrease of thermodynamic coefficients, such as the magnetic susceptibility, at low temperature (Fig. 8).

In conclusion we have shown that Van Hove singularities play an important role in HTSC, and that by taking them into account, we may explain many of their normal and superconducting properties. We argue that the observed topological transition, occurring around $p = 0.2$ hole per copper atom, is a crucial feature to explain the physical properties of the cuprates, in particular the Hall effect. The presence of VHs is now well established and almost all theoretical models take them into account under various names, VHs, saddle points, hot spots, etc.

References

1. Bednorz JG, Müller KA (1986) Z A Phys B 64:189
2. Van Hove L (1953) Phys Rev 89:1189
3. Labbé J, Bok J (1987) Europhys Lett 3:1225
4. Ino A, Kim C, Nakamura M, Yoshida Y, Mizokawa T, Fujimori A, Shen ZX, Takeshita T, Eisaki H, Uchida S (2002) Phys Rev B 65:094504

5. Bok J, Bouvier J (2007) to be published in Progress in Superconductivity Research, Nova Science Publisher
6. Bouvier J, Bok J (1995) Physica C 249:117
7. Bouvier J, Bok J (1997) Physica C 288:217
8. Bok J, Force L (1991) Physica C 185:1449
9. Force L, Bok J (1993) Solid Stat Comm 85:975
10. Bok J, Bouvier J (1997) Physica C 274:1
11. Bouvier J, Bok J (1997) J of Superc 10:673
12. Pattnaik PC, Kane CL, Newns DM, Tsuei CC (1992) Phys Rev B 45:5714
13. Bok J, Bouvier J (2004) Physica C 403:263
14. Altshuler BL, Aronov AG (1985) In: Efros AL, Pollak M (eds) Electron–Electron Interaction In Disordered Systems. Elsevier Science Publishers B.V.
15. Bouvier J, Bok J, Kim H, Trotter G, Osofsky M (2001) Physica C 364:471
16. Koïke et al. (1989) Physica C 159:105
17. Cooper R, Loram JW (1996) J Phys I France 6:2237
18. Hwang HY et al. (1994) Phys Rev Lett 16:2636

Bozovic I (2007) In: Bussmann-Holder A, Keller H (eds)
High T_c Superconductors and Related Transition Metal Oxides.
Springer-Verlag Berlin Heidelberg, 43–55
DOI 10.1007/978-3-540-71023-3

Possible Jahn–Teller Effect and Strong Electron–Phonon Coupling in Beryllium Hydride

I. Bozovic

Brookhaven National Laboratory, Upton, NY 11973, USA
bozovic@bnl.gov

Abstract Possibility of Jahn–Teller effect in extended systems, such as polymers and quasi-one-dimensional metals, is discussed using beryllium hydride as an example. If this polymer compound could be made metallic by doping (e.g., with Lithium), we conjecture that it should show strong-electron phonon coupling and possibly superconductivity with a relatively high critical temperature.

Introduction: K.A. Mueller and Jahn–Teller Bipolarons

As we all know, the search and the discovery of high-temperature superconductivity (HTS) in cuprates [1] was guided by Karl Alex Mueller's thinking about Jahn–Teller effect (JTE) and strong electron-phonon coupling. Imagine an isolated ionic cluster with a Cu^{2+} cation sitting at the center of an octahedral cage comprised of six O^{2-} anions. If the highest occupied one-electron state of this complex can be identified with the half-filled Cu $3d^9$ state, one would expect the octahedron to distort as the result of JTE. This indeed could cause strong coupling between the electrons and the ionic displacements. Strong electron-phonon coupling has been traditionally related to a "high" T_c according to the standard BCS theory. Whether this reasoning really applies to cuprates is debatable; but (to quote D. Pavuna) what will be remembered after all these debates about HTS have long faded out is that Alex and George discovered the phenomenon in 1986. My understanding is that Alex actually still believes that in cuprates JTE is operational and leads to formation of bipolarons which Bose-condense into the HTS state [2, 3]. Thus it seems appropriate in the Festschrift honoring his 80th birthday to talk about JTE, strong electron-phonon coupling, and superconductivity, from my personal angle. I hope this will find at least one interested reader – although knowing him (i.e., Alex) I am certainly ready for criticism. In any case, here is my offering.

Superconductivity in Hydrides

The quest for high-temperature superconductivity (HTS) has been a passion of numerous experimentalists and theorists for many years. One of the earliest candidate materials has been metallic hydrogen. Already in 1935, Wigner predicted [4] that molecular hydrogen, which at low temperature is an insulator, would turn metallic (and mono-atomic) under high-enough pressure, $p > p_c = 25$ GPa. This prediction has been revisited subsequently by other prominent theorists including A. Abrikosov, N. Aschcroft, Yu. Kagan, etc., who calculated [5–7] much higher critical pressure – as high as 1500 GPa. In 1968, Aschcroft raised the stakes further by predicting [6] boldly that metallic hydrogen would turn superconducting at $T_c \sim 200$ K. I was not around at that time, but I guess that this must have caused consternation. This was an order of magnitude larger than the highest T_c found in any superconductor known at that time. It was also much higher than what was widely believed to be the theoretical limit, largely under the influence of W. McMillan and some other leading theorists. One of their key arguments was that very strong electron-phonon coupling should inevitably cause lattice instability and distortion. On Aschcroft's side was the fact that hydrogen has the smallest mass and hence the highest Debye temperature – and his calculations. So far, we only know with certainty that his prediction about a very high critical pressure was most probably correct – as far as I know p_c has never been reached experimentally. (At least not for *solid* hydrogen – liquid hydrogen apparently can become metallic under shock compression [8].)

This makes one turn attention to metallic hydrides. Here, experimentalists had better luck. Palladium hydride, PdH, was found to be superconducting at a respectable $T_c = 8.8$ K, even though pure elemental Pd is not superconducting, not even under high pressure. Even more interesting, PdH showed a negative hydrogen isotope effect: $T_c = 10.7$ K in PdD. Still further increase was achieved by Cu doping: $T_c = 16.6$ K in $Pd_{1-x}Cu_xH$. This was rather interesting *per se* but it did not look like a big step in the quest for HTS.

However, in 1986, Overhauser predicted [9] that $LiBeH_3$ and/or Li_2BeH_4 may be metallic and show HTS such as it was envisioned for metallic hydrogen. This stirred a lot of interest. One of his statements, that these two compounds have modified-perovskite crystal structure, has been questioned and criticized [10], primarily because his conclusions were based on rather pour experimental diffraction data showing just a dozen powder reflections [11]. Much more work followed, theoretical and experimental [12–19], and now it appears that neither $LiBeH_3$ nor Li_2BeH_4 are metallic, much less superconducting. There was one glitch in late 1996, when all the daily news disseminated the Reuters announcement of a sensational discovery of superconductivity at $T = 350$ by K.S. Contreras and J.-P. Bastide at the National Institute of Applied Science in Lyon, France. The alleged room-temperature superconductor was powder composed of Li, Be, and H. However, this result [20] could not be reproduced and was eventually dismissed by the scientific community [21]. So, the quest for HTS in hydrides is still open to anyone with courage (and resources) to try. Additional motivation can be found in more recent work of Aschcroft [22–25], who has persisted at the forefront of theoretical research on possible hydrogen superconductivity till this days.

Vibronic Coupling in 0D: The Jahn–Teller Effect

The Jahn–Teller theorem has the unfortunate reputation of being the most misinterpreted result in physical sciences. (For the clearest account see e.g. Englman's book [26].) Actually, in its simplest version, the statement is almost trivial. If one assumes a molecule (or an ionic cluster) to have some symmetric structure, determines its electronic spectrum by a quantum-chemical calculation, and finds that the highest occupied one-electron level is degenerate and partially occupied, the original structure actually must be unstable with respect to a distortion that removes the degeneracy.

To see this, let us denote by $|e\rangle$ and $|e'\rangle$ the two degenerate molecular orbitals that correspond to the highest occupied one-electron level, and let the molecule be subject to a small asymmetric distortion of magnitude Q. In the first-order perturbation theory, the level will split into two, one level going down in energy and the other one going up. If the original degenerate level was occupied by a single electron – I am ignoring the spin for simplicity – the total electron energy will be reduced by $\Delta E = \langle e'|Q\partial V/\partial Q|e\rangle$ which is linearly proportional to Q. The elastic deformation of the molecule will cost some energy proportional to Q^2. The total energy will thus have the minimum at some $Q_{\min} \neq 0$. Jahn and Teller actually went one step further: for every possible molecular symmetry (described by the point group **P**), they identified the (symmetry of) active distortions. Now, contrary to what has been stated even in many textbooks, this does not mean that the molecule will simply distort statically. Rather, the total energy will have multiple (symmetry-related) minima with equal probability of occupancy, and the molecule will tunnel from one to another [26]. The overall **P** symmetry of the (vibronic) Hamiltonian must be preserved!

Vibronic Coupling in 1D: The Peierls Instability

Let us turn now to the simplest translation-symmetric system, a periodic linear chain of atoms, each contributing a single atomic orbital $|e\rangle$. Let us assume that these orbitals have a significant overlap $O = \langle e'|e\rangle$, so that the hopping integral $t = \langle e'|H|e\rangle$ is large enough (specifically, $t \gg kT$). This gives the electron band $\varepsilon(k) = 2t * \cos(ka)$. The electron states are extended Bloch waves, $\psi(k) = u(k) * \exp(ikx)$. If the band is partially occupied, the chain is metallic. The simplest case occurs when each atom contributes one electron. The band is half-filled; the Fermi level is found at $k_F = \pi/2a$. In this case, the matrix element of linear vibronic coupling, $\langle\psi(-k_F)|\partial U/\partial Q|\psi(k_F)\rangle$, is nonzero for $Q = 2k_F = \pi/a$. The chain is unstable against dimerization, since $Q = \pi/a$ corresponds to the wavelength $a' = 2a$. This is the well-known Peierls instability.

The above phenomenon, the Peierls Effect (PE), is closely related to JTE, since both are electronic instabilities driven by the degeneracy of the relevant one-electron states, namely the HOMO levels $|e\rangle$, $|e'\rangle$ in JTE and the Fermi level states $\psi(k_F)$, $\psi(-k_F)$, in the Peierls model. However, there is a profound difference: JTE is a strong instability in the sense that the total electron energy reduction is linear in Q. In contrast, Peierls instability is a much weaker one, because in that case for a small distortion Q only the few states very close to E_F (within some range $\Delta E \ll E_F$) are perturbed, since the Fermi statistics and Pauli principle do not allow the electrons to crowd very much. The total electron energy reduction in PE is $\Delta E \sim Q^2 * \ln Q$. The point is that while both JTE and PE originate in linear vibronic coupling, they are qualitatively different in the sense that JTE is a strong (linear) instability while PE is a weak (logarithmic) one.

The (Absence of) JTE and PE in Higher Dimensions

In real 3D crystals, one in general does not expect JTE instability because systematic degeneracy only occurs at some special (high-symmetry) k-vectors, which however form a "subset of measure zero" within the set of all possible k-vectors, and generally don't count for much at all [26]. The asymmetric ones, having no non-trivial point group symmetry, grossly outnumber the special ones. To say it more technically, irreducible representations of the relevant symmetry group (the crystallographic space group that leaves the k-vector invariant) are one-dimensional for every general k-vector. So there is simply no band degeneracy – except perhaps at few irrelevant points.

If the crystal is metallic, there will be so-called *star* degeneracy, i.e., there will be several states at the Fermi level connected by point-group symmetry of the corresponding wave-vectors; this is so-called *star degeneracy*. For example, in a tetragonal crystal, in the plane perpendicular to the rotation symmetry axis C_4, the four states $\psi(\mathbf{k}_F)$, $\psi(C_4\mathbf{k}_F)$, $\psi(C_2\mathbf{k}_F)$, $\psi(C_4^{-1}\mathbf{k}_F)$ must be degenerate. This degeneracy is similar to what we had in the PE case described in Sect. 3. The difference here is that for a general shape of the Fermi surface, if we allow the crystal to distort along $\mathbf{Q} = 2\mathbf{k}_F$, this will reduce the energy of the electrons at these four $\mathbf{k}_F$ vectors but will raise the energy of all the other electron pairs, since these other ones are not connected by the same $\mathbf{Q}$. For this reason, the Peierls instability does *not* occur in 2D or 3D. (The clearest exposition of this point was written by D. Thouless in a chapter of a book that I read many years ago but which I cannot trace now.)

An exception to this rule – i.e., a vibronic instability – can occur if there is extensive nesting of the Fermi surface, so that there are many pairs of electron states at the Fermi level connected by the same **Q**. But this is not highly likely to just happen by accident, and even if it does in some sense it would be not too different from saying that the dimensionality is effectively reduced.

There has been some confusion about the above point because the so-called cooperative JTE indeed occurs in some real (3D) crystals. The resolution of the conflict is that if the crystal contains molecules or ionic clusters that are sufficiently isolated from one another and from the crystal matrix, JTE can indeed occur in these individual molecules or clusters. If an assembly of such JTE-active molecules or clusters interact weakly with one another, this interaction breaks the original **P** symmetry of the molecule and can align the distortions, leading to collective behavior. Note, though, that cooperative JTE still originates in individual molecules or clusters, while weak inter-molecular coupling merely phase-locks the distortions [26].

Another possibility for JTE to occur in a crystal is if an electron gets trapped and localized by a defect or self-trapped due to small polaron formation. In this case, again the local **P** symmetry may become relevant and JTE may occur; this possibility will be discussed later in the last section.

JTE in Polymers and Quasi-1D Conductors

For the reasons that should be clear from the previous section, most experts believe that JTE is a "point" or "0D" phenomenon, restricted to molecules and small atomic or ionic clusters, and that it can not occur in periodic systems with extended Bloch-wave states. Nevertheless, I think that there might be an exception to this rule that even the experts are unaware of. Namely, in certain *quasi*-1D systems, such as polymers, extensive *band* degeneracy can occur [27, 28]. In this case, the number of (JTE-active) one-electron states that contribute to the vibronic instability can be very large – of the order of the number of atoms, like in the standard JTE [29, 30].

The difference between the Peierls model, which one can classify as 1D, and a polymer, which is *quasi*-1D, is that apart from the translational symmetry along the chain axis common to both cases, the later may have nontrivial, discrete point-group symmetry. Examples are: rotation C_n around the (vertical) polymer axis, order-two rotation U around a horizontal axis, vertical and horizontal mirror planes. Also possible are screw axes and glide planes. The spatial symmetry groups of stereo-regular (periodic) polymers are the line groups [31–33]. They give rise to quantum numbers such as quasi-momentum, quasi-angular momentum, and parities with respect to mirror symmetry planes.

As a very simple model example, consider a periodic array of benzene rings stacked along the z-axis and take into account only a single π orbital per C atom. Sixfold-rotation symmetry axis C_6 provides the quantum number $m = 0, \pm 1, \pm 2, 3$, the quasi-angular momentum. Any vertical symmetry plane makes $m = +1$ equivalent to $m = -1$ and $+2$ to -2, hence there will be two doubly-degenerate levels, which we can denote as $E_{1,-1}$ and $E_{2,-2}$. If we allow for some hopping between the neighboring rings, these states will evolve into electron energy bands that carry the same quantum numbers. Hence, there will be two bands, $E_{1,-1}$ and $E_{2,-2}$, doubly-degenerate throughout the Brillouin zone (BZ).

As a real physical example, consider e.g. a single-wall, zig-zag (4,0) carbon nanotube. It has $L8_4/mcm$ line group symmetry, and every k-vector has the C_{8v} point group symmetry [34]. The latter group has some two-dimensional irreducible representations, and as a consequence, in this nanotube some electron bands are two-fold degenerate throughout the Brillouin zone [34].

Band degeneracy in principle opens a possibility for the *band JTE* – instability of the polymer with respect to a distortion that would split and separate the two bands [29, 30]. For this to occur in reality, apart from the band degeneracy it is also necessary that the band is relatively flat, so that $\Delta E = Q\langle\psi(-k_F)|\partial U/\partial Q|\psi(k_F)\rangle$ is larger that the bandwidth. However, if the band is too narrow (i.e., if the bandwidth is less than $k_B T$), localization will occur and we are back to 0D.

$(BeH_2)_x$ Band Structure, *p*-JTE

Long ago, already in late seventies, I have identified a candidate compound for the *band JTE* – a bizarre electron-deficient polymer, $(BeH_2)_x$. It has been a favorite of quantum chemists at that time (of punching cards and Kbyte computers) because it only contains few lightest atoms – and because it actually exists, in contrast to other simple model chains frequently studied then.

The $(BeH_2)_x$ polymer structure is illustrated in Fig. 1. The translational repeat unit of $(BeH_2)_x$ polymer contains 2 Be and 4 H atoms. The polymer is invariant with respect to the line group $L4_2/mcm$ which can be generated by an order-four screw axis $(C_4|1/2)$, two vertical mirror-symmetry planes $(\sigma_x|0)$ and $(\sigma_y|0)$ and a horizontal symmetry plane $(\sigma_z|0)$. The relevant subgroup of the line group $L4_2/mcm$ that leaves the k-vector invariant is $L4_2mc$ and hence its irreducible representations can be used to label the Bloch states and the electron energy bands. Only three of these are relevant here: ${}_kA_0$, ${}_kA_2$, and ${}_kE_{1,-1}$. In the tight-binding model, it is sufficient to consider only two atomic orbitals, $|\psi_H\rangle$ and $|\psi_{Be}\rangle$ and two hopping integrals, $t = \langle\psi_H|\mathbf{H}|\psi_{Be}\rangle$ and $t' = \langle\psi_H|\mathbf{H}[C_4|1/2]|\psi_H\rangle$. Using the proper symmetry-adapted linear combinations of atomic orbitals (LCAOs) [27], one can easily calculate the relevant (i.e., the lowest three) bands; these are shown in Fig. 2. (More detailed and accurate *ab initio* SCF Hartree-Fock calculations give quite similar bands, plus others above and below [35, 36].)

Apparently, in this model, the highest occupied band $E_{1,-1}$ is two-fold degenerate throughout the BZ. This may change if we allow the structure to

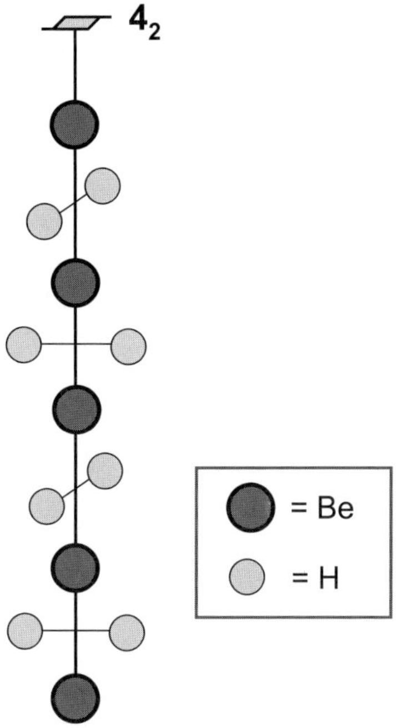

Fig. 1 A model of $(BeH_2)_x$ polymer. The line group is $L4_2/mcm$

distort. Consider, for example, libration of H_2 units around the z-axis, one clockwise and the next one counter-clockwise by the same angle. This breaks the screw axis $(C_4|1/2)$ symmetry, and the line group reduces to L2/mcm which has no two-dimensional irreducible representations. Hence, the entire $E_{1,-1}$ band splits into two non-degenerate bands. However, if the $E_{1,-1}$ band was originally full, this would provide no savings in energy – one band shifts down, the other up by the same amount $\Delta\varepsilon$, and there is no vibronic instability. However, if the $E_{1,-1}$ band is partially occupied, the situation is different – after the distortion, the lower band is full and the upper one is partially filled, hence there is a net energy gain. This will be offset by the elastic restitution force, but for librations this can be relatively weak since there is no change in the length of chemical bonds. So, this would be an example of the band JTE, with very strong *linear* vibronic (libronic!) coupling, because all the electrons in the $E_{1,-1}$ band would participate [29, 30].

In reality, one could reach this situation by p-type doping, e.g., by replacing some Be by Li. (This has actually been accomplished [37], as we will expound below.) So, Li-doped $(BeH_2)_x$ was my prime candidate for "high-temperature" superconductivity; note that this was happening long before the cuprate revolution and hence to me that meant something like 20 K. I was aware of a major theoretical hurdle, which H. Thomas, who was also interested in this problem, pointed out to me first at some conference in late

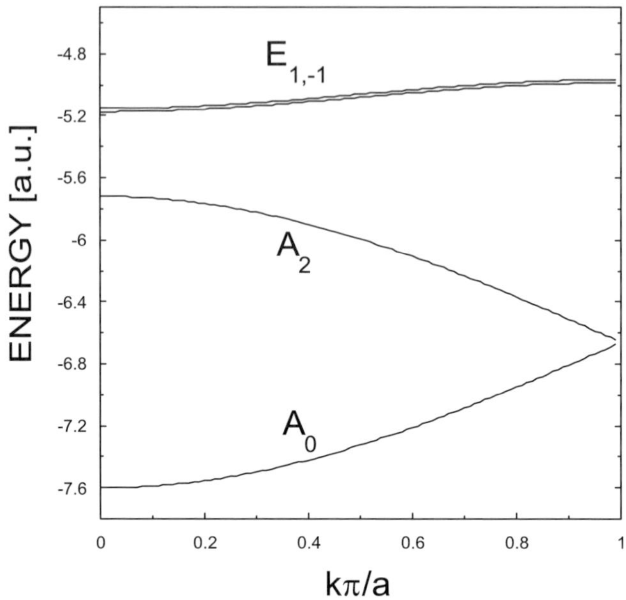

Fig. 2 Simplified tight-binding band structure of $(BeH_2)_x$ polymer. Note that the $E_{1,-1}$ band is twofold degenerate throughout the Brillouin zone. It is fully occupied in $[BeH_2]_x$ but could be made partially occupied by e.g. Li doping

seventies. (I have learned from Alex that his own thinking was inspired by an early paper of H. Thomas and coworkers [38], of which I have learned only much later, since it was published in a journal that was not on my reading list.) His killer question was whether indeed the above situation is at all compatible with metallic conduction, or else the carriers would necessarily be localized and immobile. It should be possible to answer this question theoretically, determining more accurately the real electron spectrum of $(BeH_2)_x$, the width of the $E_{1,-1}$ band, the elastic constants, etc. Experimentally, the question is simply whether Li-doped $(BeH_2)_x$ can be made metallic, or it will stay insulating at all doping levels.

Experiments: $(BeH_2)_x$ Crystal, Transport

Indeed, after my theoretical "insight" that $(BeH_2)_x$ polymer may be a candidate HTS material, I tried hard to find out what is known experimentally about this material. I was rather surprised (at that time) to find exactly nothing in physics journals. After a substantial effort – at that time I did all my literature search by manually browsing through numerous issues of Physics Abstracts and the Russian Referativnii Zhournal – I found *one* interesting paper published in "Inorganic Chemistry", in a journal that I certainly did not read regularly. The paper, by Brendel et al. [37], reported a recipe how to prepare crystalline $(BeH_2)_x$. They mixed amorphous polymeric solid BeH_2 with some Lithium (e.g. 0.5–2.5 mol % of LiH) and then applied compaction-fusion, i.e., exposed the powder to high temperature and pressure for a fixed time interval, then releasing both p and T. They explored the phase diagram up to $p = 12$ kbar and $T = 250$ °C (during compaction) and found several crystalline phases, depending on p and T. The densest phase (0.77 g/cm^3) was stable, white to light gray in color. However, in some narrow (p, T) range the product turned out to be "glassy and black", even though it was indistinguishable from the normal compacted material by chemical analysis, X-ray diffraction, and infrared spectroscopy. There was no evidence of free Be metal, either. The authors speculated about electron delocalization in this metastable phase, which presumably meant that it was indeed metallic. The primary role of Li in this work was to terminate and shorten the $(BeH_2)_x$ chains and promote crystallizations. However, in principle this doping could also introduce carriers (holes) in the topmost, two-fold degenerate $E_{1,-1}$ band.

So, after reading the Brendel et al. paper [37], my interest in Li:$(BeH_2)_x$ increased further. Indeed I would love to try to synthesize this and other light-metal hydrides in thin film form, using powerful COMBE technique and high-throughput testing and characterization that now I have in my hands [39]. I have tried repeatedly to get funding for such experiments, but so far without success. Nevertheless I am optimistic that the climate in US research administration circles has changed for better and that now there may be a much more benevolent look at efforts to search for superconductivity in new classes of materials hitherto unexplored. Also, in the meantime I have found (thanks to Google and the Web of Science) a little more information on beryllium hydride. It turned out that beryllium hydride is also attractive to people interested in (i) solid rocket propellants, (ii) very efficient moderators (e.g. for making very small nuclear bombs), and (iii) hydrogen storage materials [40–43]. Learning this, I suspect that physical properties of beryllium hydride may have indeed been studied in much detail, even though so little has been published.

Vibronic Coupling in Cuprates?

In relation to superconductivity, clearly we are interested in crystals that are metallic. This implies that there must be sufficient overlap between the relevant atomic orbitals. As we have seen, JTE proper cannot occur in higher dimensions (2D and 3D) but only in molecules or small isolated atomic clusters [26]. The usual story about crystal field splitting of a fivefold-degenerate d level into a threefold-degenerate t_{2g} and twofold-degenerate e_g levels, and subsequent further splitting of e_g due to JTE distortion may apply to an isolated CuO_6 octahedron, but not to a metal in which electrons occupy extended Bloch orbitals. In fact, in cuprates strong Cu – O hybridization produces a relatively broad ($\sim$ 1 eV) band, judging from Angle-resolved photoemission spectroscopy (ARPES) and Angle-resolved magneto-resistance oscillations (AMRO); in optimally doped and overdoped samples both techniques show nice (although essentially 2D) Fermi surfaces [44–47]. On this side of the phase diagram, there are other signs of decent Fermi-Liquid, such as proper Wiedeman–Franz–law behavior.

At low doping levels, the situation may be different; specifically below the metal-to-insulator crossover at $x_c \approx 0.05$ one could imagine that electron localization may occur, in which case one should not be surprised to detect concomitant local lattice distortion (cuprates are essentially ionic crystals!), and even perhaps JTE. If this indeed happens, polaron or bipolaron formation may play the key role [48–51]. But to say the truth, I have not understood yet what causes this crossover in the first place – the jury is still out. In manganites the situation may be a bit clearer, with some typical signatures of both large and small polaron behavior (at low and high temperature, respectively) [52–54].

As for the immediate future, in my group at Brookhaven National Laboratory we have mastered the technique of growing quite reproducibly atomically smooth and perfect HTS films, in particular of $La_{2-x}Sr_xCuO_4$ with doping level ranging from $x = 0$ to $x = 0.50$. This indeed opens the door to many interesting experiments some of which are already well on track. I hope that before long, on the scale of months, we will be able to announce results that will shed some new light on the great cuprate puzzle, and Alex will rejoice.

Appendix

Karl-Alex Mueller: A Personal Touch

I first met Alex at Ted Geballe's beautiful estate in Woodside, CA (not far from Stanford) exactly 20 years ago, in the spring of 1987. The party started early in afternoon and lasted till late in the night. Pierre-Gilles de Gennes, whom I already knew from his earlier visits to (what was then) Yugoslavia, was also present. He and several other guests succumbed to the temptation of the sunny California day to swim in the large pool. We did not know that two in the company were to become Nobel laureates subsequently, Alex by the end of the same year, and Pierre-Gilles four years later. But at this party they were still mortals and I used the chance to bug them with some physics. I found soon enough that Alex nourishes strong opinions (to borrow Nabokov's phrase). I guess I do as well, so our conversation appeared more like a debate. We have continued discussing in the same spirit on several other occasions – I remember well our long walks and talks in snow (which didn't seem to bother Alex at all) in Closters, Switzerland, as well as several private dinners in warmer places such as Los Alamos and Santa Fe, and elsewhere. Even though our opinions are rarely identical (as I am afraid this article testifies), Alex has never ceased to impress me with his lucidity and logic – apart from his physics intuition, which no one can question. And then there is his infectious enthusiasm – he cares about physics a lot, it is his great passion.

Another seems to be fast cars. My favorite K.-A. Mueller anecdote is that he actually *drove* from Zurich to Bremen and back – I guess about 10^3 kilometers one way – just to visit my laboratory, see my MBE system, and discuss physics. (True, he was still young then – only 73!) Actually, he tried – and succeeded, his enthusiasm was infectious – to convince me to try to reproduce the result of a group from Japan who claimed to have seen the giant proximity effect (GPE) in Josephson junctions with $La_{1.85}Sr_{0.15}CuO_4$ (LSCO) electrodes and $La_{0.7}Sr_{0.3}MnO_3$ barrier [55, 56], both polaronic materials according to Alex. I indeed tried – but failed; we saw no supercurrent whatsoever, even with very thin manganites barriers. This of course did not prove much – there are many ways in which one can make a bad junction, and hence I have never published this. We got one possibly interesting result from our attempts to reduce the lattice mismatch and improve the epitaxy by reducing the La content. Eventually I tried using pure $SrMnO_3$, although I could not find any information about its bulk lattice constants. I e-mailed to several expert friends

Fig. 3 April 21, 2006 at the house of Leilani and Steve Conradson (*standing left*, with their son in between): the celebration of Alex's 79th birthday. *Seated, left to right*: Davor Pavuna, Takeshi Egami, Alex, and Hugo Keller. *Standing right*: Natasha and Ivan Bozovic. Missing in the picture is Dragan Mihailovic who took this photo

and two, Ted Geballe and Darrell Schlom, responded promptly that $SrMnO_3$ is actually hexagonal. But thanks to the nine-hour time-zone difference the experiment was already underway and I actually observed excellent hetero-epitaxy. We synthesized $SrMnO_3$ in perovskite structure, pseudomorphic to LSCO, by virtue of epitaxial strain and stabilization. However, this perovskite $SrMnO_3$ did not transmit supercurrent, so we never published this result either, although I did show it at the MRS Meeting later the same year and couple other conferences. Subsequently other groups published the same result. We followed this line of investigation by trying other candidate compounds and eventually succeeded to demonstrate GPE using underdoped LSCO as the barrier material [57, 58].

Recently, Alex has suggested to me another exciting idea that I cherish and indeed intend to put to experimental test as soon as possible. Maybe this time I will be luckier – future will tell.

Acknowledgements This work was supported by DOE (Contract MA-509-MACA).

References

1. Bednorz JG, Mueller KA (1986) Z Phys B 64:189
2. Mueller KA (1999) J Supercond 12:3
3. Mueller KA (2002) Physica Scripta 102:39
4. Wigner E, Huntington HB (1935) J Chem Phys 3:764
5. Abrikosov AA (1954) Astron Zh 31:112
6. Aschcroft NW (1968) Phys Rev Lett 21:1748
7. Browman EG, Kagan Yu, Kholas A (1972) Sov Phys JETP 35:783
8. Nellis WJ (2006) Rept Prog Phys 69:1479
9. Overhauser AW (1987) Phys Rev B 35:411
10. Selvam P, Yvon K (1989) Phys Rev B 39:12329
11. Bell NA, Coates GE (1968) J Chem Soc A 1968:628
12. Press MR, Rao BK, Jena P (1998) Phys Rev B 38:2380
13. Yu R, Lam PK (1988) Phys Rev B 38:3576
14. Martins JL (1988) Phys Rev B 38:12776
15. Khowash PK, Rao BK, McMullen T, Jena P (1997) Phys Rev B 55:1454
16. Seel M (1991) Phys Rev B 43:9532
17. Seel M, Kunz AB, Hill S (1989) Phys Rev B 39:7949
18. Pauling L (1990) Proc Natl Acad Sci USA 87:244
19. Bastide JP (1990) Solid State Comm 74:355
20. Contreras S, Lucas R, Bastide JP, Claudy P, Escorne M, Comptes Rendus de L (1997) Academie des Sciences Serie II Fascicule B-Mecanique Physique Chimie Astronomie 324:641
21. Souw V et al. (2002) Phys Rev B 65:094510
22. Edwards B, Ashcroft NW (1997) Nature 388:652; see also Edwards PP, Hensel F, ibid, p 621
23. Ashcroft NW (2004) Phys Rev Lett 92:187002
24. Feng J, Grochala W, Jaron T, Hoffmann R, Bergara A, Ashcroft NW (2006) Phys Rev Lett 96:017006

25. Babaev E, Sudbo A, Ashcroft NW (2004) Nature 431:666
26. Engelman R (1972) The Jahn–Teller Effect in Molecules and Crystals. Wiley, London
27. Bozovic I, Delhalle J (1984) Phys Rev B 29:4733
28. Bozovic I (1984) Phys Rev B 29:6586
29. Bozovic I (1985) Phys Rev B 32:8136
30. Bozovic I (1985) Mol Cryst Liq Cryst 117:475
31. Vujicic M, Bozovic I, Herbut F (1977) J Phys A 10:1271
32. Bozovic I, Vujicic M, Herbut F (1978) J Phys A, p 2133
33. Bozovic I, Vujicic M (1981) J Phys A 14:777
34. Bozovic I, Bozovic N, Damnjanovic M (2000) Phys Rev B 62:6971
35. Karpfen A (1978) Theor Chim Acta 50:49
36. Armstrong D, Jamieson J, Perkins PG (1979) Theor Chim Acta 51:163
37. Brendel GJ, Marlett EM, Niebylski LM (1978) Inorg Chem 17:3589
38. Hock KH, Nickisch H, Thomas H (1983) Helv Phys Acta 56:237
39. Bozovic I (2001) IEEE Trans Appl Superconduct 11:2686
40. Job PK, Rao KS, Srinivasan M (1983) Nucl Sci Eng 84:293
41. Subbarao K, Srinivasan M (1983) Nucl Technol 49:315
42. Vandam H, Deleege PFA (1987) Annals of Nuclear Energy 14:369
43. Zaluska A, Zaluski L, Strom-Olsen JO (2000) J Alloys and Compounds 307:157
44. Damascelli A, Hussain Z, Shen Z-X (2003) Rev Mod Phys 75:473
45. Plate M et al. (2005) Phys Rev Lett 95:77001
46. Hussey NE et al. (1996) Phys Rev Lett 76:122
47. Hussey NE et al. (2003) Nature 425:814
48. Alexandrov AS, Mott N (1995) Polarons and Bipolarons. World Scientific
49. Alexandrov AS (2003) Theory of Superconductivity: From Weak to Strong Coupling. IOP publishing
50. Andreev AF (2004) JETP Lett 79:88
51. Andreev AF (2006) J Supercond and Novel Magnetism 19:181
52. A Shengelaya et al. (1996) Phys Rev Lett 77:5296
53. Zhao GM et al. (1996) Nature 381:676
54. Coey JMD, Viret M, von Molnar S (1999) Advances in Physics 48:167
55. Bozovic I et al. (2002) Phys Rev Lett 89:107001
56. Bozovic I et al. (2003) Nature 422:873
57. Kasai M et al. (1990) Jap J Appl Phys 29:L2219
58. Kasai M et al. (1992) J Appl Phys 72:5344
59. Bozovic I et al. (2004) Phys Rev Lett 93:157002

Chu CW (2007) In: Bussmann-Holder A, Keller H (eds)
High T_c Superconductors and Related Transition Metal Oxides.
Springer-Verlag Berlin Heidelberg, 57–73
DOI 10.1007/978-3-540-71023-3

The Competing Interactions in Multiferroics and Their Possible Implications for HTSs

C. W. Chu

TCSUH, University of Houston,
Hong Kong University of Science and Technology
& Lawrence Berkeley National Laboratory, USA
cwchu@ust.hk

Abstract Multiferroics exhibit multiple competing interactions and display concurrently various ground states, e.g. the coexistence of magnetism and ferroelectricity. Studies by us and others show that ferroelectricity in multiferroics can be induced by magnetism, pressure or an external magnetic field. It has also been shown recently that superconductivity can coexist with ferromagnetism and magnetic field can induce superconductivity. This raises the question if superconductivity can evolve from ferroelectricity directly or indirectly through magnetism. Since multiferroics belong to the class of highly correlated electron material systems which posses transition temperatures up to above room temperature, it is conjecture that it may not be impossible to achieve superconductivity with a very high T_c by optimizing the various interactions present.

Introduction

In late February 1987, after the Symposium on the Frontier of Science at the National Academy of Sciences in Washington DC, Alex Mueller mentioned to me that it was ferroelectricity that started him on the exciting and fruitful journey of high temperature superconductivity after attending an international conference on ferroelectricity in Porto Rico in 1974. The seminal discovery of high temperature superconductivity (HTS) in perovskite-like copper oxides in 1986 by Alex Mueller and George Bednorz [1] has demonstrated that HTS exists in oxides with crystal structures similar to those where ferroelectricity often occurs. It is known that ferroelectricity is characterized by the presence of an electric polarization, resulting from the absence of spatial inversion symmetry or the existence of a noncentrosymmetry, a character that a metal and a superconductor cannot support. Therefore the exclusion of superconductivity from ferroelectricity usually is expected, as was evidenced in 1964 by the observation of superconductivity in WO_3 upon doping with alkaline metals after making it metallic and suppressing the ferroelectric state [2]. The failure of subsequent studies to detect ferroelectricity in the metallic HTS cuprates appears to be not surprising, but the absence of ferroelectricity in the insulating parent compounds of HTS does not seem to be so obvious [3]. The interplay between superconductivity and ferroelectricity remains an open question. The low temperature semiconducting behavior detected in the 214 LSCO after the suppression of superconductivity by a strong magnetic field [4] and the absence of spatial inversion crystal symmetry recently observed [5] in metallic compounds such as $CePt_3Si$, UIr, $CeRhSi_3$, and Li2(Pd1-xPtx)B suggest that the exclusion of superconductivity from ferroelectricity may be worth re-visiting. Study of it can yield important insight into the interplay of the two phenomena. Multiferroics may be the proper material system for the study. It should be noted that back in 1964, Ginzburg already predicted [6] a possible transition directly from an insulating (or more exactly semiconducting) state to a superconducting state.

In this paper, I would like to present some of our resent observations on multiferroics that show a possible similarity between the ferroelectricity-magnetism coexistence and the superconductivity-magnetism coexistence by the application of pressure. Whether such a possible similarity can be translated to the coexistence of ferroelectricity and superconductivity under pressure (at least above the superconducting transition temperature T_c) or in a magnetic field is yet to be determined. If it is proven, to achieve a higher T_c does not appear to be impossible.

Multiferroics

The discovery of high temperature superconductivity (HTS) in perovskite-like copper oxides in 1986 by Alex Mueller and George Bednorz [1] rekindled immense interest in the study of other perovskite and perovskite-like oxides, the most abundant inorganic compounds on earth, leading to the discovery or rediscovery of colossal magnetic resistance, magnetic-field induced metal-insulator transition and magnetoelectric effect, among others. The oxides where the above effects, including the high temperature superconductivity, are detected consist invariably of transition metal elements with unfilled d-shells. These transition metal oxides thus form an interesting material class of their own that exhibits a myriad of ground states, different kinds of interactions and vast types of orders with a wide range of transition temperatures.

The study of the competition and coupling of various interactions and different orders in single phase solids has long fascinated condensed matter physicists in exploring the nature of the ground states. Perovskite and perovskite-like transition metal oxides provide an ideal environment for such studies. For example, competitions between and coexistence of superconductivity with ferromagnetism, antiferromagnetism, charge density waves, spin density waves, Mott insulator, charge ordered insulators etc. have extensively been carried out in this class of materials in recent years. As mentioned earlier that the interplay between superconductivity and ferroelectricity is still an open question. Knowing that a metal cannot support the required electric polarization for a ferroelectric and that superconductivity is synonymous with an ideal metal with zero resistivity, we have decided to study the superconductivity-ferroelectricity interplay indirectly by examining possible coupling first between ferroelectricity and magnetism (and later couplings) in the special group of perovskite-like transition metal oxides under high pressures (and later between magnetism and superconductivity).

This special group of perovskite-like transition metal oxides is known as multiferroics, a name coined by Schmid [7] for materials that possess simultaneously two or more of the ferroic properties: namely ferromagnetic, antiferromagnetic, ferrimagnetic, ferroelectric, antiferroelectric, ferrielectric ferroelastic, ferroplastic and ferrotoroidic, all in one phase. The simultaneous occurrence of a spontaneous magnetization and a spontaneous polarization may give rise to the magnetoelectric effect which allows the control of the polarization by en external electric field or the alteration of the maganetization by an external electric field and thus adds an extra degree of freedom to memory device design. The study of multiferroics is of both fundamental and device interests. However, magnetism is known to arise from the exchange interaction between the *d*-electrons in the transition element which tends to prevent the cooperative atomic displacements needed for ferroelectricity from taking place. Therefore multiferroic materials are few in nature. Initial search for multiferroic materials and the magnetoelectric effect associated was done mainly in the former Soviet Union in the 50 and 60s [8]. Unfortunately, the interest waned due to the limitation of materials and the smallness of the effect. However, the situation has drastically changed in recently years and a startling revival of interest in multiferroics is taking place. This is due to the observation of giant magnetoelectric effect, the detection of complex phase diagrams induced by magnetic and/or electric field, the discovery of more multiferroics, the emergence of spintronics, the progress made in compound and composite material synthesis and characterization, and advancement in theoretical understanding. As a result, intensive multiferroics have become one of the emerging areas of research in the last few years [9]. In what folloes, I would like to show some of our recent results on multiferroics that are related to later discussions concerning their implications for HTS.

The Manganites $RMnO_3$ and RMn_2O_5 (R = Y or Rare Earth)

Manganites $RMnO_3$ (RMO113) and RMn_2O_5 (RMO125) with R = Y or rare earth account for the majority of the known multiferroics to date and are the most studied multiferroics. The recently discovered Kogame salt $Ni_3V_2O_8$ has also been extensively investigated.

$RMnO_3$

$RMnO_3$ forms in two crystal structures: hexagonal with $P6_3$ cm symmetry and orthorhombic with Pnma symmetry as the ionic radius of R increases. Both structure phases may exist for R lies in the region between Er and Y. The hexagonal phase usually becomes ferroelectric with a Curie temperature $T_c \sim 600-1000$ K and on further cooling antiferromagnetical with a Neel Temperature $T_N < 100$ K. Some Mn-ions undergo a spin-reorientation at $T_{sr} \sim 30-40$ K before they order antiferromagnetically below $T'_N \sim 10$ K. On the other hand, the orthorhombic phase exhibits an orbital order or Jahn-Teller distortion at a temperature above ~ 700 K and orders antiferromagnetically at a $T_N \sim 40-150$ K. At or below T_N, the compound can display incommensurate magnetic order, lock-in magnetic transition, and ferroelectricity. The orthorhombic phase, when doped with alkaline metal of Ca or Sr, becomes ferromagnetic. It shows the colossal magnetoresisitive (CMR) effect and exhibits an insulator-metal transition or charge order upon cooling.

Hexagonal $YMnO_3$ Showing the Definitive Indirect Coupling Between the Ferroelectric and Magnetic Orders: Although it had been realized that the magnetoelectic coupling between the magnetic and ferroelectric properties was of great interest both scientifically and technologically and experiments were carried out in the 60s [8], definitive evidence for such coupling was sparse at best and interest in the topic waned in the ensuing decades. Inspired by the observations of CMR effect [10] in the metallic orthorhombic manganites doped with alkaline metals and the expanding knowledge on manganites in general, we decided to search for evidence of the magnetoelectric coupling in the insulating hexagonal $YMnO_3$, which displays a ferroelectric $T_c \sim 914$ K and a low antiferromagntic $T_N \sim 80$ K. We have measured the dielectric and magnetic properties as a function of temperature both in the absence and presence of an external magnetic field. As shown in Fig. 1, an inverse *S*-shape anomaly in the dielectric constant ε is detected near T_N, an unambiguous indication of the coupling between the ferroelectric and magnetic orders [11]. A small but distinctive magnetoelctric effect is also detected in the presence of a 5 T field, as evidenced by a slight suppression ofε.

In the absence of an externally magnetic or electric field, the ferroelectric and magnetic orders in a multiferroic material compound may couple via the secondary lattice effect. The observed ε- anomaly can be attributed to the lattice change associated with the antiferromagnetic transition at T_N. Phenomenologically, the coupling between the polarization (*P*) and magnetization (*M*) in the presence of total electric (*E*) and magnetic (*H*) fields can be described by a magnetoelectric susceptibility tensor X, i.e. $P = X^{ee}E + X^{em}H$ and $M = X^{me}E + X^{mm}H$. It is clear that the off-diagonal tensor elements ($X^{me} = X^{em}$) provide the coupling. By assuming that T_c and T_N are close, the Landau theory gives a smooth change in ε at T_N in contrast to the observed *S*-anomaly. Since $M = 0$ for an antiferromagnet, the coupling observed must

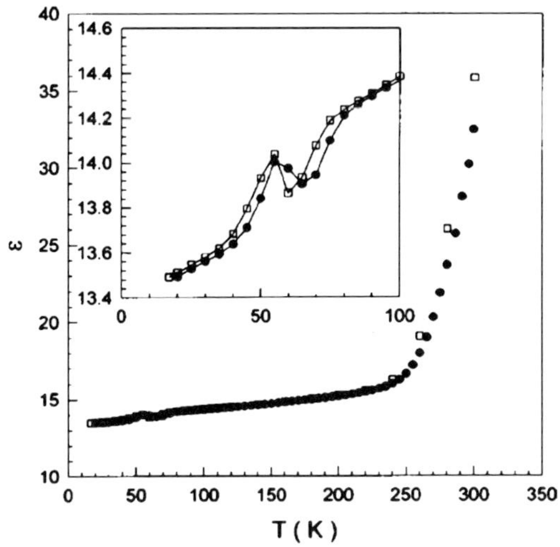

Fig. 1 $\varepsilon(T)$ at 75 ***MHz*** near T_M at (□) 0 ***T*** and (•) 5 ***T***

have arisen from the magnetostrictive effect as was demonstrated later by us through thermal expansion measurements as shown in Fig. 2 [12].

Hexagonal $HoMnO_3$ showing a strong magnetoelectric effect and a complex magnetic *T–H* phase diagram rich in physics: As was pointed out in the phenomenological analysis above, the ferroelectricity-antiferromagnetic coupling can happen only indirectly through the magnetostrictive effect associated with the antiferromagnetic transition at T_N [11]. Later the symmetry argument also demonstrates the impossibility of a direct coupling of the two orders [13]. Hexagonal $HoMnO_3$ is ferroelectric below 830 K and its Mn-sublattice exhibits an antiferromagnetic order below $T_N \sim 76$ K. A sharp Mn-

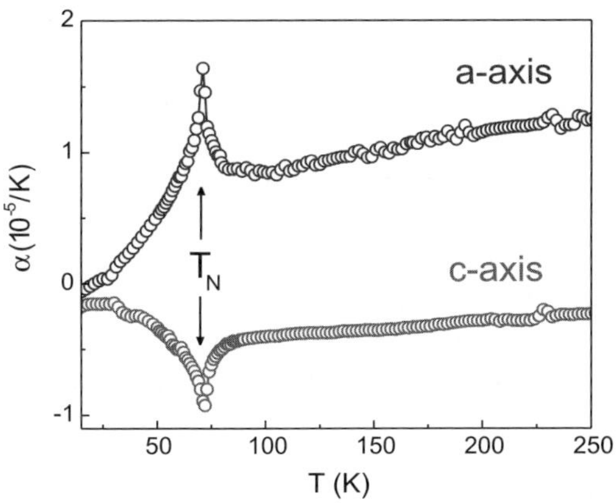

Fig. 2 Thermal expansivities of $YMnO_3$ along the hexagonal *a*- and *c*-axes

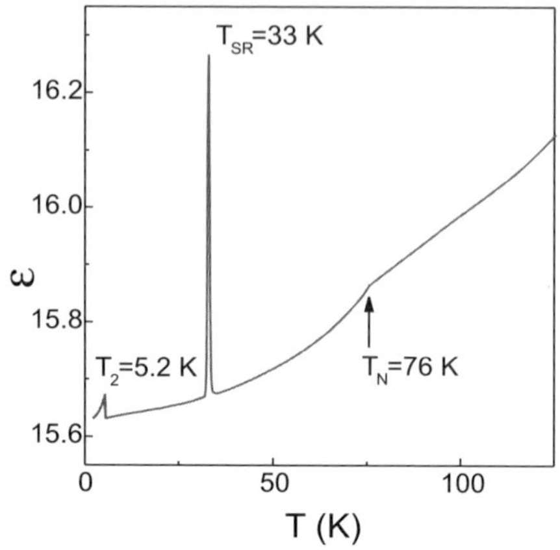

Fig. 3 ϵ(T) of $HoMnO_3$. The three magnetic phase transitions are labeled T_N, T_{SR}, and T_2

spin reorientation transition takes place at $T_{SR} \sim 33$ K together with another Mn-spin reorientation close to $T'_{SR} \sim 5$ K. It displays [13] distinct ε-anomalies at the various magnetic transient temperatures, T_N, T_{SR} and T'_{SR}, respectively, as shown in Fig. 3, suggesting a strong magnetic-ferroelectric coupling. The compound therefore offers an excellent opportunity for the investigation of the magnetoelectric effect. We have therefore measured [13] the magnetic field effect on the ε-anomalies up to 10 T down to 1.8 K. Figure 4 shows that the unexpected and complicated evolution of ε-peak at T_{SR} with H. Above 10 K, as H increases, the ε-peak at T_{SR} varies in the following ways: 1. its overall magnitude decreases and moves toward lower temperatures continuously; 2. a slanted plateau develops as defined by two Temperatures T_1 and T_2 with $T_{SR}(H = 0) > T_1(H) = T_{SR}(H) > T_2(H)$ at $H > 0$; 3. an additional slanted plateau appears between T_1 and T_2 with $T_2 > T_1$ (reason for the labelling will be evident in Fig. 5 later); 4. above ~ 3.5 T, the two plateaus merge into one and 5. above ~ 4 T, the ε-anomaly disappears above 10 K. Below 10 K, similar but even more complex ε-behavior is observed [14]. Based on these $\varepsilon(T, H)$ results, a complex phase diagram rich in physics is constructed [14] as shown in Fig. 5. The phase diagram has also been confirmed by anomalies in our magnetic, calorimetric and dilatometric measurements. It shows several multicritical points wherever different phase boundaries meet. For example, two tricritical points are located at the intersection of $T_5(H)$ with $T_2(H)$ and $T_3(H)$, respectively. Below 3 K and near 2 T, three phase boundaries $T_1(H)$, $T_3(H)$ and $T_4(H)$ may merge and give rise to a tetracritical point close to 0 K. The hexagonal HMO113 indeed provides a rich environment for the study of the coupling of the various interactions involved in the multiple phase transitions in the compound.

Orthorhombic $YMnO_3$ and $HoMnO_3$ demonstrating the induction of ferroelectricity by magnetism, the direct coupling between the ferroelectric

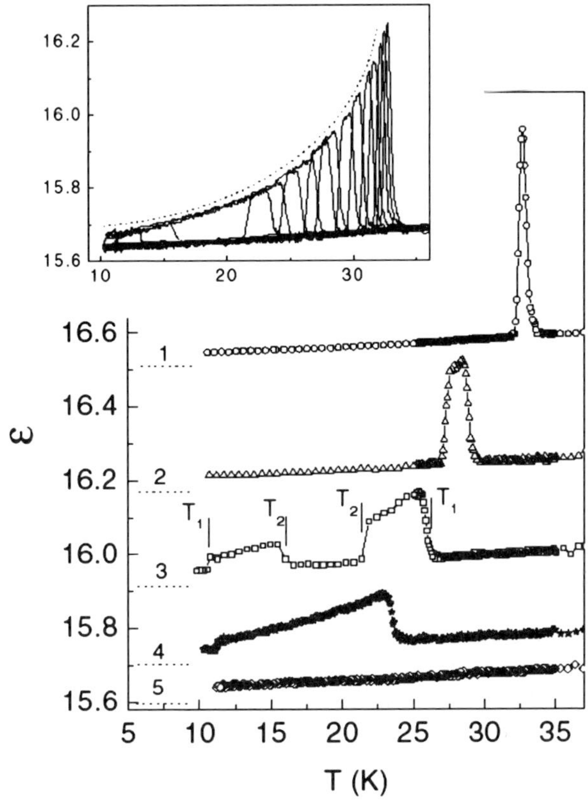

Fig. 4 $\varepsilon(T)$ for selected external magnetic fields H, (1) $H = 0$, (2) $H = 2.6$ T, (3) $H = 3.3$ T, (4) $H = 3.7$ T, (5) $H = 4.1$ T. Different curves are offset by a constant (indicated by *dotted lines*). T_1 and T_2 are marked by *vertical bars* next to curve 3. *Inset*: all data plotted on the same scale

and the magnetic orders, and the rare-earth dependent magnetic field effect on ferroelectricity: Due to the ionic size of Y and Ho, $YMnO_3$ (YMO113) and $HoMnO_3$ (HMO113) crystallize in the hexagonal form when they are synthesised at ambient pressure. They become ferroelectric well above room temperature and order antiferromagnetically at $T_N \sim 75$ K. However, the metastable orthorhombic YMO113 and HMO113 can be stabilized by high pressure synthesis. Both compounds undergo an antiferromagnetic transition with $T_N \sim 42$ K to an incommensurate phase, followed by another magnetic transition at a lower temperature $T_L = 26–28$ K to another incommensurate state with a temperature-independent modulation vector for YMO113 or to a commensurate E-type order for HMO113 [15]. It will be interesting to determine the role of symmetry in the occurrence of ferroelectricity if ferroelectricity is indeed detected in the orthorhombic $YMnO_3$ and $HoMnO_3$ albeit at a much lower temperature than their hexagonal allotropes. Earlier,

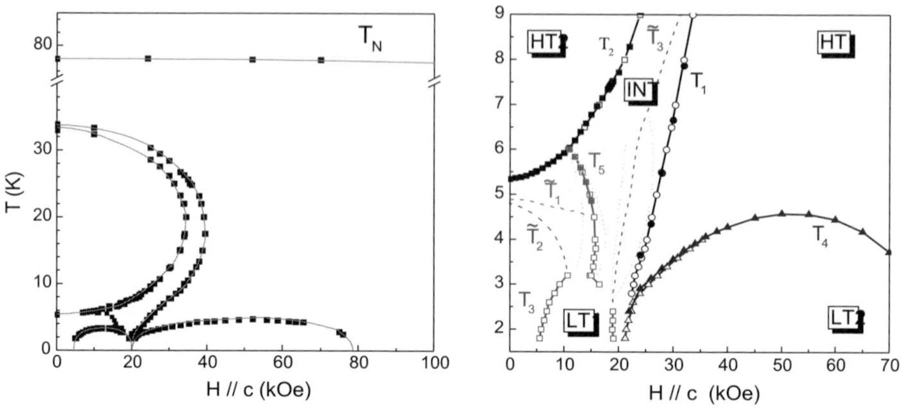

Fig. 5 Complex magnetic high-field phase diagram of $HoMnO_3$. *Right panel*: details at low temperature. Phase boundaries are labeled T_1 to T_5. Additional anomalies of physical properties are observed at $\widetilde{T}_1$ to $\widetilde{T}_3$

we found [16] a rapid $\sim 60\%$ ε-increase accompanying the antiferromagnetic transition of the two compounds at $T_L \sim 26$–28 K, as shown in Fig. 6, suggesting the existence of a strong magnetoelectric effect and a possible concurrent ferroelectric transition in these compounds. The figure also shows a large field influence on the ε-anomaly of HMO113 but not YMO113. It should be noted that the insulating orthorhombic RMO113 when doped by alkaline element becomes metallic below room temperature and exhibits the CMR effect [10]. Aside from learning the role of crystal symmetry, the study will also

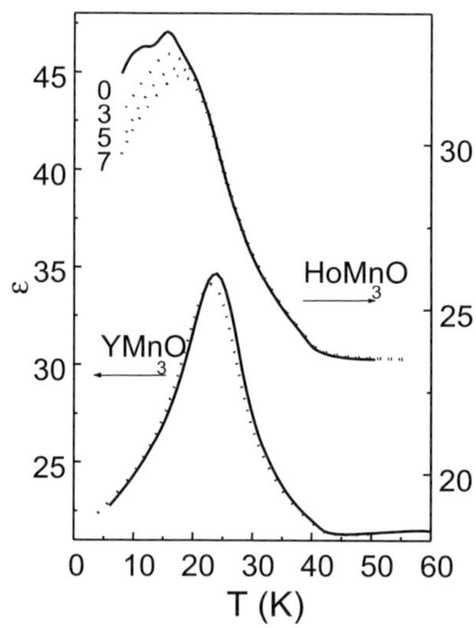

Fig. 6

provide insight to the magnetic effects of the Mn-spin and magnetic moment associated with R in the occurrence of ferroelectricity and the evolution of ferroelectricity to metal via doping. We have therefore decided to search for ferroelectricity in YMO113 and HMO113 by measuring their polarization by integrating the pyroelectric current with time in different magnetic fields.

The temperature dependences of the spontaneous polarization P of polycrystalline YMO113 and HMO113 are displayed [17] in Fig. 7. It is evident that the P of YMO113 starts to rise rapidly at ~ 28 K, signally the onset of a ferroelectric transition, P is independent of H up to 7 T, consistent with the ε-data shown in Fig. 6. The transition coincides with the incommensurate-to-incommensurate antiferromagnetic transition at T_L. For HMO113, the P starts to grow but slowly initially at $T_L \sim 26$ K, the temperature for the incommensurate-to-commensurate antiferromagnetic transition. However, it increases rapidly below $T_{NC} \sim 15$ K, where the effective Ho-moment increases drastically due to its rotation in the *a*–*b* plane and formation of a noncolinear magnetic structure. A magnetic field suppresses P drastically, especially below T_{NC} reflecting the similar ε-behavior in the presence of an external field. The results clearly demonstrate that ferroelectric transition is triggered by the onset of the antiferromagnetic transition from incommensurate to incommensurate or commensurate of the wave vector, and helped by the effective magnetic moment of R. The alignment of the magnetic moment by the external field also has a positive influence on P. All of the above point to con-

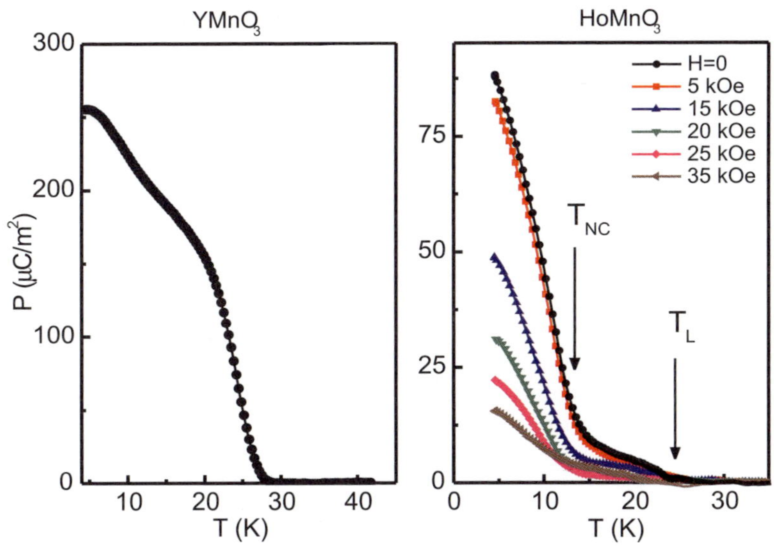

Fig. 7 Ferroelectric polarization of orthorhombic $YMnO_3$ (left) and $HoMnO_3$ (right) as obtained from pyroelectric current measurements. The polarization of $HoMnO_3$ decreases quickly in external magnetic fields

clusion that magnetism can create ferroelectricity under proper condition. While several recent microscopic models based on the symmetry argument of the frustrated Mn-ions can account for the simultaneous appearance of the ferroelectric and the antiferromagnetic transitions [18], detailed mechanism has yet to be unraveled.

RMn_2O_5

Another class of rare-earth manganite multiferroics that has attracted great attention recently is RMn_2O_5 (RMO125). RMO125 displays a complex orthorhombic structure (Pbam) with edge-sharing $Mn^{4+}O_6$ octahedra forming ribbons along the *c*-axis and a pair of $Mn^{3+}O_5$ pyramids connected at a common edge of their bases, bridging the adjacent ribbons of octahedra in the *a–b* plane [19]. The complex structure gives rise to competitions of various magnetic interactions associated with the spin frustration. As a result, RMO125 exhibits a complex magnetic phase upon cooling. Some of these magnetic transitions remove the inversion symmetry and lead to accompanying ferroelectric transitions [20]. We have therefore decided to examine the spin lattice couplings in and the magnetic field effects on these transitions in RMO125 with R = 0 y, H_0 and T_b, and especially on the ferroelectric transition.

The cascade of phase transitions: The sequence of transitions is exemplified by that of $DyMn_2O_5$ as evidenced by the ε-anomalies [21] in Fig. 8: on cooling, an antiferromagnetic order develops at $T_{N1} = 43$ K with an incommensurate magnetic modulation, followed by a lock-in transition at $T_{C1} = 39$ K into a commensurate phase; on further cooling, a spin reorientation takes place at $T_{N2} = 27$ K, accompanied by another spin reorientation at $T_{C2} = 13$ K; and the sequence is completed by a transition back to the incommensurate magnetic modulation at $T_{C3} = 6.5$ K when Dy orders antiferromagnetically. A ferroelectric order occurs concurrently with the incommensurate-commensurate magnetic transition at $T_{C1} = 39$ K and the simultaneous po-

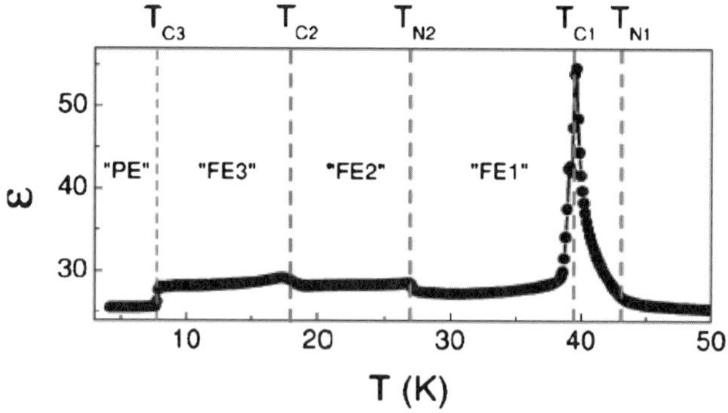

Fig. 8

larization changes as one crosses the other phase boundaries at T_{N2}, T_{C2} and T_{C3}. Below T_{C3} when Dy orders antiferromagnetically, the polarization drops to close to zero or DMO125 may have re-entered to the paraelectric state.

The strong spin-lattice anisotropic coupling: It has been proposed [20] that ferroelectricity observed in RMO125 originates from the competing interactions of the Mn^{3+}/Mn^{4+} spins, the magnetic moments of R and the lattices. It is conjectured that the geometric magnetic frustration among the Mn^{3+}- Mn^{4+} spins gives rise a degeneracy of the magnetic ground state that can be lifted by a distortion of the lattice. A strong spin-lattice coupling is therefore expected for the ionic displacement involved in the creation of a ferroelectric state and or the change of the ferroelectric polarization. To search for the effect, we have measured the thermal expansion coefficients along the three crystal axes of RMO125 with R = Ho, Dy and Tb by a home-made high resolution dilatometer. Also determined are the ε along the b-axis and the specific heat. All measurements share similar but not identical features for the three compounds [23]. The subtle difference in changes of the thermal expansion coefficients along different directions depends on the detailed magnetic structure of the magnetic phases. Typical data are shown [23] in Fig. 9 for

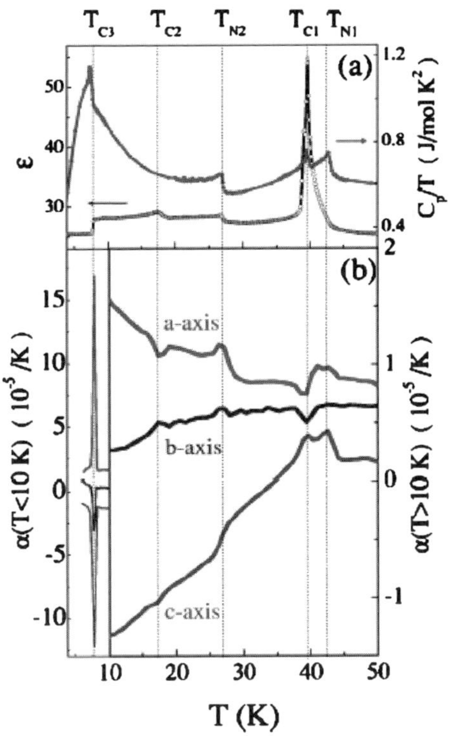

Fig. 9

DMO125, consistent with our magnetic and calorimetric data. Strong lattice anomalies clearly are evident, confirming the important role of spin-lattice coupling present in RMO125 multiferroics.

The large magnetoelastic effect creating or destroying ferroelectricity: Once the strong spin-lattice coupling arising from the internal magnetic field associated with specific spin configurations in RMO125 is proven by the above thermal expansion measurements, it is reasoned that a strong magnetoelastic effect caused by an external field should exist. We have therefore investigated the magnetoelastic strain on the lattice along the three crystal directions of DMO125 at low temperature and found [23] large changes in the lattice parameters as a function of the field and abrupt striction effects crossing the phase boundaries as shown in insets a–c of Fig. 10. The low temperature magnetic phase diagrams based on the magnetoelastic anomalies detected are constructed and shown in insets d–f of the same figure. Due to the anisotropic magnetic structures of DMO125, the H–T phase diagram clearly is different from the others. The ferroelectric transition temperatures at T_{C2} and T_{C3} are

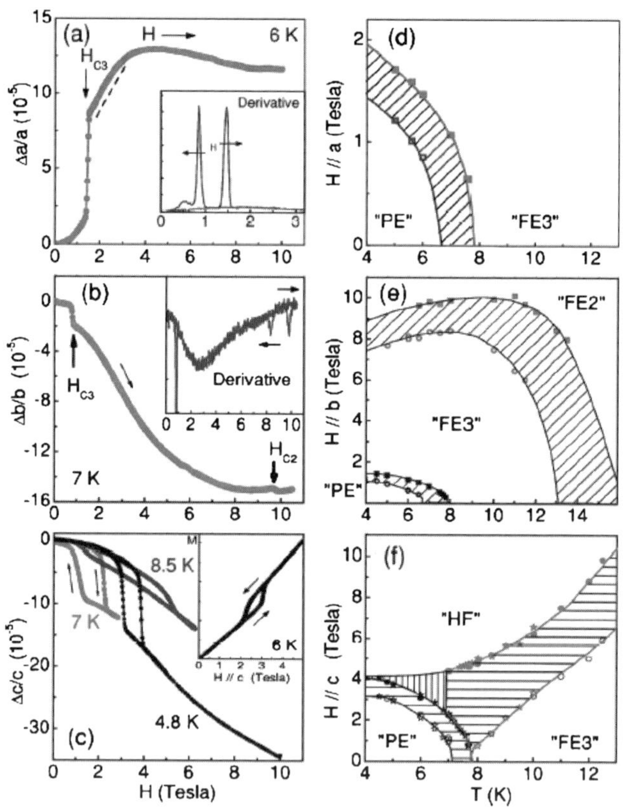

Fig. 10

suppressed completely by magnetic fields along the *a* and *b*-directions, while a new high field phase (HF) is created but only in fields along the *c*-direction. Recently our polarization measurements show that the both the low field phase labeled as PE and the high field phase (HF) are paraelectric. The results demonstrate that while magnetism can create ferroelectricity, it can also destroy ferroelectricity. The observation shows the intricate interaction between magnetism and ferroelectricity.

The unusual pressure effect enhancing or suppressing ferroelectricity: In the study of high temperature superconductivity, pressure has played an important role. While the pressure effects on magnetism has been studied extensively and on ferroelectricity to a certain extent, pressure effects on the ferroelectric and magnetic orders in multiferroics have only been investigated sparsely at best. Recently, we have examined the influence of pressure on the stability and the polarizations of different ferroelectric phases of RMO125 with R = Ho, Dy and Tb. All T-P phase diagrams obtained are similar except that DMO125 shows an extra low temperature phase [24]. Figure 11 shows only the phase diagram of DMO125 for visual clarity. The phase diagrams for all RMO125 studied share several common features, i.e. the pressure effects on T_{N1} and T_{C1} are positive, on T_{N2} is negative, these transition temperatures change only slightly with pressure but the ferroelectric polarization varies greatly, e.g. the FE3 phase is suddenly removed beyond a critical pressure resulting in a large pressure-induced increase of the ferroelectric polarization. For DMO125, a new phase was induced at low temperature by pressure. The sudden change of the polarization without large variation in the transition temperature suggests that the magnetic structure effect on the crystal symmetry of RMO215 must be very subtle and the creation or destruction

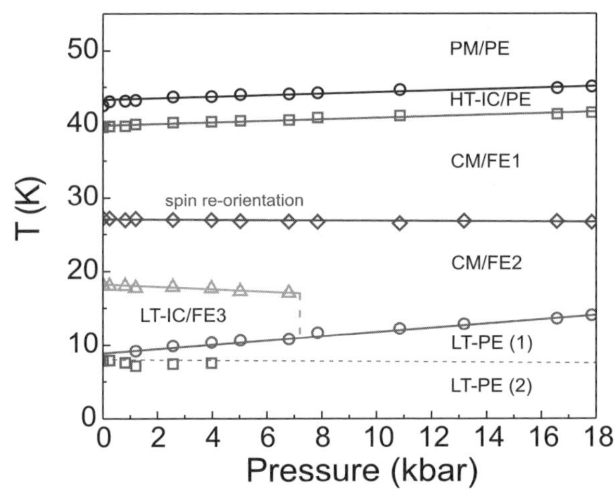

Fig. 11 Pressure-temperature phase diagram of $DyMn_2O_5$

of ferroelectricity can be achieved more easily by modifying the magnetic structure through physical or chemical means.

Superconductivity from Ferroelectricity Directly or Indirectly via Magnetism?

The study of multiferroics by us and others demonstrated that there exist multiple ground states in the materials through the delicate interplay of the magnetism and ferroelectricity. In contrast to original thinking that magnetism acts against ferroelectricity, multiferroics show that both orders can coexist and couple to each other, creating many interesting ground states. Pressure and magnetic field are found to be able to tune-in or tune-out a specific ground state of ferroelectricity. The situation is similar to the early belief that magnetism excludes superconductivity, and now magnetism is found to play an important positive role in high temperature superconductivity. Additionally, pressure and magnetic field have been used successfully to create or destroy superconductivity [25]. It has been common knowledge that the electric dipole moment associated with ferroelectricity can only be supported by an insulator with a non-centrosymmetry and not by a metal due to Coulomb interaction. However, in 1995 localization of electrons has been observed [4] in high temperature superconductors at low temperature when superconductivity is suppressed by a magnetic field (Ginzburg already predicted a direct semiconductor-to superconductor transition in 1964) and recently superconductors with a non-centrosymmetry have been discovered [5]. In view of the above, perhaps it is the right time to explore the possibility to achieve superconductivity or to enhance superconductivity from ferroelectricty directly or indirectly through magnetism, utilizing physical means such as pressure, electric field and magnetic field or chemical means such as doping. To turn a ferroelectric WO_3 into a superconductor has long been achieved [2] by Matthias in 1964 by alkaline element doping, unfortunately, the maximum transition temperature obtained is 6 K. A systematic study on effects of doping on the symmetry, ferroelectricy and transport properties of WO_3 may provide important insights for searching for a new way for superconductivity.

As mentioned at the beginning, multiferroics, similar to high temperature superconductors, belong to the interesting material class of transition metal oxides which exhibit a myriad of ground states with transition temperatures close to 0 K to above 1000 K. It seems that competition between any two leads to the stabilization of one at the expense of the other. However, in a real material, more than two interactions exist and can act cooperatively. By adjusting the individual interactions, one may be able to obtain an effective interaction optimal for the occurrence of a desired ground state. For example, according to the BCS theory, strong electron-phonon interaction and a large electron density of states are needed for a high T_c. However, a strong electron-phonon interaction and high density of states often result in lattice instabilities, leading to a structural deformation or a collapse of the lattice [26] by adjusting the two, one may obtain a maximum T_c. This is exemplified by the T–P phase diagrams of V_3Si [27, 28] and Nb_3Sn [29], the conventional high temperature intermetallic superconductors, shown in Fig. 12. The high electron-phonon interaction together with the relatively high density of states triggers a lattice instability manifested by a martensitic transformation at $T_L > T_c$. Pressure enhances the T_c while suppresses T_L for V_3Si. The opposite is true for Nb_3Sn. Since the superconducting state benefits from the soft phonon effect close to

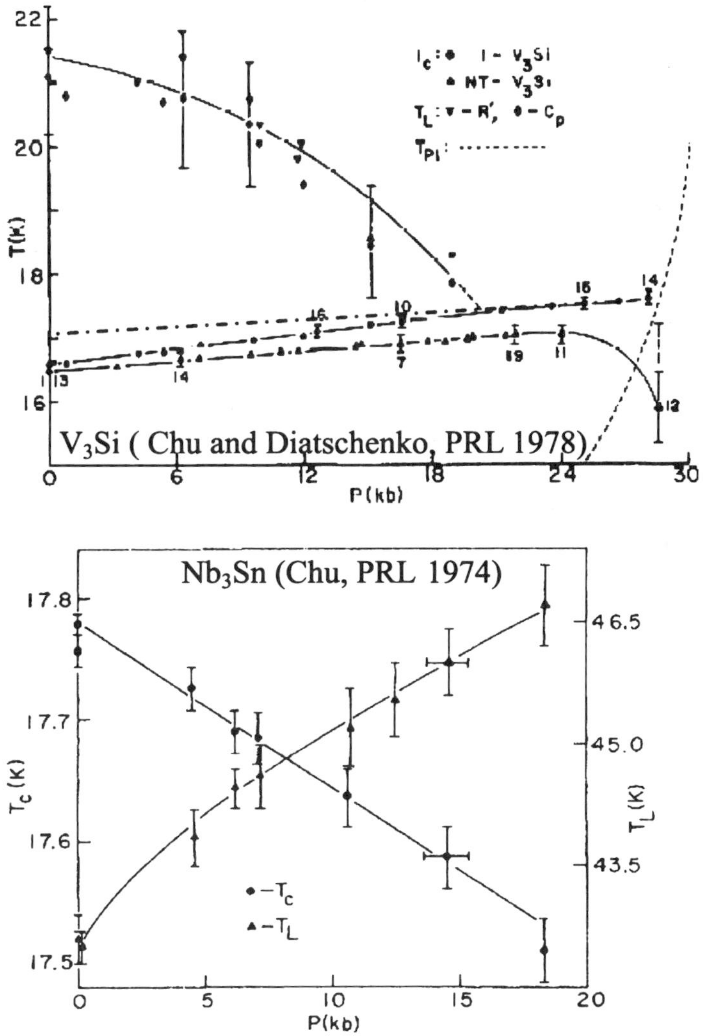

Fig. 12 A15 conventional high T_c superconductors

the lattice transformation, T_c is enhanced as T_L is pushed closer to T_c for V_3Si and T_c is suppressed as T_L is pushed away from T_c for Nb_3Sn, demonstrating the important role of optimization, instead of maximization, of the interactions involved in raising T_c. T_c is maximum when it crosses or ends with T_L at a critical pressure. The overall effect can be represented by a phase diagram that resembles, to a certain extend, the generic T_c-hole concentration phase diagram for the cuprate HTSs with the pseudogap temperature T_{ps} replacing T_L, if the T_{ps} crosses or ends with T_c (a point presently remains unsettled), one may conclude that whatever the driving mechanism for the opening of a pseudogap is the same as that for HTS. However, if T_{ps} merges with T_c and never ends, different conclusion needs to be drawn.

Given the wide range of transition temperatures of the different ground states in the highly correlated electron oxide systems of transition metal elements, it may not be impossible to achieve superconductivity though a novel channel by optimizing the various interactions present in the materials and to obtain a T_c high than the present record of 134 K at ambient and 164 K under pressure in $HgBa_2Ca_2Ca_3Ca$.

Acknowledgements The author would like to thank his colleagues Bernd Lorenz, Clarina dela Cruz, Fei Yen and Yaqi Wang for the work and valuable discussions on multiferroics. This work in Houston is supported in part by AFOSR Award No. FA9550-05-1-0447, the T. L. L. Temple Foundation, the John J. and Rebecca Moores Endowment, the Strategic Partnership for Research in Nonotechnology through AFOSR Award No. FA9550-06-01-0401 and the State of Texas through TCSUH; and at Lawrence Berkeley National Laboratory by the Director, Office of Science, Office of Basic Energy Sciencesm Division of Materials Sciences and Energy of the Department of Energy.

References

1. Bednorz JG, Müller KA (1986) Z Phys B64:189
2. Raub C et al. (1964) Phys Rev Lett 13:746
3. Chu CW et al. unpublished
4. Ando Y et al. (1995) Phys Rev Lett 75:4662
5. Bauer E et al. (2004) Phys Rev Lett 92:097001
6. Ginzburg VL, Kirzhnitz DA (1964) Sov Phys JETP 19:269
7. Schmid H (1994) Ferroelectrics 136:95
8. Bokov VA, Myl'ikova IE, Smolensk GA (192) Sov Phys JETP 15:447
9. Hill NA (2000) J Phys Chem B 104:6694
10. Huang HY et al. (1995) Phys Rev Lett 75:914
11. Huang ZJ et al. (1997) Phys Rev B 56:2623
12. dela Cruz CR et al. (2005) Phys Rev B 71:060405(R)
13. Lorenz B et al. (2004) Phys Rev Lett 92:087204-1
14. Lorenz B et al. (2005) Proc ACERS Annual Meeting, April 2005
15. Muñoz A et al. (2001) Inorg Chem 40:1020
16. Lorenz B et al. (2004) Phys Rev B 70:212412
17. Lorenz B et al. preprint

18. Sergienko IA et al. (2006) cond-mat/0608025
19. Prellier W et al. (2005) J Phys Cond Matt 17:R803
20. Blake GR et al. (2005) Phys Rev B 71:214402
21. dela Cruz CR et al. (2006) Phys Rev B 73:100406(R)
22. Hur N et al. (2004) Phys Rev Lett 93: 107202
23. dela Cruz CR et al. (2006) Phys Rev B 74:180402(R)
24. dela Cruz CR (2006) PhD Thesis
25. Mackenzie AP, Grigera SA (2005) Science 309:1330
26. Chu CW (1978) In: Chu CW, Woollam JA (eds) High Pressure and Low Temperature Physics. Plenum Press, New York, p 359
27. Chu CW, Testardi L (1974) Phys Rev Lett 32:766
28. Chu CW, Diatchenko V (1978) Phys Rev Lett 41:572
29. Chu CW (1974) Phys Rev Lett 33:1283

Conder K (2007) In: Bussmann-Holder A, Keller H (eds)
High T_c Superconductors and Related Transition Metal Oxides.
Springer-Verlag Berlin Heidelberg, 75–84
DOI 10.1007/978-3-540-71023-3

Site-Selective Oxygen-Isotope Exchange in $YBa_2Cu_3O_{7-x}$

K. Conder

Laboratory for Developments and Methods, Paul Scherrer Institute, 5232 Villigen PSI, Switzerland
kazimierz.conder@psi.ch

Abstract Kinetic aspects of the oxygen-isotope exchange in $YBa_2Cu_3O_{7-x}$ (Y123) are discussed. In the structure of Y123 the CuO chains are the "fastest paths" for the ^{18}O diffusion, from which further exchange with the apical and planar sites takes place. Based on isothermal oxygen-isotope exchange experiments and an assumed diffusion model, oxygen-tracer diffusion coefficients in Y123 have been estimated. Total and site-selective oxygen-isotope effects on transition temperature T_c were measured. It was found that T_c in Y123 is mainly affected due to the oxygen-isotope substitution in CuO_2 planes.

Introduction

The discovery of superconductivity by K.A. Müller and G. Bednorz [1] in the La – Ba – Cu – O system in 1986, started an exceptional scientific activity in the new field of high-T_c superconductivity. Shortly after the first discovery, series of other superconducting compounds were synthesized. $YBa_2Cu_3O_{7-x}$ (Y123) obtained by Wu et al. [2] in 1987 was the first material having a transition temperature T_c higher than the boiling point of liquid nitrogen. For this compound, the highest transition temperature $T_c = 93$ K is reached for the composition $YBa_2Cu_3O_{6.92}$, whereas for $7 - x < 6.4$ the material is not superconducting. As the oxygen-content has a decisive influence on the properties, studies of the oxygen thermodynamics, oxygenation kinetics, and the oxygen ion ordering in the structure have been of the great interest. There is also another aspect which attracted our attention to the oxygen chemistry in the cuprates. For the conventional superconductors it is known, that the transition temperature T_c depends on the isotope mass of the elements building a crystal lattice. This relationship, ($T_c \propto M^{-\alpha}$, with $\alpha = 0.5$) was a strong experimental evidence for an electron-phonon interaction in the electron pairing mechanism within the framework of the BCS theory [3, 4]. Therefore, it was of particular interest to investigate an isotope effect in the cuprates. Especially, a replacement of the ordinary oxygen-isotope (^{16}O) with a stable isotope-18 results in a large relative change of the mass. Further, as the structures of the superconducting cuprates are relatively complex, there are different sites available for oxygen. Thus, in the Y123-structure there are three oxygen sites: in CuO_2 planes, in CuO chains and so known apical oxygen between planes and chains (Fig. 1a). To probe the role of the different oxygen

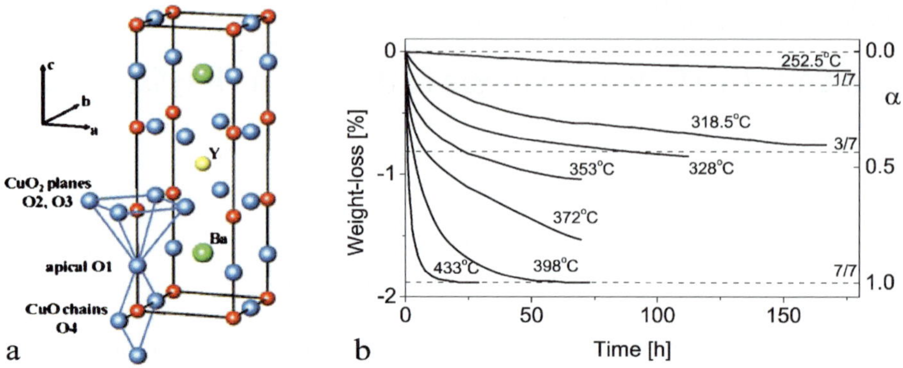

Fig. 1 **a** Crystal structure of $YBa_2Cu_3O_{7-x}$. In the unit cell, there are four oxygen sites in the CuO_2 planes (O2, O3), two apical sites (O1) and one CuO chain site (O4); **b** Weight-loss curves obtained during isothermal oxygen back-exchange in $YBa_2Cu_3{}^{18}O_{7-x}$. Parameter α ($0 \leq \alpha \leq 1$) describes a progress of the isotope exchange process

sites in the superconductivity mechanism, it was proposed by K.A. Müller in 1990 that one should measure oxygen-isotope effects (OIE) arising from diverse oxygen sites. The work presented here summaries our studies made in attempt to manufacture site-selectively oxygen-isotope substituted samples.

Oxygen Exchange Kinetics in $YBa_2Cu_3O_{7-x}$

The oxygen-content in a mixed valence oxide phase depends both on the thermodynamics (equilibrium) and kinetics of the reaction of the oxygen gas with the solid material. In an equilibrium situation, temperature and the oxygen partial pressure determine the oxygen-content in the sample. However, the equilibrium can sometimes be reached only after a very long time, and therefore, kinetic studies illuminating the reaction rate and mechanism are of a great interest. The experimental success of a site-selective substitution of the oxygen-isotope in Y123 required exact measurements of the tracer (isotope) oxygen diffusion. Such measurements are made at equilibrium conditions i.e. at constant temperatures and oxygen partial pressures so that, the overall oxygen-content in the sample stays constant and only one stable oxygen-isotope is exchanged with another one. Generally, in an oxygen exchange process, a gas (oxygen) transport to the surface of the solid, a surface reaction and an oxygen diffusion within the solid material have to be considered by an evaluation of the mechanism. It is commonly assumed that in the case of the oxygen exchange between the gas atmosphere and the solid Y123 the diffusion of oxygen in the solid phase is the step determining the rate of the whole process. The diffusion progress can be described by the Fick's laws. Typical solutions of the Fick's differential equations for various diffusion conditions and resulting concentration profiles of the diffusing specimen, can be found in the literature [5, 6]. In case of the oxygen diffusion during the isotope exchange the process is characterized by the tracer (self diffusion) coefficients. The oxygen-tracer diffusion in Y123 was intensively studied in many research groups [7–17]. Routbort and Rothman [18] reviewed the work on Y123, Y124, $La_{2-x}Sr_xCuO_4$ and $Bi_2Sr_2Ca_{n-1}Cu_nO_{2n+4}$. Different approaches can be found in the literature concerning investigation of the oxygen-tracer diffusion coefficient in Y123. Both tracer and chemical (oxygen diffusion during oxidation) coefficients have been calculated assuming pair interactions between oxygen sites in the structure [7, 8]. Among the experimental methods, the SIMS (Secondary Ion Mass Spectroscopy) technique was mostly used for the profiling of the tracer (isotope ^{18}O) distribution in the samples and further determination of the diffusion coefficient [9–11, 16, 17]. The same technique was used in investigations of the diffusion anisotropy in Y123 single crystals [9–15]. The oxygen self-diffusion coefficient was also determined from the measurements of internal friction [12, 13], weight change using thermogravimetry [14] and ionic conductivity [16, 19]. Unfortunately, the reported results of the oxygen-tracer diffusion coefficient in Y123 scatter by several orders of magnitude. One of the reasons for that huge discrepancy in the reported values is a strong anisotropy of the oxygen-tracer diffusion as reported by Rothmann et al. [9], Tsukui et al. [20], and Bredikhin et al. [11]. In these studies performed on single crystals, the tracer diffusion coefficients measured along the *b*-axis were 10^3–10^5 larger than those along the *c*-axis.

As we now know, the large oxygen-tracer diffusion anisotropy provides an amazing possibility of a site-selective oxygen-isotope substitution in Y123 structure. An exchange selectivity of the apical and chain oxygen sites was reported for the first time by Cardona et al. [21] in a partially exchanged $TmBa_2Cu_3{}^{18}O_{0.93}{}^{16}O_{6.07}$. The exchange process has been performed at 520 °C in Ar-$^{18}O_2$ atmosphere. For this material very small (≤ 0.1 K) oxygen-isotope effect was found. Nickel et al. [22] reported studies of synthesis and OIE measurements of oxygen-isotope site-selective substituted Y123 by an exchange in $^{18}O_2$ gas at 320 °C (32 h) on a thermobalance. Unfortunately, not sufficient reproducibility of the samples' properties has been obtained and therefore, the reported OIE's are not credible.

In order to explore a possibility of a site-selective oxygen-isotope substitution in Y123, we decided to perform systematic kinetic studies of the exchange process [23]. These were carried out during a back-exchange i.e. using already ^{18}O-substituted Y123 samples. Powder samples (average grain size about 30 μm) have been annealed on thermobalance at different temperatures in a stream of natural oxygen. Weight changes of the sample resulting from a replacement of the heavy isotope-18 with the light isotope-16 have been recorded. Thermogravimetric curves obtained during isothermal exchange experiments performed at selected temperatures for the same batch of the material are shown in Fig. 1b.

For the experiments performed at temperatures close to 400° and higher, a complete isotope exchange could be achieved in a reasonable time. At lower temperatures (see curves for 318.5 and 328 °C) the process slows down after the weight-loss reaches the value corresponding to the exchange of 3 of the total 7 oxygen sites. It is straightforward to believe that the remaining four are those from the CuO_2 planar sites (Fig. 1a). At the lowest investigated temperature (252.5 °C), it looks that only one oxygen site from each unit cell (O4, CuO chains) can be exchanged. Thus, depending on the temperature, all the oxygen sites, only apical and chain sites, or perhaps only chain sites (at the lowest temperature) can be substituted with an isotope. This can be understood assuming that the chains in the structure are the "fastest paths" for the ^{18}O diffusion, from which further exchange with the apical and planar sites takes place.

To verify the kinetic results, Raman measurements of the samples exchanged at different conditions have been performed [24]. In the Raman spectra (Fig. 2) the characteristic modes for the apical oxygen O1 at 502 cm^{-1} (O_a) and the planar oxygen O2, O3 at 340 and 435 cm^{-1} (O_p) are observed. Upon substitution with ^{18}O these bands are shifted to lower frequencies – compare spectra for sample **1** ($^{16}O_{pac}$ – with ^{16}O in planar, apical and chain sites) and sample **4** ($^{18}O_{pac}$ – fully substituted with ^{18}O). From the fully substituted samples **1** ($^{16}O_{pac}$) and **4** ($^{18}O_{pac}$) the site-selective substituted samples **2** ($^{16}O_p{}^{18}O_{ac}$) and **3** ($^{18}O_p{}^{16}O_{ac}$) have been obtained by annealing (320 °C, 200 h) carried out in the atmosphere of $^{18}O_2$ and $^{16}O_2$, respectively (Fig. 2b). For these samples only the shift of the band characteristic for the apical oxygen without a change of the bands for the planar sites can be observed (compare pairs **1** → **2** and **4** → **3**).

Analyzing the isothermal exchange experiments as shown in Fig. 1, it was assumed that oxygen-isotope diffuses along every CuO chain. Then, the isotope from the chains can be exchanged with oxygen from the apical and further the planar sites. The Fick's equation describing the above proposed diffusion model, could be solved [25] using numerical computation methods [26, 27]. Figure 3 shows temperature dependences of the oxygen-tracer diffusion coefficients D_{ab}, D_{c1} and D_{c2} for the diffusion along the chains and between chain-apical, and the apical-planar sites, respectively. As it can be seen, differences between the values of the diffusion coefficients along *b*- and *c*-crystallographic axes are larger than eight orders of magnitude. However, in the assumed model the diffusion lengths are very different in crystallographic direction *b* (diffusion through the whole crystallite i.e. 15 μm-half of the linear size) and *c* (adjacent oxygen sites in the unit cell). Therefore, in the bottom part of Fig. 3, the diffusion coefficients divided by the square of the diffusion length are presented, i.e. values which are proportional to the effective diffusion fluxes [25]. As can be seen, below 230 °C the diffusion flux along the chains dominates, whereas the diffusion fluxes to the apical and the planar sites are much smaller. This should assure a possibil-

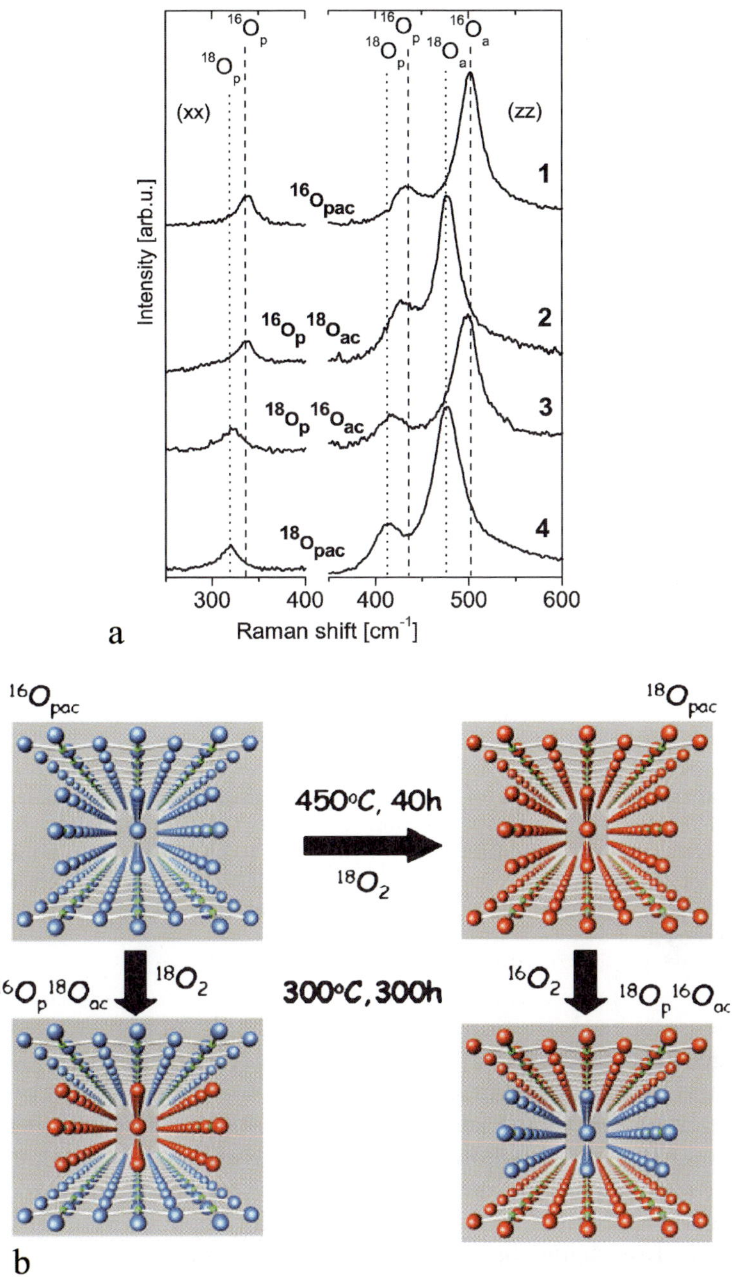

Fig. 2 **a** Raman spectra for the total and site-selective oxygen-isotope exchanged Y123. O_a and O_p show frequency bands for apical and planar oxygen sites, respectively; **b** Experimental procedure used in order to obtain oxygen-isotope substituted samples. ^{18}O- and ^{16}O-isotopes are depicted with red and blue colors, respectively. Copper cations are shown in green color; Y and Ba are omitted

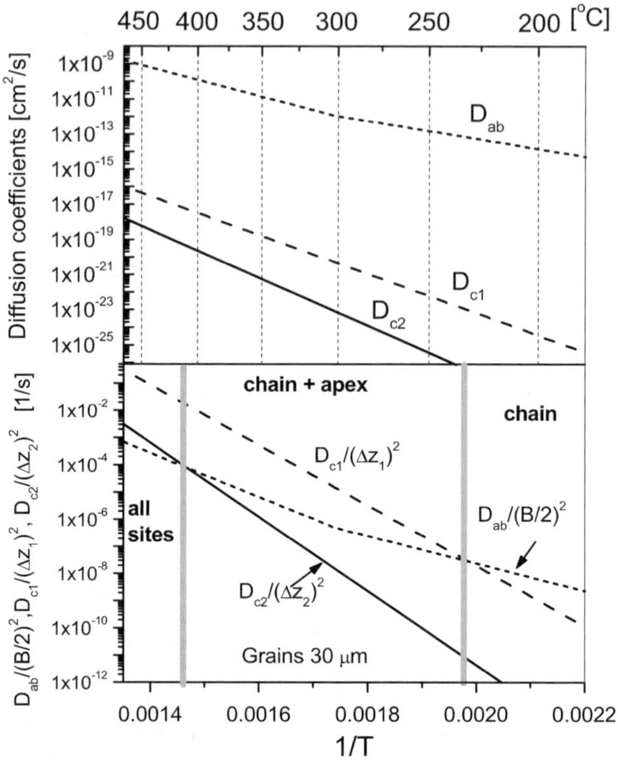

Fig. 3 Temperature dependence of the tracer-oxygen diffusion coefficients determined in this work. The *lower figure* shows diffusion coefficients D_{ab}, D_{c1} and D_{c2} divided by the second power of the average diffusion legth. These are: $B/2 = 15\ \mu m$ – half of the assumed crystallite length, $\Delta z_1 = 1.83$ Å and $\Delta z_2 = 2.32$ Å are distances between chain-apical and apical-plane sites in the unit cell, respectively. Temperature ranges of the exchange selectivity are shown

ity of the isotope substitution of the chain sites only. At higher temperatures the oxygen-isotope transport to the apical sites (see the line $D_{c1}/(\Delta z_1)^2$) increases and becomes comparable to that along the chains, so that both these sites are exchanged at the same time. Above 400 °C the chain diffusion is the limiting step of the whole process and the exchange is no longer site-selective. The temperature ranges for the selective exchange marked in Fig. 4 are in a good agreement with annealing conditions applied in order to obtain site-selectively exchanged samples (compare Fig. 2b).

As the isotope-exchange is non-equilibrium, dynamic process, the progress of the exchange α and the exchange selectivity depend not only on temperature but also time of the process and the size of the material crystallites. This is illustrated by the calculations presented in Fig. 3. At low temperatures selectivity of the exchange can be reached, but the process has to be per-

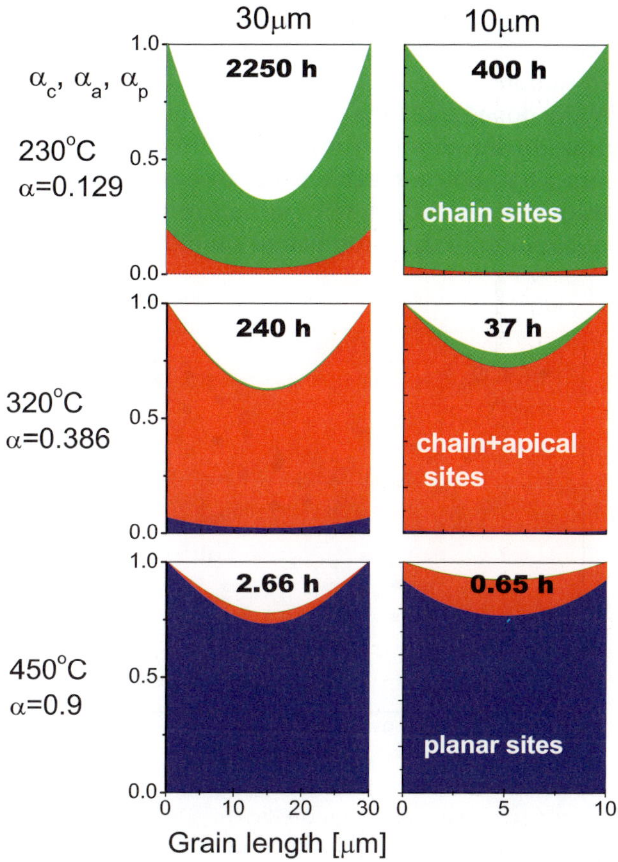

Fig. 4 Calculated isotope distributions in 30 and 10 μm grains illustrating a possibility of a site-selective oxygen-isotope exchange at 230, 320 and 450 °C. The isotope distributions in the chain (α_c), apical (α_a) and planar (α_p) sites are shown with green, red and blue colors, respectively. The α-parameter indicate the total isotope exchange progress. The times necessary to reach the desired exchange progress are given for each diagram

formed during a certain time. If this time is too short, insufficient exchange progress will be obtain. Opposite, the exchange selectivity will continuously diminish. It is evident from the Fig. 4 that the exchange selectivity is better and the exchange-process times are shorter for smaller crystallites (grains) of a material used.

Oxygen-Isotope Effect in Site-Selective Substituted Samples

Magnetization measurements have been performed for totally and site-selective oxygen-isotope exchanged samples [24]. The samples were obtained using an annealing procedure as shown in Fig. 2a. Figure 5 shows a temperature dependence of the (normalized to 10K) magnetization for two sets of the ^{18}O-substituted samples.

In the optimally doped Y123 the OIE is very small (about 0.2 K) and originates mainly due to the oxygen sites in the CuO_2 planes. The apical and the chain oxygen sites show a much smaller or even vanishing OIE. The results demonstrate that in anticipated mechanisms of superconductivity in cuprates, phonon modes of the CuO_2 planes should be considered.

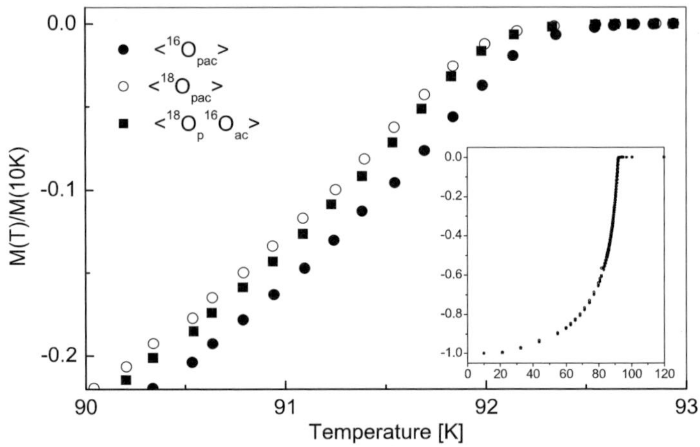

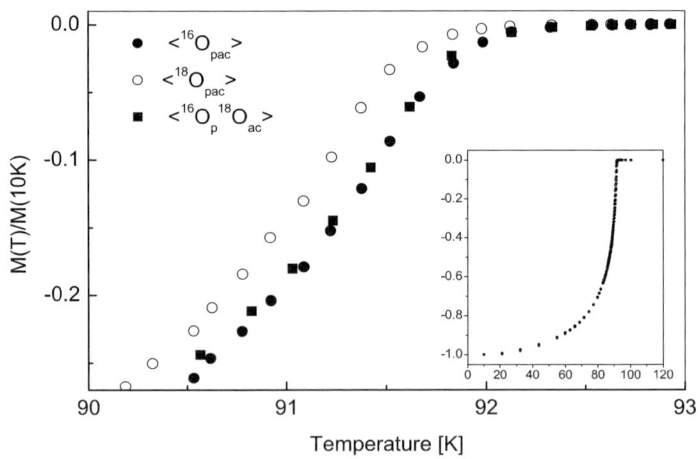

Fig. 5 Magnetization curves obtained for the total and site-selective oxygen-isotope substituted Y123 for two sets of the samples. Note that the curves in the entire temperature range (see *inserts*) are identical indicating excellent reproducibility of the samples' properties

Conclusions

Measurements of the oxygen-isotope exchange in Y123 powder samples have been performed using thermogravimetry. The results of the measurements were interpreted using a three-dimensional diffusion model. It was found that the tracer (^{18}O isotope) diffuses along the CuO chains in the structure. From the chains further exchange with the apical and planar sites takes place. Tracer diffusion coefficients in the a–b plane (in a macroscopic scale, through a whole crystallite) and two microscopic coefficients for the oxygen-isotope exchange between chain-apical and apical–planar sites have been evaluated using numerical calculations. The determined values of the diffusion coefficients allow calculation of the isotope distribution at all three oxygen sites during an exchange process in dependence on temperature, time and grain size. It has been shown that the OIE is mainly due to the oxygen sites in the CuO_2 planes.

References

1. Bednorz JG, Müller KA (1986) Z Phys B 64:189
2. Wu MK, Ashburn RJ, Torng CJ, Hor PH, Meng RL, Gao L, Huang ZJ Wang YZ, Chu CW (1987) Phys Rev Lett 58:908
3. Buckel W (1990) Supraleitung. VCH Verlagsgesellschaft, mbH Weinheim, FRG
4. Cyrot M, Pavuna D (1992) Introduction to Superconductivity and High-T_c Materials. World Scientific Publ. Co. Pte. Ltd
5. Sharma BL (1970) Diffusion in Solids. Trans Tech Publications, Germany
6. Ghez R (1988) A Primer of Diffusion Problems. Wiley, Inc,
7. Salomons E, de Fontaine D (1990) Phys Rev B 41:11159
8. Choi J-S, Sarikaya M, Aksay IA, Kikuchi R (1990) Phys Rev B 42:4244
9. Rothman SJ, Routbort JL, Baker JE (1989) Phys Rev B 40:8852
10. Rothman SJ, Routbort JL, Welp U, Baker JE (1991) Phys Rev B 44:2326
11. Bredikhin SI, Emel'chenko GA, Shechtman V, Zhokhow AA, Carter S, Chater RJ, Kilner JA, Steele BCH (1991) Physica C 179:286
12. Xie XM, Chen TG, Wu ZL (1989) Phys Rev B 40:4549
13. Tallon JL, Staines MP (1990) J Appl Phys 68:3998
14. Ikuma Y, Akiyoshi S (1988) J Appl Phys 64:3915
15. Tsukui S, Yamamoto T, Adachi M, Shono Y, Kawabata K, Fukuoka N, Nakanishi S, Yanase A, Yoshioka Y (1991) Jpn J Appl Phys 30:L973
16. Turillas X, Kilner JA, Kontoulis I, Steele BCH (1989) J Less-Common Met 151:229
17. Sabras J, Peraudeau G, Berjoan R, Monty C (1990) J Less-Common Met 165:239
18. Routbort JL, Rothman SJ (1994) J Appl Phys 76:5615
19. Vischjager DJ, van der Put PJ, Schram J, Schoonman J (1988) Solid State Ionics 27:199
20. Tsukui S, Yamamoto T, Adachi M, Shono Y, Kawabata K, Fukuoka N, Nakanishi S, Yanase A, Yoshioka Y (1991) Jpn J Appl Phys 30:L973
21. Cardona M, Liu R, Thomsen C, Kress W, Schönherr E, Bauer M, Genzel L, König W (1988) Solid State Commun 67:789
22. Nickel JH, Morris DE, Ager JW (1993) Phys Rev Lett 70:81

23. Conder K, Kaldis E, Maciejewski M, Steigmeier EF, Müller KA (1993) Physica C 210:282
24. Zech D, Keller H, Conder K, Kaldis E, Liarokapis E, Poulakis N, Müller KA (1994) Nature 371:681
25. Conder K (2001) Mater Sci Eng R32:41–102
26. Wong SSM (1992) Computational Methods in Physics, Engineering, Prentice Hall, New Jersey
27. Crank J (1983) The Mathematics of Diffusion, Oxford University Press

Dalal NS (2007) In: Bussmann-Holder A, Keller H (eds)
High T_c Superconductors and Related Transition Metal Oxides.
Springer-Verlag Berlin Heidelberg, 85–97
DOI 10.1007/978-3-540-71023-3

Coexistence of Order–Disorder and Displacive Behavior of KH_2PO_4 and Analogs from Electron Paramagnetic Resonance

N. S. Dalal

Department of Chemistry & Biochemistry and NHMFL, Florida State University, Tallahassee, FL 32306-4390, USA
dalal@chemmail.chem.fsu.edu

Abstract An early report by Müller et al. [Phys Rev Lett 36:1504 (1976)] on the electron paramagnetic resonance (EPR) studies of slow (100 MHz-range) molecular dynamics in ferroelectric crystals has proven to be pioneering in the field of phase transitions. Subsequent studies over the last 30 years have extended this methodology and utilized it for unraveling microscopic mechanism(s) of ferroelectric and antiferroelectric phase transitions, especially for the KH_2PO_4 (KDP) family. We here discuss these EPR as well as some related electron nuclear double resonance data, utilizing the conjecture that an anomaly in the isotropic part of the electron Zeeman tensor (g_{iso}) or of hyperfine coupling, a_{iso}, implies an electronic instability. It is concluded that the phase transitions of the KDP-family exhibit an order–disorder character as well as instability (displacive behavior). This is in conformity with recent high-resolution NMR data, but in contrast with the notion that these materials are model order–disorder systems.

Introduction

The present undertaking discusses the utilization of electron paramagnetic resonance (EPR) spectroscopy for investigating dielectric solids that are diamagnetic in their native state, but can be made slightly paramagnetic by lightly doping them with transition ions, such as Fe^{3+}, as pioneered by Müller [2–7], or by using molecular probes such as the $CrO_4{}^{3-}$ (Cr^{5+}) ion [1, 8–13], $SeO_4{}^{3-}$ radical [14–22] and the $AsO_4{}^{4-}$ center [23–40]. While Müller's studies using Fe^{3+} as a probe in $SrTiO_3$ and $LaAlO_3$ are considered as some of the classic examples of the power of EPR technique in probing microscopic mechanisms of phase transitions in the perovskites [2–7], it is perhaps less well-known that he and his coworkers have also made pioneering contributions to EPR studies on other classes of compounds, not counting the field of superconductivity. A good example that discussed here is his pioneering EPR studies on the KH_2PO_4 (KDP) family, using $CrO_4{}^{3-}$ ion as a paramagnetic probe. The first major publication appeared in 1976 [1], and several other significant ones followed well into the 1980s [8–11, 41], on the way to his discovery of high-temperature superconductivity [42]. Since the present author was intimately involved in the initiation of the KDP work [1] and has continued over the last three decades [12, 13, 16, 19–22, 25, 27–40], this current undertaking is considered an opportunity to highlight Müller's strong impact in this field as well.

It seems timely to provide here a brief overview of the relevant historical developments. In the late 1970s the important issue was the detection of the slow dynamics of lattices near their phase transitions, in relation to the central peak (CP) phenomenon [43]. CP refers to a peak observed at unusually low energies that was unexpectedly found in neutron inelastic scattering studies of lattices such as $SrTiO_3$ in the vicinity of their (cubic-tetragonal) structural phase transitions, around 105 K for $SrTiO_3$ [44]. Interestingly, it was noted that as the sample was cooled towards the phase transition temperature, T_c, the CP peak intensity grew at the expense of the soft mode peak [2–5, 44], thereby suggesting that CP should be a fundamental characteristic of such phase transitions. In KDP also, CP peaks were observed, mainly by light scattering [45–49]. In particular, KDP was the first compound for which Cowley had predicted the occurrence of such a peak, well before the peak's discovery [50]. In general, however, the time scale of CP was too long to be well studied by scattering techniques, but Muller and others [30] realized that this time-scale was optimally suited for this problem. His studies using Fe^{3+} in $SrTiO_3$ was a classic example of such work, and showed that the CP width was of the order of a few MHz, much narrower than could be measured by scattering techniques [7]. EPR spectroscopy enables easy and direct access to such slow dynamics because the peak positions as well as lineshape are directly affected by the atomic fluctuations at the EPR probe's local site [7, 30]. It turned out that the CP peak was related in part to crystal strains and impurities, and also to the slow dynamics related to dynamic cluster formation in a lattice as it approached its T_c, and EPR provided some of the most detailed data on the CP phenomenon [1–11, 30].

While the origin of the CP phenomenon still remains intriguing, and not fully understood, the focus has now shifted from "dynamics" to "electronic structure" point of view; in particular whether a transition can be considered as a pure order–disorder type or whether it is also accompanied by electronic instabilities or displacive characteristics. The KDP-family again acquired renewed interest, because these compounds have traditionally been considered as model order–disorder types, the order–disorder relating to the displacement of the hydrogens (H's) in their double minimum potential wells along the $O-H...O$ bonds connecting the adjoining PO_4 groups [51, 52]. Earlier ^{31}P NMR studies of KDP and its deuterated analog KD_2PO_4 (DKDP) have been interpreted to provide direct evidence for the order–disorder mechanism, since the position of the signal in the paraelectric phase was judged to be the average of those in the ferroelectric phase, albeit within the large error bars set by the broadness of the peak widths [54]. However, repetition of these ^{31}P NMR studies carried out recently using the modern high-resolution techniques (magic angle spinning using single crystals and rf decoupling) showed that this is not the case. Specifically, the high resolution data enabled direct measurement of the isotropic part of the chemical shift tensor, δ_{iso}, (average peak position), and δ_{iso} exhibited a clear anomaly at the phase transition, thereby providing definitive evidence of displacive character [54–60]. This follows from the fact that (a) δ_{iso} is a quantity that involves contributions from all the orbitals of the entire molecular fragment containing the nucleus under study, and (b) δ_{iso} is invariant to any rotational or translational change in a molecular fragment. Thus an anomalous variation in δ_{iso} would indicate a change in the electronic structure of the local environment of the site being probed, and hence a direct evidence for the existence of a displacive component in the transition mechanism. Similar conclusions have been reported also from ^{13}C high resolution NMR studies of squaric acid ($H_2C_4O_4$), a related hydrogen-bonded antiferroelectric compound [55–60]. These observations prompted us to re-examine the earlier spin-probe EPR studies on the KDP family, which have long been interpreted in terms of an order–disorder behavior [30, 51–54]. In analogy to the above-mentioned high resolution NMR work, we have looked for an anomaly in the isotropic component of the electron Zeeman tensor component, g_{iso}, which is the EPR equivalent of δ_{iso} in NMR. As discussed below, we have indeed found anomalies in this parameter in the EPR data on the KDP-family of ferroelectrics as well as antiferroelectrics, pointing to the coexistence of order–disorder and displacive behavior in these and related materials [30, 61].

Experimental Details

As discussed in detail elsewhere [30], the major caution in planning an EPR spin-probe study is to match the electronic charge, size and, in particular, the local site symmetry that the probe settles into a lattice being investigated. The three most common probes used in the EPR studies related to KDP are AsO_4^{4-}, SeO_4^{3-} and CrO_4^{3-} radical ions. All three of them are easily introduced, respectively, via doping a lattice lightly (usually 0.1 mol %) with CrO_4^{2-}, SeO_4^{2-} and AsO_4^{3-} ions, and then x-irradiating the doped material for about an hour or so. Too much radical concentration is not desirable because the radical-radical dipolar interaction causes an unwanted line broadening and thus poorer spectral resolution. Too little concentration results in an undesirably poor spectral intensity.

The KDP-family of compounds easily grow brick-shaped crystals, with their polar (tetragonal) axes along the longest dimension. They are thus easy to align in a magnetic field in the three mutually perpendicular planes, *ab*, *bc* and *ca*. Most work has been carried out with commercially available X-band (8–10 GHz), K-band ($\sim$ 16–22 GHz; much used by Müller and coworkers [1–11, 41]), and occasionally [26] Q-band (35 GHz). The temperature dependence of the EPR spectral lineshape yields direct evidence for the order–disorder motion of the H's coupled with the heavier nuclei. However, the observed change in the isotropic part of the g-value demonstrates the role of the displacive behavior.

Much more detailed information has been obtained through electron nuclear double resonance (ENDOR) measurements mainly by Dalal and coworkers, initially using a self-designed spectrometer [16, 20, 25, 27–31], and later a Bruker system [33–39]. One significant ENDOR study on protons around the CrO_4^{3-} center was reported by Gailard, Gloux and Müller [9]. As will be shown here for the first time, the change in the magnitude of the proton superhyperfine coupling measured via ENDOR complements the g-values in providing evidence for the displacive character.

Experimental Results with Different Probes

(a) Studies with AsO_4^{4-}

The AsO_4^{4-} center was first identified by McDowell and coworkers in x-irradiated KH_2AsO_4 in 1966 [23]. The center is formed by the capture of an X-ray generated photoelectron by the AsO_4^{3-} unit in the crystal according to:

$$AsO_4^{3-} + e^- \rightarrow AsO_4^{4-} .$$

The unpaired electron enters an a_1-type molecular orbital of the AsO_4^{3-} unit, and it spreads over the entire H_4AsO_4 fragment. Its EPR spectrum thus consists of a main quartet of peaks due to the hyperfine interaction of the As ($I = 3/2$) nucleus. At room temperature, each of these peaks is further split into a $(1:4:6:4:1)$ quintet due to the superhyperfine interaction of the 4 surrounding protons; a feature that helps its identification directly [23]. As the next important step, Blinc, Cevc and Schara [24] noted that the quintet changes to a $(1:2:1)$ triplet as the sample is cooled down to about 220 K, as can be noted also from Fig. 1.

Blinc et al. [24] thus demonstrated that the four H's undergo a hopping type dynamics in their O – H...O potential wells, at the slow rate of 100 MHz. Subsequently, Dalal et al. carried out precise electron nuclear double resonance (ENDOR) measurements and showed that the proton and As motional correlation times are different at high temperatures but they become the same as the temperature approached the phase transition [25, 29–39], which has now become a text-book example of the use of EPR for studying phase transitions [51].

Order-disorder vs Displacive Behavior: The observation of the quintet-triplet transition (Fig. 1) provides direct evidence of the order–disorder type motion of the H's in the KDP-family, at the MHz time scale as evident from Fig. 2. On the other hand, the ENDOR measurements clearly show that there

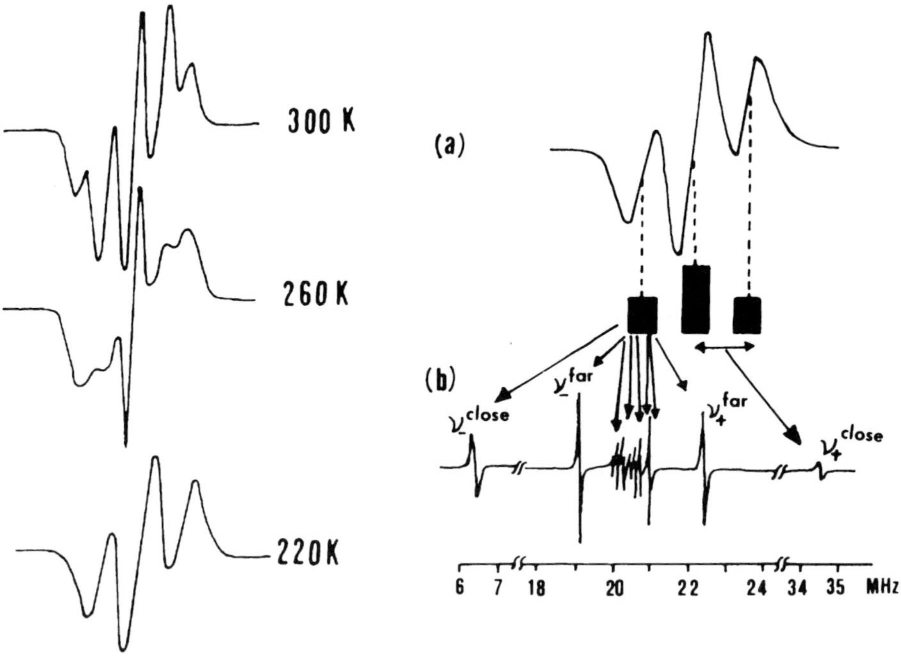

Fig. 1 **a** Proton superhyperfine structure on an As peak from the $AsO_4{}^{4-}$ center in KH_2AsO_4 [23, 24]. **b** ENDOR spectrum at 4.2 K, resolving many shells of protons around an AsO_4 unit [25, 27, 30]

is a redistribution of charge density on the four oxygens as the crystal goes through the phase transition. This has been discussed in detail by Rakvin and Dalal [37]. They show that the ENDOR measured superhyperfine couplings for KH_2AsO_4 are: $a_{close} =- 30$ MHz, and $a_{far} =- 4$ MHz cannot be reconciled by any calculation of a simple distance dependence of the "close" and "far" protons. The large variation can be easily understood by recognizing that the proton attachment to an oxygen alters the charge distribution nonlinearly, and leads to the development of the dipole around the central AsO_4 unit, as was suggested earlier by Bystrov and Papova [38].

(b) Evidence of the Coexistence using SeO_4^{3-}

Extensive EPR [14–22, 30] and ENDOR [16, 30, 34, 38] studies have also been carried out on the KDP family using the $SeO_4{}^{3-}$ probe. In short, the results obtained are fairly similar to those described above for the $AsO_4{}^{4-}$ center. One observes a well defined quintet-triplet transition in the EPR spectra [14–16, 30], similar to that shown in Fig. 1 above, *that provide direct evidence for the order–disorder behavior*. Again, complementary ENDOR studies provide the evidence for the strong electronic charge redistribution at the phase tran-

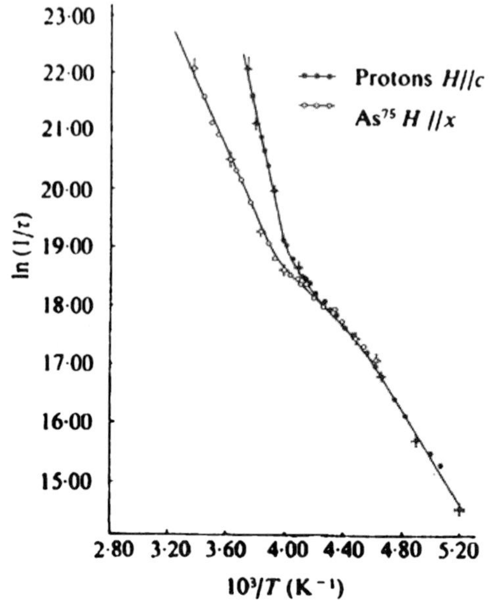

Fig. 2 Correlation times for the motion of ^{75}As and of protons calculated from the temperature dependence of ^{75}As and proton superhyperfine structure (Fig. 1) in KH_2AsO_4 [27]

sition. For example, the isotropic component of the proton superhyperfine couplings are found to be $a_{close} =- 17.7$ MHz, and $a_{far} =- 1$ MHz [38]. Using the various theoretical models of deducing spin density, Rakvin and Dalal obtain an unpaired electron density of 0.3 on the oxygens to which the proton is attached, and only 0.05 on the oxygens from which the proton has moved away to the "far" position, showing that *"displacive" character is also involved in the transition mechanism.*

(c) Studies using CrO_4^{3-}

As mentioned in the introduction, this probe has yielded perhaps the most decisive results on the slow dynamics as well as electronic instabilities related to the phase transitions in the KDP-family. It was first introduced by Müller, Dalal and Berlinger in their study of the slow dynamics in KH_2PO_4, KD_2PO_4, KH_2AsO_4 and KD_2AsO_4 [1]. Figure 3 shows a schematic how the $CrO_4{}^{3-}$ probe is located in the lattice: $Cr^{(5+)}$ substitutes for the $As^{(5+)}$ or the $P^{(5+)}$ ion in the original lattice.

This probe remains a favorite because its spectrum can be interpreted by the very simple spin Hamiltonian, that for an $S = 1/2$ system, without any complication from hyperfine structure, since the most dominant isotope, ^{52}Cr (92% natural abundance) has the nuclear spin $I = 0$. The remaining stable iso-

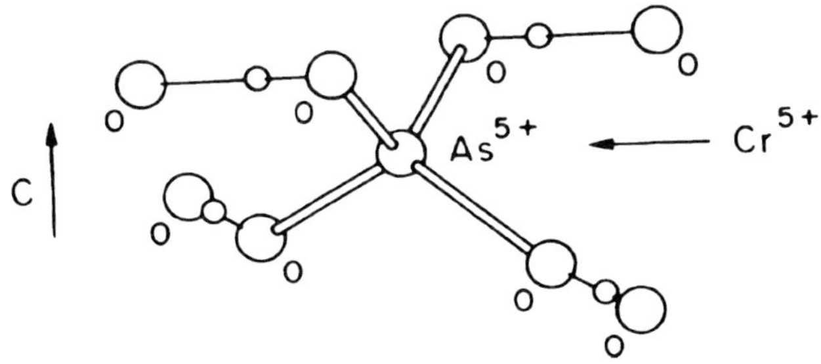

Fig. 3 Structure and location of the $CrO_4{}^{3-}$ probe in KH_2AsO_4. In KDP, the Cr^{5+} ion substitutes for the P^{5+} atom. The small circles are the four H's in the O – H...O bonds surrounding the AsO_4 unit; they exhibit superhyperfine couplings with the unpaired electron of the $CrO_4{}^{3-}$ center (or equivalently, the Cr^{5+} ion) (from Müller et al. [1])

tope, ^{53}Cr, does have a nuclear spin $I = 3/2$ and thus its spectrum is a quartet with 8% (total) intensity compared to that of the strong main peak (arising from the ^{52}Cr isotope). Below we discuss the salient features of the EPR properties of this probe, as well as how EPR spectroscopy is used in studying phase transitions.

Figure 4 shows a typical EPR spectrum of the $CrO_4{}^{3-}$ probe doped in the antiferroelectric hosts $NH_4H_2AsO_4$, and its deuterated analog $ND_4D_2AsO_4$; the former chosen to explain how the protons in the O – H...O bonds show up as the superhyperfine spiltting on the main peak, see Fig. 1 for the $AsO_4{}^{4-}$ radical, and Figs. 4 and 5 for the $CrO_4{}^{3-}$ center, respectively [12, 13, 30].

Direct evidence for the order–disorder dynamics can be noticed from the temperature dependence of the proton superhyperfine structure for H along the *c*-axis, as shown for KDP in Fig. 5. The analysis of the spectra on the left panel using the formalism of the Modified Bloch Equations yield the correlation times for the hopping motion of the protons in the O – H...O bonds, and shows directly that the protons undergo hopping dynamics in their O – H...O bonds. Figure 6 shows additional data on the temperature dependence of polarization fluctuations in the KDP-family; from the domain related g-tensor splittings in KH_2AsO_4 (KDA) and KD_2AsO_4 (DKDA), as reported by Muller et al. [1] in their seminal paper. The right panel in Fig. 6 shows the polarization fluctuation times obtained from the motional narrowing analysis of the data in Fig. 6. As the authors noted, the temperature dependence of the correlation times could be fitted to a power law of the form A $(T - T_c)^m$, with $m \sim 2.2 \pm 1$, which is of the order of unity, but a thermally activated process could not be ruled out.

On the other hand, the temperature dependence of the g-values shows that the isotropic part of the g-values changes abruptly at the phase transition. This is evident for both the antiferroelectric $ND_4D_2AsO_4$ (Fig. 7) and ferroelectric KD_2AsO_4 (Fig. 8).

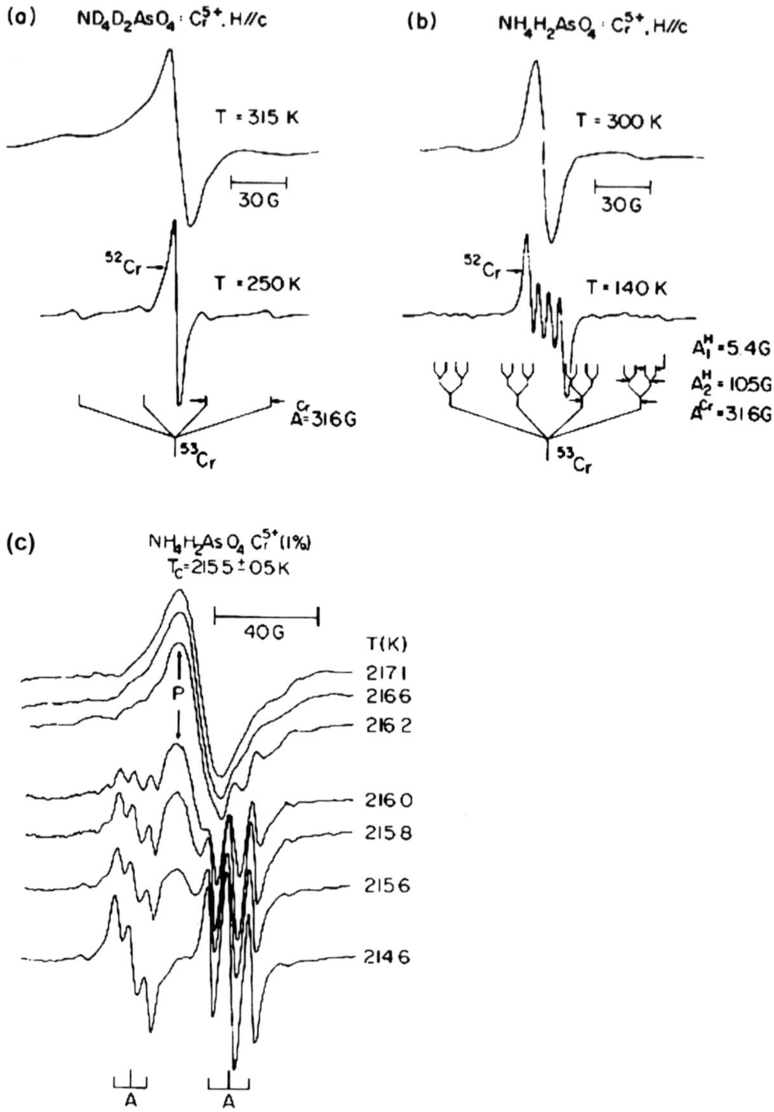

Fig. 4 EPR spectra of the CrO_4^{3-} center in $ND_4D_2AsO_4$ (**a**) and $NH_4H_2AsO_4$ (**b**). Proton superhyperfine splittings are clearly discernible in **b**. **c** shows temperature dependence of the spectra for $NH_4H_2AsO_4$. Note how the polarized clusters start to appear within a couple of degrees of the transition temperature. Also, notice that the center of the high temperature spectrum is not at the average position of the two low-temperature peaks, thereby pointing to the existence of the "displacive" component in the phase transition mechanism. See text for details [12, 13, 30]

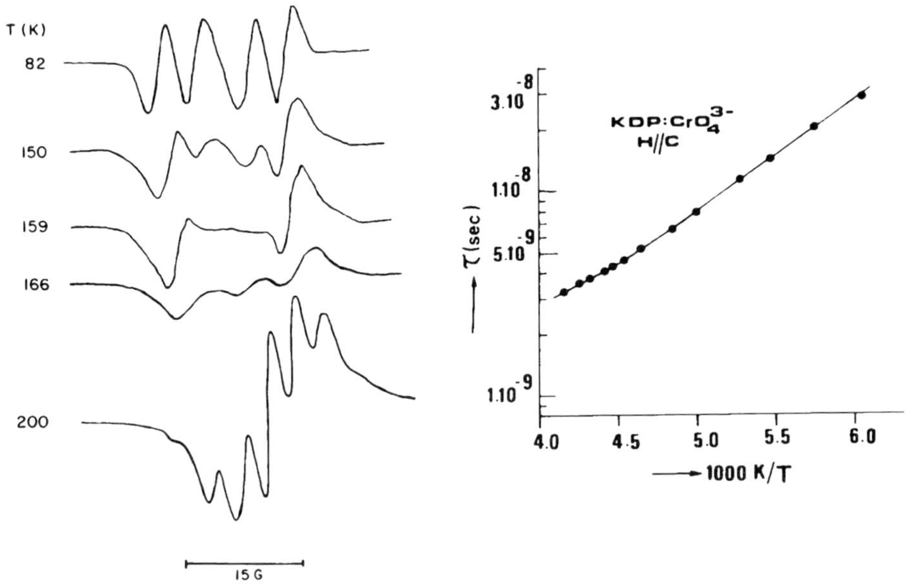

Fig. 5 *Left panel*: Temperature dependence of the proton superhyperfine splitting on the EPR spectrum of the CrO_4^{3-} center ion KDP. *Right panel*: Correlation times of the motion of protons obtained from the quintet–quartet spectra in the left panel, using the formalism of the modified Bloch Equations [12, 30]

Similar changes in the g_{iso} are also observed of the CrO_4^{3-} center in KD_2AsO_4, as can be seen from Fig. 8. Here again, the observed change in g_{iso} point directly to the presence of the "displacive" character in the phase transition in the KDP family.

Quantitatively, Dalal and Reddoch [13] have shown that the g-values decrease abruptly on entering the antiferroelectric phase from 1.9570 to 1.9557 for $NH_4H_2AsO_4$, and from 1.9565 to 1.9555 for $ND_4D_2AsO_4$. On the other hand, the ^{53}Cr couplings hyperfine increase from 27.5 to 31.5 gauss for $NH_4H_2AsO_4$ and from 28 to 31 gauss for $ND_4D_2AsO_4$. Again, these data clearly support the notion that the transition mechanism involves a displacive character.

Conclusions

We have presented a summary of three decades of EPR and ENDOR data with regard to the mechanism of the phase transition in the KDP-family of compounds that are generally considered as good models of pure order–disorder lattices [30, 41, 44, 51, 52]. Our reanalysis of especially the ENDOR data shows that in fact there is a clear "displacive" component in the transition mechanism, since there is a strong charge redistribution on the oxygens as the

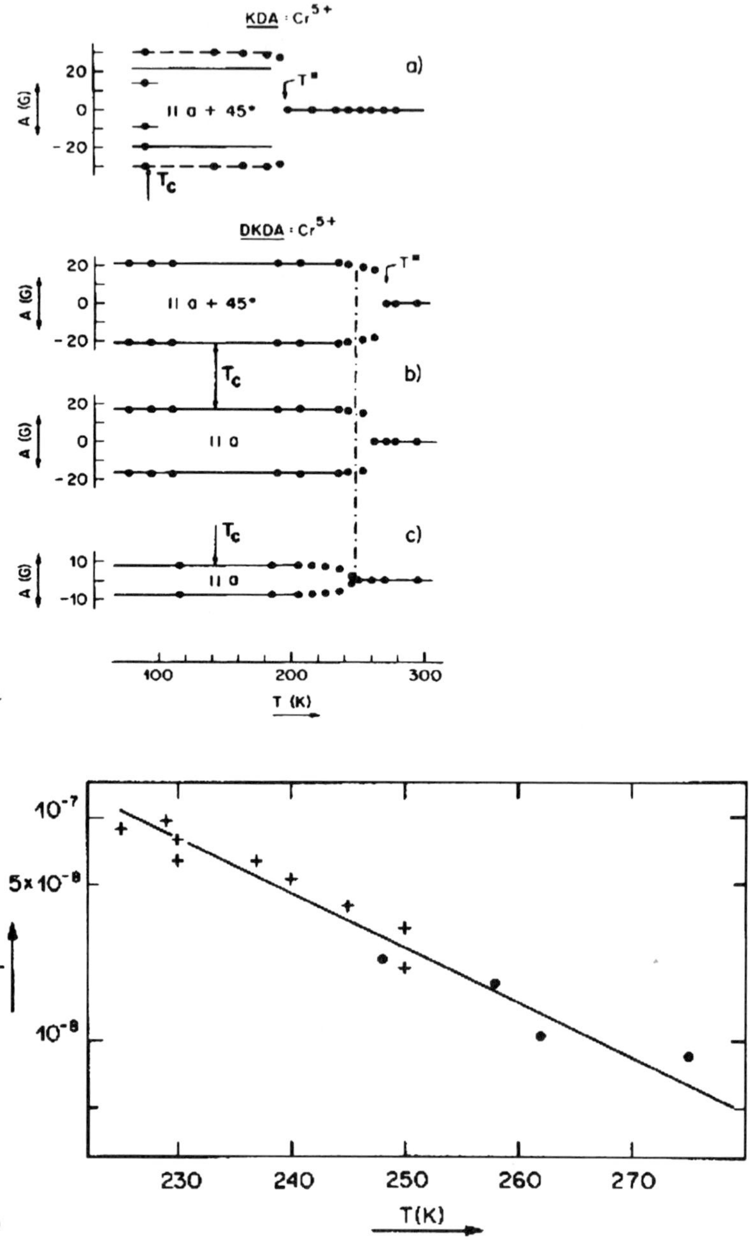

a

b

Fig. 6 **a** Temperature dependence of ^{53}Cr splittings for KH_2AsO_4 (KDA) and KD_2AsO_4 (DKDA), **b** correlation times for ^{53}Cr fluctuations [1]

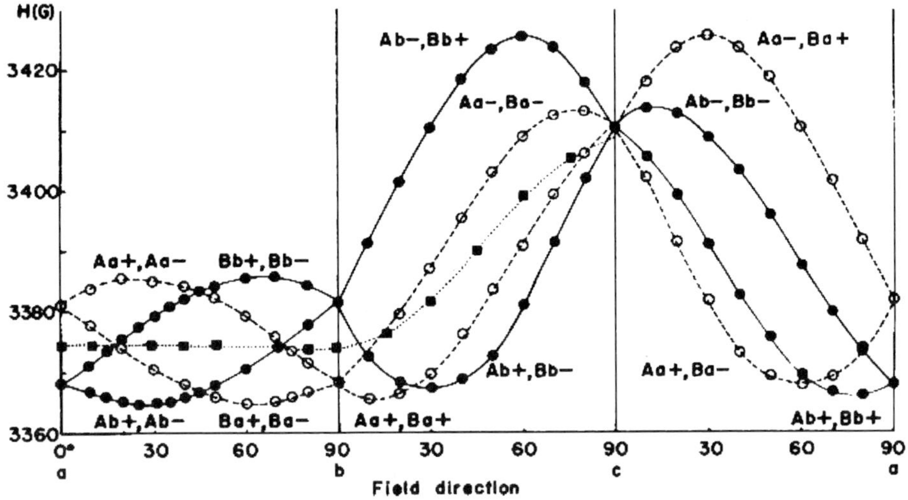

Fig. 7 Angular dependence of the g-tensor of the CrO_4^{3-} center in the paraelectric phase (*dotted line*) and antiferroelectric phase (*solid curves*) of $ND_4D_2AsO_4$. There is a clear change in the overall g-tensor, and hence in g_{iso}, the isotropic part of the g-tensor at the phase transition, thereby implicating the existence of a displacive character in the transition mechanism; see text for further details

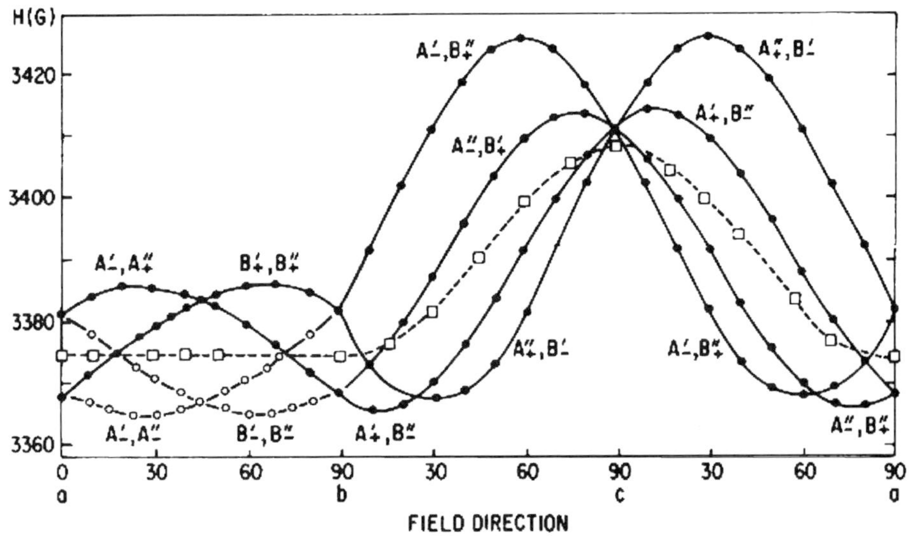

Fig. 8 Angular dependence of the EPR spectra of the CrO_4^{3-} center in KD_2AsO_4. The *dotted lines* are the data in the paraelectric phase. Clearly, the average of the g-tensor components of the low temperature (ferroelectric) phase do not coincide with the observed peaks in the high-temperature, paraelectric phase, demonstrating the displacive character of the transition mechanism

compounds undergo the phase transition. It is our impression that the studies by Muller and coworkers in the late 1970s [1] laid the foundation of this field. In addition, it seems worth adding that even in the High-T_c superconductor studies, the EPR and microwave absorption work of Müller, Dalal and their coworkers has played a significant role in the understanding of the Josephson junction formation and vortex dynamics in the YBCO and related lattices [62, 63]. It is thus clear that his influence in EPR spectroscopy goes way beyond that of most other workers in the field.

Acknowledgements I wish to thank Alex Müller for his mentorship in my early career, and Robert Blinc and Annette Bussmann-Holder for stimulating discussions over the years and the National Science Foundation for financial support.

References

1. Müller KA, Dalal NS, Berlinger W (1976) Phys Rev Lett 36:1504
2. Müller KA (1971) In: Samurelsen EJ, Anderssen E, Feder J (eds) Structural Transitions and Soft Modes. Universitetsforlaget, Oslo, p 73
3. Müller KA, Berlinger W (1972) Phys Rev Lett 29:715
4. Müller KA, von Waldkirch T (1976) In: Müller KA, Rigamonti A (eds) Local Properties at Phase Transitions. North-Holland Publ, Amsterdam, p 134
5. Müller KA (1979) Lect Notes Phys 104:210
6. Bruce AD, Müller KA, Berlinger W (1979) Phys Rev Lett 42:185
7. Reiter GF, Berlinger W, Müller KA, Heller P (1980) Phys Rev B 21:1
8. Müller KA, Berlinger W (1976) Phys Rev Lett 37:916
9. Gailard J, Gloux P, Müller KA (1977) Phys Rev Lett 38:1216
10. Müller KA, Dalal NS, Berlinger W (1977) Ferroelectrics 17:443
11. Müller KA, Berlinger W (1978) Z Phys B 31:151
12. Dalal NS, Reddoch AH, Northcott DJ (1978) Chem Phys Lett 58:553
13. Dalal NS, Reddoch AH (1988) Canad J Chem 66:2045
14. Kawano T (1974) J Phys Soc Jpn 37:848
15. Hukuda K (1975) J Phys Soc Jpn 38:150
16. Dalal NS, Hebden JA, Kennedy DE, McDowell CA (1977) J Chem Phys 66:4425
17. Hukuda K, Nakagawa Y (1978) J Phys Soc Jpn 44:1588
18. Hukuda K, Tsuoura Y, Akahoshi S (1982) J Phys Soc Jpn 51:7
19. Dalal NS (1982) J Am Chem Soc 104:5512
20. Rakvin B, Dalal NS (1991) Phys Rev B (Rapid Commun) 44:892
21. Nakagawa K, Rakvin B, Dalal NS (1991) Solid State Commun 78:129
22. Rakvin B, Dalal NS (1996) J Phys Chem Solids 57:1483
23. Hampton MA, Herring FG, Lin WC, McDowell CA (1966) Molec Phys 10:565
24. Blinc R, Cevc P, Schara M (1967) Phys Rev 159:411
25. Dalal NS, McDowell CA, Srinivasan R (1970) Phys Rev Lett 25:823
26. Lamotte B, Gaillard J, Constinescu O (1972) J Chem Phys 57:3319
27. Dalal NS, McDowell CA (1972) Phys Rev B 5:1074
28. Dalal NS, Dickinson JR, McDowell CA (1972) J Chem Phys 57:4254
29. Dalal NS, McDowell CA, Srinivasan R (1972) Molec Phys 24:403
30. Dalal NS (1982) Advan Magn Reson 10:119

31. Kahol PK, Dalal NS (1988) Solid State Commun 65:823
32. Trybula Z, Stankowski J, Szczepanska L, Blinc R, Weiss AI, Dalal NS (1988) Physica B 153:143
33. Dalal NS, Kahol PK (1989) Solid State Commun 70:623
34. Rakvin B, Kahol PK, Dalal NS (1989) Molec Phys 68:1185
35. Dalal NS, Rakvin B (1989) J Chem Phys 86:5262
36. Rakvin B, Dalal NS (1989) Phys Rev B 39:7009
37. Rakvin B, Dalal NS (1992) Ferroelectrics 135:227
38. Bystrov DS, Papova EA (1987) Ferroelectrics 72:147
39. Rakvin B, Dalal NS (1994) Phys Rev B 49:13211
40. Kahol PK, Lao X, Castello MB, Dalal NS (1994) J Phys C: Condensed Matter 6:2971
41. Muller KA (1987) Ferroelectrics 72:273
42. Bednorz JG, Muller KA (1986) Z Phys B 64:189
43. Riste T, Samuelsemn EJ, Otnes K, Feder J (1971) Solid State Commun 9:1455
44. Bruce AD, Cowley RA (1980) Advan Phys 29:219
45. Cowley RA, Coombs GJ, Katiyar RS, Ryan JF, Scott JF (1971) J Phys Chem 4:1203
46. Lagakos N, Cummins HZ (1974) Phys Rev B 10:1063
47. Mermelstein MD, Cummins HZ (1974) Phys Rev B 16:2177
48. Durvasula LN, Gamon RW (1977) Phys Rev Lett 38:1081
49. Courtens E (1977) Phys Rev Lett 39:566
50. Cowley RA (1970) J Phys Soc Jpn Suppl 28:239
51. Lines ME, Glass AM (1977) Principles and Applications of Ferroelectrics and Related Materials, Oxford University Press
52. Blinc R, Zeks B (1987) Ferroelectrics 72:193
53. Blinc R (2004) Ferroelectrics 301:3
54. Blinc R, Burgar M, Rutar V, Seliger J, Zupancic I (1977) Phys Rev Lett 38:92
55. Mehring M, Suwaleck D (1979) Phys Rev Lett 42:317
56. Mehring M, Becker D (1981) Phys Rev Lett 47:366
57. Klymachyov AN, Dalal NS (1997) Z Phys B 104:651
58. Dalal N, Klymachyov A, Bussmann-Holder A (1998) Phys Rev Lett 81:5924
59. Bussmann-Holder A, Dalal NS, Fu R, Migoni R (2001) J Phys Condensed Matter 13:L231
60. Dalal NS, Gunaydin-Sen O, Fu R, Achey R (2006) Ferroelectrics 337:153
61. Marunka D, Rakvin B (2004) Solid State Commun 129:375
62. Blazey KW, Muller KA, Bednorz JG, Berlinger W, Amoretti G, Buluggiu E, Vera A, Matacotta FC (1987) Phys Rev B 36:7241
63. Stankowski J, Kahol PK, Dalal NS, Moodera JS (1987) Phys Rev B 36:7126

Deutscher G (2007) In: Bussmann-Holder A, Keller H (eds)
High T_c Superconductors and Related Transition Metal Oxides.
Springer-Verlag Berlin Heidelberg, 99–101
DOI 10.1007/978-3-540-71023-3

The High T_c Cuprates as Nanoscale Inhomogeneous Superconductors

G. Deutscher

School of Physics and Astronomy, Tel Aviv University, Ramat Aviv, 69978 Tel Aviv, Israel
guyde@post.tau.ac.il

Because of an unforeseen event my contribution to this volume in honor of the 80th birthday of Alex Müller is much shorter than I would have liked it to be, and I regret it very much. On the other hand, this might be an opportunity to focus on one central issue of the physics of the cuprates, namely whether they are homogeneous superconductors. I will claim that they are not.

The Heat Capacity and Resistive Transitions in Granular Al

The archetype of inhomogeneous superconductivity is granular Aluminum. It is composed of nanoscale Al grains embedded in an insulating matrix, for instance Aluminum oxide. By increasing the oxide content, grains can be pulled apart until the material becomes insulating. One key point is that near the metal to insulator transition the grains size is such that the number of free electrons per grain is of the order of 1000, and therefore the splitting between the electronic levels is of the order of several meV, one order of magnitude larger than the size of superconducting gap in bulk Al. Therefore the isolated grains cannot be superconducting, and a transition to a coherent superconducting state in granular samples cannot be described as being achieved by phase locking between superconducting islands where a pairing amplitude is well established.

Heat capacity and resistive measurements of the superconducting transition in granular Al films having a normal state resistivity of the order of 1000 $\mu\Omega$ cm and a grain size of about 30A have revealed how properties evolve as the metal to insulator transition is approached [1]. Up to a resistivity of about 1000 $\mu\Omega$ cm the heat capacity transition is as predicted for a BCS superconductor with a jump that coincides with the resistive transition. Beyond that, the heat capacity jump becomes progressively broader and weaker, the resistive transition occuring at the temperature where the heat capacity transition starts. Beyond a resistivity of 10 000 $\mu\Omega$ cm the heat capacity transition is not detectable anymore but there is still a resistive transition, which becomes broader and takes place at a lower temperature. Beyond a resistivity of several 10 000 $\mu\Omega$ cm superconductivity is not observed anymore.

The Percolation Model

This experimental behavior is well reproduced by a percolation model whose basic assumption is that only grains effectively coupled to an infinite cluster do contribute to the heat capacity transition [2]. Disorder is characterized by a distribution of inter-grain resistances around an average value related to the macroscopic resistivity of the sample. It is assumed that all grains have the same mean field critical temperature T_{co} and that the mean field gap follows the BCS temperature dependence. The model uses the Ambegaokar-Baratoff expression for the Josephson inter-grain coupling energy, which is then compared to the thermal energy kT to determine if two grains are effectively coupled or not. This is a bond percolation problem. The number of effective superconducting bonds is calculated as a function of the temperature, their number increasing as it is reduced below the mean field critical temperature T_{co}. At some temperature an infinite superconducting cluster may be formed, this is the temperature where the sample's resistance will go to zero. But when the infinite cluster is just formed its mass is infinitesimally small so that a bulk measurement such as that of the heat capacity does not yet show a transition. As the temperature is lowered further, more and more grains are effectively coupled to the infinite cluster whose mass becomes a finite fraction of the total Al mass in the sample, and this shows up in the heat capacity. In spite of the simplifying assumption that only grains coupled within the infinite cluster participate in the heat capacity transition, this model fits quite well the experimental data with only one adjustable parameter which is the width of the distribution of inter-grain resistances.

Application of the Percolation Model to the Cuprates

There is a striking resemblance between the evolution of the heat capacity transition in granular Al as the oxide content is increased, and that seen in YBCO samples as oxygen doping is reduced below its optimum value. Also in that case the heat capacity jump becomes broader and weaker, and the resistive transition takes place at the beginning of the heat capacity transition [3]. This resemblance strongly suggests that a percolation model is appropriate to describe the transition in YBCO.

If a percolation model applies to the cuprates, it means that small metallic clusters, too small to be superconducting by themselves, are formed when holes are introduced before superconductivity is established. Of course it also implies that their behavior cannot be understood within any model that assumes a priori a homogeneous order parameter. The question is then what is the origin of the formation of the metallic clusters, and also what is the origin of pairing.

The formation of metallic clusters already at small doping levels, when samples are insulating on the macroscopic scale, poses no problem in a percolation model. It is in fact consistent with ESR experiments [4]. It has been suggested that they may consist of clustered bi-polarons [5].

A New Interpretation of the Phase Diagram

The origin of pairing is of course the most difficult question, but if one accepts that the percolation model applies, it follows that the origin of the high critical temperature is intimately related to the small size of the clusters that eventually get coupled to form the infinite cluster, as it is in granular Al. If we literally transpose the percolation model developed for granular Al to the cuprates, we obtain a new interpretation of their phase diagram.

Starting from very low doping levels small metallic clusters are formed, which are too far from each other to be coupled. The doping level at which superconductivity appears corresponds to the formation of an infinite cluster. Coupling occurs through the insulating matrix. The mass of the infinite cluster being at first very small, the heat capacity transition is very weak. At optimum doping most of the clusters are part of the infinite cluster, and are still coupled through the insulating matrix. In the overdoped regime an increasing fraction of the clusters get coupled by direct compact, the surface to volume ratio goes down and with it the critical temperature. This is at this stage only a very rough description that clearly needs to be elaborated further.

Acknowledgements I am profoundly indebted to Alex Müller for a number of illuminating discussions on the question of inhomogeneity in the cuprates. The percolation approach proposed here is in great part a result of these discussions.

References

1. Worthington T, Lindenfeld P, Deutscher G (1978) Phys Rev Lett 41:316
2. Deutscher G, Entin-Wohlman O, Fishman S, Shapira Y (1980) Phys Rev B 21:5041
3. Loram J (1998) J Phys Chem Solids 59:2091
4. Müller A, private communication
5. Bishop AB, Mihailovic D, Mustre de Leon J (2003) J Phys Condens Matter 15:1160

Egami T (2007) In: Bussmann-Holder A, Keller H (eds)
High T_c Superconductors and Related Transition Metal Oxides.
Springer-Verlag Berlin Heidelberg, 103–129
DOI 10.1007/978-3-540-71023-3

Essential Role of the Lattice in the Mechanism of High Temperature Superconductivity

T. Egami[1,2]

[1]University of Tennessee, Science Alliance,
Department of Materials Science and Engineering,
Department of Physics and Astronomy and the Joint Institute for Neutron Sciences,
Knoxville, TN 37996-1508, USA
egami@utk.edu

[2]Oak Ridge National Laboratory, Oak Ridge, TN 37831, USA

Abstract The historical discovery of high-temperature superconductivity by Bednorz and Müller was led by the idea of strong electron-phonon coupling in the cuprate oxides associated with Jahn-Teller distortion. Nevertheless, the majority in the field quickly converged to the view that it is a purely electronic phenomenon involving only spins and lowering of the kinetic energy, and the lattice and phonons play little or no role in the mechanism. After two decades this majority view is now seriously challenged by experimental observations of strong electron-phonon coupling and nano-scale electronic inhomogeneity. The transition metal oxides are characterized by strong competition among multiple degrees of freedom; spin, charge, orbital and lattice. It is more natural to conjecture that the high-temperature superconductivity is a synergetic phenomenon with many degrees of freedom, including phonons. We discuss the possible role the phonons may play in the synergy of multiple degrees of freedom to produce the high-temperature superconductivity.

Keywords Electron–phonon coupling · Lattice effect · Phonons · Polarons

Introduction

When the phenomenon of high-temperature superconductivity (HTSC) was discovered in 1986 [1] everybody became convinced that the phonon-based BCS mechanism [2] could not explain this phenomenon, because the critical temperature, T_c, was simply too high [3]. It did not take long for the majority in the field to come to believe that the HTSC was based upon a completely new mechanism, most likely a purely electronic one involving only the spin degree of freedom [4]. Focusing on the closeness of HTSC to the antiferromagnetic (AFM) phase in the phase diagram it has been argued that some kind of spin-fluctuations are mediating the pairing. However, in spite of the explosive level of research effort world-wide over two decades, the origin of the phenomenon remains largely a mystery. The main obstacle is the high degree of complexity, unprecedented in the field of condensed matter physics. Not only that we have to deal directly with the problem of strong electron correlation, but also the structure and interactions are very complex, involving multiple elements and many degrees of freedom.

Since the problem of strong electron correlation is so formidable it is not surprising that initially the theories focused on the simplified models, most notably the t–J model [5]. At the time of the discovery the most prevailing assumption was that the doped holes go into the z^2 orbital of Cu. In this scenario the one-band t–J Hamiltonian involving only Cu spins is a reasonable starting point, albeit it could be too simple. But it was soon discovered that the doped holes mostly enter the oxygen p-orbital, heavily hybridized with the Cu $d(x^2 - y^2)$-orbital [6]. Then the more natural minimalist model is the three-band Hubbard Hamiltonian, involving the Cu $d(x^2 - y^2)$-orbital and the oxygen p_x- and p_y-orbitals [7].

However, the three-band Hubbard model is a formidable object to work with. To circumvent this difficulty Zhang and Rice (ZR) came up with a remarkable idea, the ZR-singlet state [8]. This idea quickly became extremely popular, since it appeared to be the right solution at low doping levels, and above all it conveniently justified the one-band t–J model. On top of that the d-symmetry of the superconducting order parameter was first proposed based upon the NMR data [9, 10], and confirmed by direct measurements [11, 12]. The t–J model appears to support the d-wave superconductivity through the spin-fluctuation mechanism [13]. Thus even though the $t - J$ model was initially proposed as the minimalist model, the researchers started to believe in its reality. Some people believe that the mechanism of HTSC is already essentially known [14].

But the developments in the last decade impressed many of us that the story is not that simple. In stark contrast to early days of research the majority now recognizes that the mechanism of HTSC is very complex, and we are far from a complete resolution of the problem. The history of the research on the mechanism of the HTSC developed in a strange, preposterous way. In the beginning many theoreticians believed that they understood the principal mechanism of the HTSC, and details would be clarified in a short time. As the experimental techniques were improved and better experimental results were obtained many of these theories were proven wrong. For instance, as discussed briefly below the "spin-alone" scenario does not appear to answer the question completely, and something critical seems to be still missing from the current theory. In a sense the field developed backward, from the initial unfounded optimism and arrogance to the recent recognition that we know so little about this complex material. The right thing to do now may be to reconsider the placement of all the pieces of the giant puzzle from the beginning.

In this article we first briefly touch upon the problems of the current "spin-alone" scenario, discuss the unusual nature of the e–p interaction in the cuprates, review recent results of inelastic neutron scattering on the Cu – O bond-stretching phonons that suggest intimate involvement of phonons in

HTSC, and finally discuss the possible role of the lattice distortion and phonons in the HTSC mechanism. Since the field is still wide open, we do not claim to understand this complex phenomenon completely. But we wish to show that a pathway through the lattice effect does exist as an alternative to more standard theories, and needs to be taken seriously.

Difficulties with the Current HTSC Theories

Problems with the Basic Assumptions

As mentioned above the current quagmire in identifying the microscopic mechanism of HTSC seems to have been in part caused by confusing the theoretical expediency with the reality. In our view at least the following three basic assumptions commonly made in many theories need to be seriously re-examined:

1. Focus on the t–J model.
2. Assumption of perfect lattice periodicity.
3. Neglecting the degrees of freedom other than spins.

A. The *t–J* Model

Currently the majority of the theories focus on the one-band t–J Hamiltonian,

$$H_{t-J} = \sum_{i,j,\sigma} t_{ij} \left(c^{+}_{i,\sigma} c_{j,\sigma} + c^{+}_{j,\sigma} c_{i,\sigma} \right) + J \sum_{i,j} \left(\boldsymbol{S}_i \cdot \boldsymbol{S}_j - \frac{1}{4} n_i n_j \right) , \tag{1}$$

where t_{ij} and J are the hopping matrix and the exchange constant between the site i and j, $\boldsymbol{S}_i$ and n_i denote the spin and charge of the site i. There are at least two fundamental problems in focusing on this Hamiltonian alone. The first is the difference between the Hubbard Hamiltonian and the t–J Hamiltonian. The second is neglecting the crucial nature of the doped holes that they mainly reside on oxygen ions.

The t–J Hamiltonian is obtained by starting from the $U \to \infty$ limit of the Hubbard model and taking the effect of hopping integral t by perturbation [5, 13]. The upper Hubbard band is taken away by the Gutzwiller projection. Thus in the t–J model each atom is locally strongly spin-polarized, even when the overall interatomic spin correlation is weak. Consequently the spin fluctuation is overestimated, and it is not surprising that in the t–J model superconductivity due to spin fluctuation appears to take place [13]. However, it is still unclear that the two-dimensional Hubbard Hamiltonian supports superconductivity. Even when it appears to do so it occurs in parameter ranges that do not justify the t–J model [15].

Secondly, the t–Jmodel and one-band Hubbard model was justified by the Zhang-Rice (ZR) model, as we mentioned earlier. By being in the ZR singlet state the doped holes become spin depolarized, and gains mobility. The ZR singlet state appears to be a good model for an independent localized hole. But when the doping level is high the ZR-singlets start to overlap with each other and become delocalized. When they are delocalized it is not clear whether they still can be described as the linear combination of the local bound ZR state or not. It is quite possible that they are better described by a more complex model, such as the three-site model by Emery [7, 16]. Also they may couple to other degrees of freedom, such as the lattice, and this could fundamentally change the symmetry of the ZR-state. If the t–J model is not the right model for the cuprates, we have to go back to square one in the discussion of the mechanism of HTSC.

B. Perfect Lattice Periodicity

There have been a large number of reports that question the basic assumption that the HTSC can occur in the perfectly periodic lattice without phonons, as discussed below [17]. While these early reports were largely dismissed as by-products of strong coupling, even the staunch majority could not ignore the spin-charge stripe state [18] and the obvious nano-scale electronic inhomogeneity observed by STM/STS [19–22]. A complex state with the unit cell larger than the square cell of the CuO_2 plane being the underlying ground state is a real possibility.

C. Coupling to other degrees of freedom

Recent observations of strong e–p coupling by angle-resolved photoemission spectroscopy (ARPES) [23–25] are seriously challenging the majority belief that phonons are irrelevant to the HTSC. The argument that the quantum-criticality related to the charge degree of freedom is also gaining support [26]. More detailed arguments will be given below.

The "Spin-alone" Mechanism

There are at least two different views in the "spin-alone" community. The first group believes in exotic mechanisms, such as the resonating valence bond (RVB) mechanism, and argues that the lowering of the kinetic energy by releasing the AFM constraint, rather than the potential energy gain by pairing, provides the driving force for the HTSC [3, 4, 27]. The reality of the RVB state in the two-dimensional square lattice has never been proven by experiment or simulation, and theories have never been sufficiently developed to test the validity of the predictions. As for the idea of lowering the kinetic energy, the experimental results are mixed [28–30], and it appears to be untrue for overdoped samples [30].

The second group considers spin-fluctuations as the "glue" for pairing, instead of phonons for the BCS theory [31–33]. In terms of the theoretical framework, still the conventional strong coupling theory is used. In this sense these ideas are "conventional". The problem of this approach is that the strength of spin fluctuations strongly depends on the doping level, and decreases quickly with doping. In the $YBa_2Cu_3O_{6+\delta}$ system, for instance, the intensity of the spin fluctuations below 50 meV extraporates to zero at the barely overdoped state of $\delta = 1$, where T_c is still sizable [34]. Recent evaluation of the absolute value of the total spin correlation through the careful examination of the inelastic neutron scattering data suggests that the total driving force due to spin fluctuation can amount to 2 meV per Cu for $YBa_2Cu_3O_{6.95}$ and is much larger than the condensation energy estimated from the specific heat data [35]. However, this energy is still much smaller than kT_c. If we consider that there are additional energy costs of pair formation, such as the increase in the kinetic energy and the Coulomb repulsion energy, it does not appear to be sufficient in accounting for the HTSC phase transition. Thus it may be fair to conclude that the "spin-alone" scenario is far from being proven correct.

Early Evidences of Lattice Involvement

It is well known that Bednorz and Müller focused on the cuprates, thinking that the involvement of the Jahn-Teller distortion will enhance the electron-phonon (e–p) coupling [36]. For this mechanism to work the holes must enter the z^2 orbital of Cu, but as mentioned above it was soon discovered that the doped holes mostly enter the oxygen p-orbital heavily hybridized with the Cu $d(x^2 - y^2)$-orbital [6]. Unfortunately this minor misstep and the near absence of the isotope effect [37] caused the massive short-circuit effect for the majority to abandon the phonon mechanism entirely. But the isotope effect is absent in many BCS superconductors [38], and strong isotope effects have been observed for the penetration depth [39, 40] and the pseudogap temperature [41, 42]. In addition the z^2 orbital Jahn-Teller effect is not the only source of the strong e–p coupling.

Actually the most important essence of the Bednorz-Müller idea is that because of the strong e–p coupling the doped holes may lead to polaronic superconductivity. The BCS mechanism is appropriate for weak e–p coupling. As the e–p coupling becomes stronger the Cooper pairs become smaller and polaronic. While it is usually assumed that the large polaronic mass will bring the critical temperature, T_c, down [43], details of the polaronic superconductivity is still not well understood, and hence there is hope that the polaronic HTSC may be possible. In fact Alexandrov proposes a superlight bipolaron mechanism [44, 45] in which the mass of the bipolaron remains low in spite of strong coupling. However, the theory includes many layers of approximations, and the validity of these approximations is not yet fully tested. The theory of polaronic superconductivity was first proposed by Schafroth [46] before the BCS theory. But mathematical difficulties prevented this theory from being developed, and the same difficulties still hamper the effort. Consequently the idea of polaronic HTSC still has not gained the full citizenship in the theoretical physics community.

On the other hand there are a large number of experimental observations that suggest the polaronic nature of the Cooper pairs in the cuprates [16]. Probes of local structure, such as EXAFS [47, 48], pulsed neutron pair-density function (PDF) [49, 50], neutron resonance [51], and ion-channeling [52, 53], all show the evidence of anomalous local lattice distortion or lattice dynamics at T_c or the pseudo-gap temperature, T_{PG}. An example is seen in Fig. 1 that shows the PDF peak height at 3.4 Å for $Tl_2Ba_2CaCu_2O_8$ [49]. The PDF describes the distribution of distances between atoms, and is obtained by Fourier-transforming the structure function, $S(Q)$, ($Q = 4\pi \sin\theta/\lambda$, where θ is the diffraction angle and λ is the wavelength of the probe) determined by neutron or X-ray diffraction, including both Bragg peaks and diffuse scattering [54]. Since it includes the diffuse scattering it can describe aperiodic as well as periodic structure. This technique has long been used in the study of liquids and glasses, but owing to the advances in the synchrotron based X-ray or neutron sources that offer probes with short wavelengths it became possible to apply this to crystals [54]. In obtaining the PDF the Fourier integral of $S(Q)$ has to be carried out to infinity in Q, but Q is bound by the wavelength of the probe, since $Q < 4\pi/\lambda$. Terminating the integral at values of Q not large enough for $S(Q)$ to be constant results in the unwanted oscillations called termination errors. Thus the availability of high energy probes that allows for a wider Q range was critical for the success of this method.

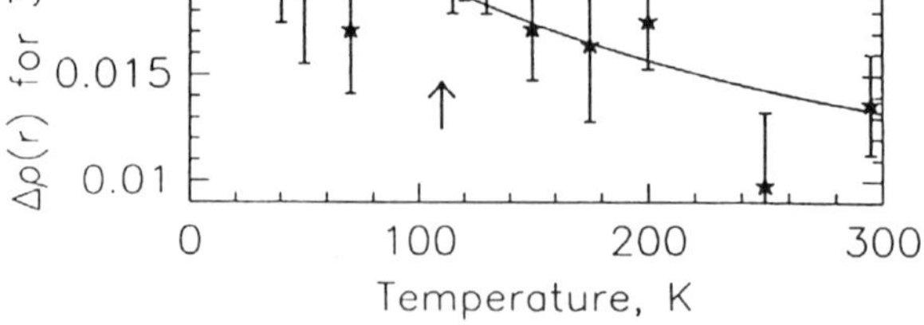

Fig. 1 The pulsed neutron PDF peak height of $Tl_2Ba_2CaCu_2O_8$ at 3.4 Å as a function of temperature [49]. The *arrow* indicates superconducting transition temperature, T_C

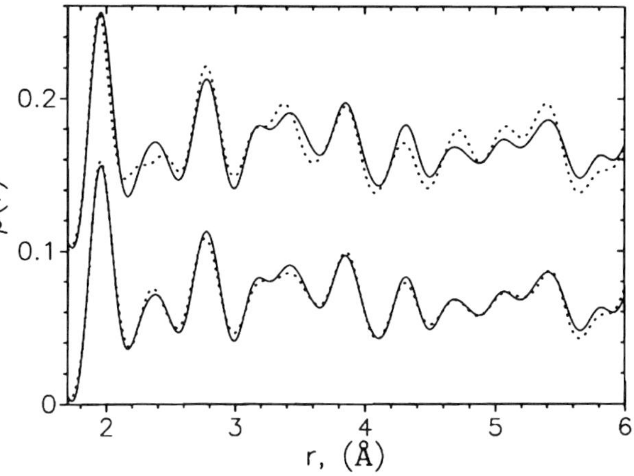

Fig. 2 The PDF of $Tl_2Ba_2CaCu_2O_8$ at $T = 120$ K (*solid line*, *above* and *below*), and model PDF with only Tl and O(3) (oxygen in Tl plane) adjusted (*dotted line*, *above*), and model PDF with O(1) and O(2) also adjusted (*dotted line*, *below*) [49]

The PDF of $Tl_2Ba_2CaCu_2O_8$ at $T = 10$ K is shown in Fig. 2 [49]. The peak at 1.95 Å corresponds to the in-plane Cu – O distance, while the out-of-plane Cu – O distance forms a subpeak at 2.35 Å. The in-plane O – O distance is a major part of the peak at 2.8 Å, and the distance between the apical oxygen (O2) and in-plane oxygen (O1), 3.2 Å, forms another subpeak. Modeling suggests that the unusual increase of the PDF intensity at 3.4 Å around $T_c \approx T_{PG}$ is due to the O1 – O2 distance splitting into two distances, at 3.0 and 3.4 Å. Such a bifurcation of distances because of local lattice distortion is a strong evidence for the formation of polarons. The fact it shows up only in the vicinity of T_c and its dependence on the detector angle suggest the dynamic nature of the distortion. Only near T_c the dynamic slow-down makes the distortion visible to the PDF, and far from T_c its vibration is too fast so that the PDF shows only the averaged distance [55].

However, these observations were dismissed by the majority merely as byproducts of strong coupling. Indeed as the pairing force becomes strong the spread of the Cooper pair becomes small, and they naturally assume a polaronic character. It took more direct evidence of strong *e–p* coupling through the angle-resolved photoelectron spectroscopy (ARPES) by Lanzara [22] for the field to wake up to the real possibility of the phonon contribution.

Unconventional Nature of Electron-Phonon Coupling in the Cuprates

The *e–p* coupling in the cuprates as estimated by the density functional theory (DFT) calculations is only of the order of 0.3, much too small to account for the HTSC [56]. But a rather convincing evidence of strong *e–p* coupling was provided by the ARPES measurements. The electron dispersion observed by the ARPES is known to have a "kink" just below the Fermi level [57]. Initially this kink was interpreted in terms of the coupling to the spin resonance mode observed by inelastic neutron scattering [57, 58]. But Lanzara argued that the kink has to be due to phonons, based upon the independence of the kink position in energy of the superconducting gap [22] and the isotope effect [23]. Lanzara also pointed out that the isotope effect extends up to several hundreds of meV, far above the energy of the phonons, indicating a strong multi-phonon effect [23]. She argued that this multi-phonon effect is the evidence of polaron formation.

The phonon mode that was proposed to couple strongly to the "nodal" quasiparticles, nearly along the $[\pi, \pi]$ direction, are the half-breathing Cu – O bond-stretching LO phonons shown in Fig. 3 [59, 60]. The most important characteristic of this mode is that it modifies the Cu – O distance, and thus the Cu-d/O-p overlap. Since the undoped cuprates are charge-transfer insulators [61], modifying the Cu-d/O-p overlap means creating virtual charge transfer between Cu and O, across the Fermi level, as shown in Fig. 4. This

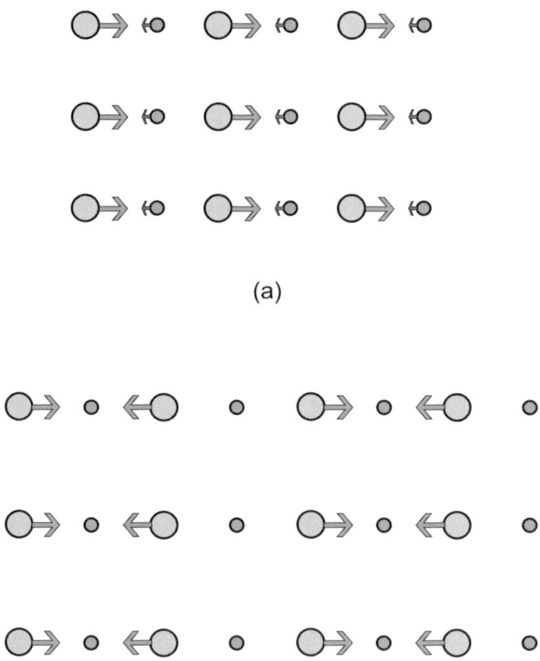

Fig. 3 Cu – O bond-stretching LO phonon modes. **a** Ferroelectric mode at $Q = 0$, and **b** Half-breathing mode at $Q = [0.5, 0]$

phonon-induced charge transfer is the source of an unconventional strong *e–p* coupling, and that is the reason our group has continuously focused on this mode [60, 62–66].

The *e–p* coupling of a covalent solid such as the cuprate is best described by the Su-Schrieffer-Heeger (SSH) model [67],

$$H_{e\text{-}p} = \frac{\alpha}{a} \sum_{i,j,\sigma} (u_i - u_j) \left(c^{+}_{i,\sigma} c_{j,\sigma} + c^{+}_{j,\sigma} c_{i,\sigma} \right) , \tag{2}$$

where u_i is the displacement of the i-th atom, a is the interatomic distance. In the SSH coupling phonons modify the hopping integral t, so that its effect extends much beyond the phonon energy from the Fermi surface. As an example the optical conductivity (Fig. 5) calculated for a one-dimensional two-band Hubbard model,

$$H_{\text{Hubbard}} = \sum_{i,j,\sigma} t_{ij} \left(c^{+}_{i,\sigma} c_{j,\sigma} + c^{+}_{j,\sigma} c_{i,\sigma} \right) + U_1 \sum_{i} n_{i,\uparrow} n_{i,\downarrow} + U_2 \sum_{j} n_{j,\uparrow} n_{j,\downarrow} \tag{3}$$

with the SSH coupling (Eq. 2) to a frozen Cu – O bond-stretching optical phonon [66]. The optical conductivity shown in Fig. 5 is modified up to sev-

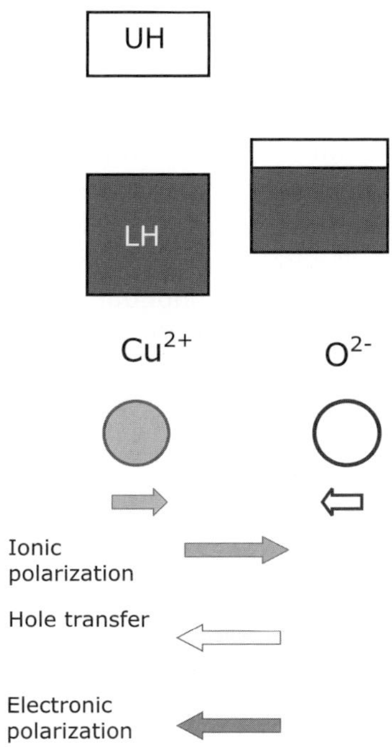

Fig. 4 Phonon-induced charge transfer. As Cu and O ions move closer by phonon the *d–p* overlap increases and doped holes on O are transferred to Cu. This produces electric polarization *pointing left*

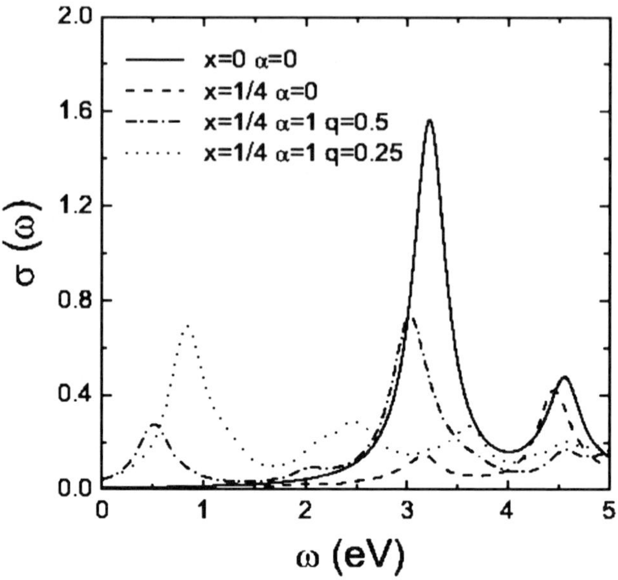

Fig. 5 Optical conductivity of two-band Hubbard model [66]. *Solid line* for undoped system ($x = 0$), without phonon ($\alpha = 0$), *dashed line* for doped ($x = 0.25$) but without phonon ($\alpha = 0$), and for doped ($x = 0.25$) with ($\alpha = 1$) for $q = 0.5$; zone-boundary (*dash-dot*), and for $q = 0.25$ (*dotted*)

eral eV by phonon, and phonons produce the in-gap states. This illustrates the strong effect of multi-phonon processes in this doped charge-transfer insulator. While the optical conductivity being modified up to high energies upon superconductive transition was argued as evidence for an electronic mechanism [27, 68, 69], the result above shows that the phonon mechanism is also consistent with the observation, since phonons are strongly coupled to charge excitations. This is also consistent with the strong isotope effect on the electronic band structure recently observed by the ARPES measurement [23] and the isotope effect on the effective mass [39, 40].

The phonon-induced charge transfer is well known in ferroelectric titanates [70]. In the titanates, such as $BaTiO_3$, lattice distortion produces not only ionic polarization but also electronic polarization due to deformation-induced charge transfer. The total polarization is the sum of these two terms,

$$P = uZ + a\Delta Z\,, \tag{4}$$

where u is the ionic displacement, a is the inter-ionic distance and ΔZ is the charge transfer due to displacement. The Born effective charge Z^* is defined as

$$Z^* = Z + \frac{a}{u}\Delta Z\,. \tag{5}$$

In ferroelectric titanates the nominal charge on Ti is + 4, but the effective charge can be twice as much. An elegant Berry phase theory was developed to calculate the electronic polarization and the Born effective charge [70, 71]. In the doped cuprates as it turned out the effective charge strongly depends on the phonon wavevector q, and the induced electronic polarization is opposite in direction compared to the ionic polarization.

The effective charge calculated for the one-dimensional two-band Hubbard model with a Cu – O bond-stretching optical phonon is shown in Fig. 6 [66]. Unlike the ferroelectric titanates the charge transfer effect at $q = 0$ is very small, and the effect is pronounced to large q vectors. This means while the charge transfer between oxygen and titanium is important for the ferroelectricity of titanates, in the cuprates the relevant charge transfer processes are those among the copper ions, instigated by the displacements of oxygen ions. For instance in the case of the zone-boundary "half-breathing" mode, oxygen ions move between the neighboring copper ions which by themselves are not moving. Since the local charge on oxygen does not change much during the motion, Cu – O charge transfer does not appear to be important. But actually charge transfer between Cu ions is induced by Cu – O charge transfer, and induced charges mostly go through the oxygen ion rather than staying there. This crucial role of oxygen in the Cu – Cu charge transfer is expressed in terms of a large electronic contribution to the effective charge of oxygen shown in Fig. 2. It was also found that the phonon-induced charge transfer

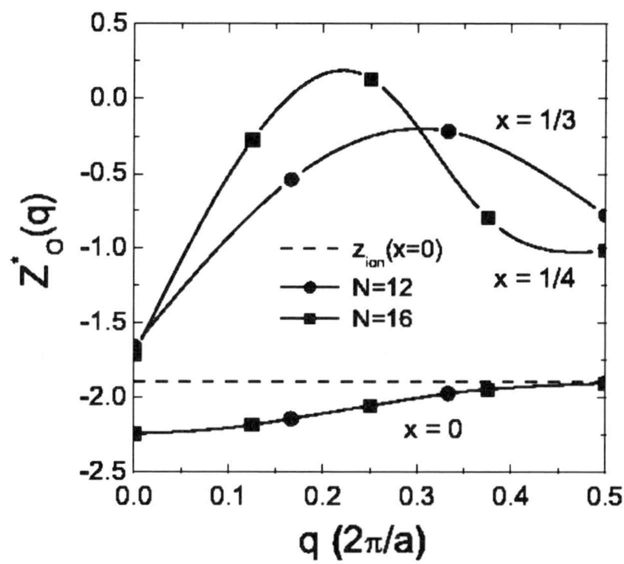

Fig. 6 Effective charge of oxygen calculated for the two-band Hubbard model for various doping levels ($x = 0, 1/4, 1/3$) [66]. The ionic value is – 1.9, but the large electronic contribution makes the effective charge much less negative, or even positive

is spin-dependent, and the phonon affects the spin correlations and dynamics [66]. Thus in the cuprates spin, charge and phonons all interact, rendering an unconventional character to the *e–p* coupling in the cuprates.

Evolution of the Cu – O Bond-Stretching LO Phonons with Temperature

While most of the phonon modes in the cuprates show little temperature dependence, the Cu – O bond-stretching LO mode shows strong and unusual dependence on temperature. As shown in Fig. 7 for $La_{1.85}Sr_{0.15}CuO_4$ [60], at $T = 300$ K the mode shows a usual sinusoidal dispersion. But at $T = 10$ K the dispersion becomes discontinuous at $[\pi/2, 0]$ in the unit of $1/a$, or [0.25, 0] in the unit of $2\pi/a$, where a is the tetragonal unit cell size, or the in-plane Cu – Cu distance, as though there is a zone boundary at $[\pi/2, 0]$ [60]. Recently it was proposed to explain this discontinuity in terms of the spin-charge stripe formation [72]. However, the stripe formation should influence the transverse mode more strongly [73], while there is no discontinuity in the transverse mode at all [65]. An alternative explanation is the sub-Brillouin zone formation as discussed below.

The same discontinuity is seen for $YBa_2Cu_3O_{6+\delta}$ (YBCO-123) [59, 65]. But upon a closer look more unusual temperature dependence was found. Figure 8 shows the temperature dependence of the inelastic neutron scattering intensity for $YBa_2Cu_3O_{6.95}$ corrected for the Bose factor, at $Q = [0.25, 0]$,

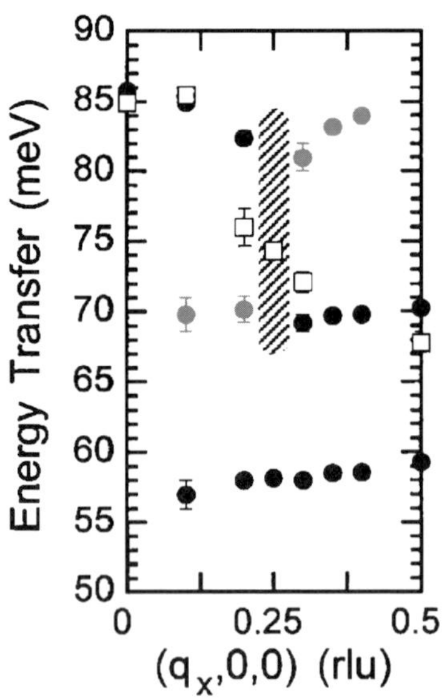

Fig. 7 Dispersion of the oxygen LO phonon mode of LSCO at 10 K (*filled square*) and 300 K (empty square) [60]. The branch shown with *filled circles* is the Cu – O – Cu bonding mode

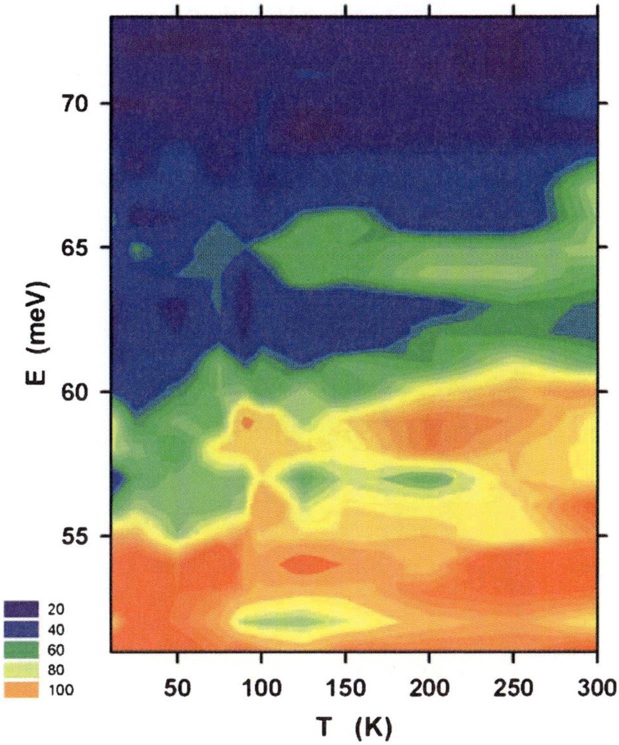

Fig. 8 Temperature dependence of $S(Q, E)$ at $Q = 3.25$ for $YBa_2Cu_3O_{6.95}$, corrected for the Bose-Einstein factor [65]

measured at the HFIR of Oak Ridge National Laboratory [65]. The mode at 64 meV shifts to 53 meV below about 100 K. The similar plots for $YBa_2Cu_3O_{6.8}$ and $YBa_2Cu_3O_{6.6}$ are shown in Figs. 9 and 10 [74]. In all cases the high temperature mode around 64 meV becomes softened at low temperatures. The softening appears to happen around the pseudogap temperature (= 93 K for $YBa_2Cu_3O_{6.95}$, 160 K for $YBa_2Cu_3O_{6.8}$, and ~ 300 K for $YBa_2Cu_3O_{6.6}$). Figure 11 [65] shows the difference in the average intensity from 51 to 55 meV, and that from 56 to 68 meV for $YBa_2Cu_3O_{6.95}$. The change is apparently happening around $T_c = T_{PG}$.

Now the softened energy at low temperature is dependent on composition. Figure 12 [74] shows the phonon energy of this mode at 10 K against T_c (= 93 K for $YBa_2Cu_3O_{6.95}$, 80 K for $YBa_2Cu_3O_{6.8}$, and ~ 60 K for $YBa_2Cu_3O_{6.6}$), and the phonon energy at high temperature for $T_c = 0$. Interestingly the amount of softening is linearly related to T_c [74]. Furthermore, this relationship is virtually identical with the one observed by the STM/STS for $Bi_2Sr_2CaCu_2O_8$ (BiSCCO-1211) [22]. In the STM/STS study by the group of Davis they locally determined the superconductive gap, Δ, through the first

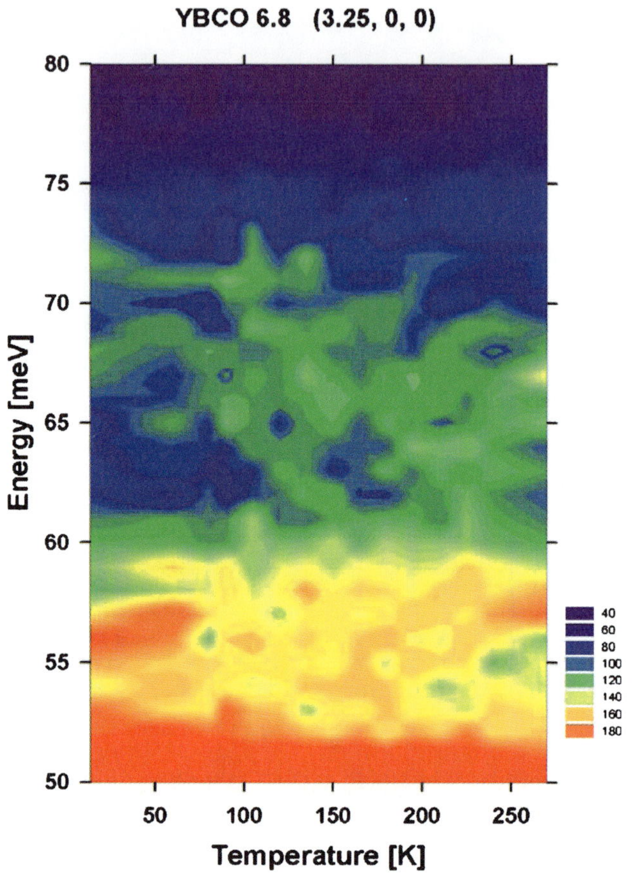

Fig. 9 The same as Fig. 8, but for $YBa_2Cu_3O_{6.8}$ [73]

derivative of the tunneling current, dI/dV, and the energy of the coupled boson mode, Ω, through the second derivative, d^2I/dV^2. They found that Δ and Ω are linearly related, and this relationship is very close to the one shown in Fig. 12. The STM/STS study and the neutron scattering were done on different materials, BiSCCO for STM and YBCO for neutron. But the difference in the compounds may not be so important here, since the high-energy portion of the phonon density of states and the in-plane lattice constants for the two are very similar. It is therefore tempting to equate the interacting boson mode observed by the STM/STS to the Cu – O bond-stretching LO mode.

In addition it was found that the high-temperature mode at 64 meV is a localized mode. Figure 13 shows the difference in the neutron inelastic scattering intensity for $YBa_2Cu_3O_{6.6}$ between $T = 290$ K and 10 K. The extra mode at 64 meV has almost no dispersion, indicating that it is a local mode. At low temperatures this mode is no longer local, and shows a normal disper-

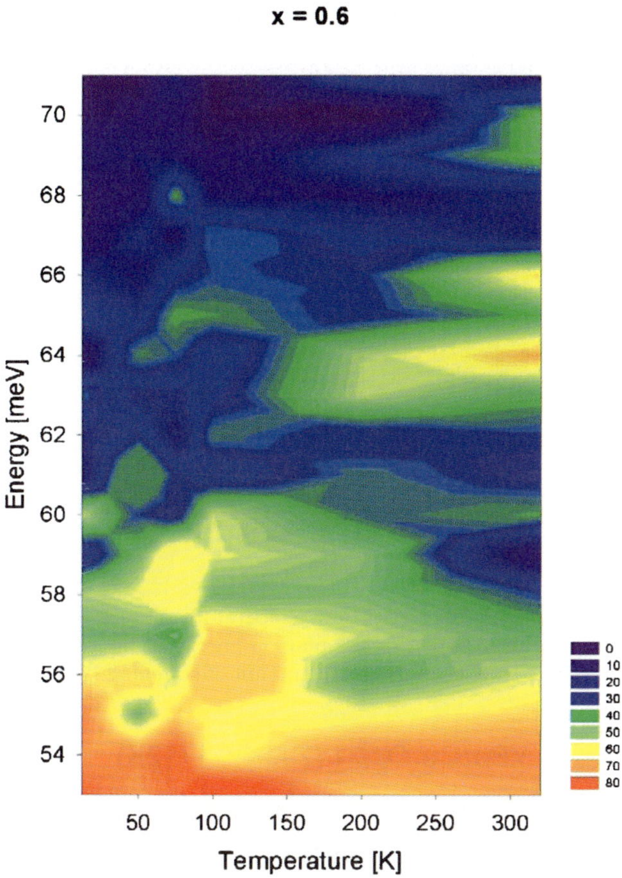

Fig. 10 The same as Fig. 8, but for $YBa_2Cu_3O_{6.6}$ [73]

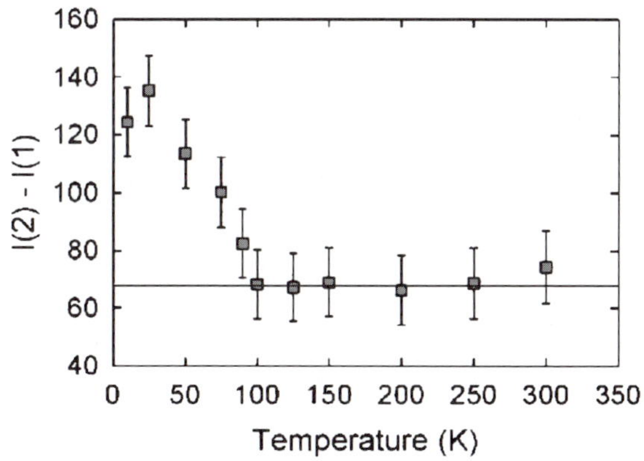

Fig. 11 Difference in the average of $S(3.25, E)$ from 51 to 55 meV, I(2), and $S(3.25, E)$ from 56 to 68 meV, I(1), as a function of temperature [65]

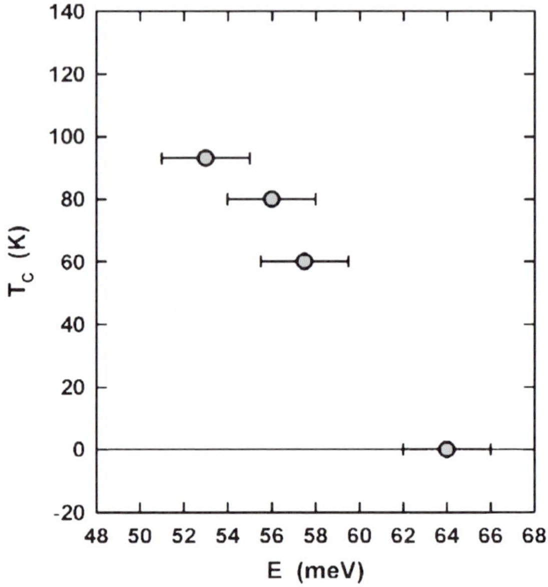

Fig. 12 The softened phonon energy at low temperatures (10 K) plotted against T_C for $YBa_2Cu_3O_{6+\delta}$ ($\delta = 0.6$, 0.8, 0.95). Also the phonon energy at high temperature ($> T_{PG}$), 64 meV, is plotted for $T_C = 0$, to indicate the amount of softening [74]

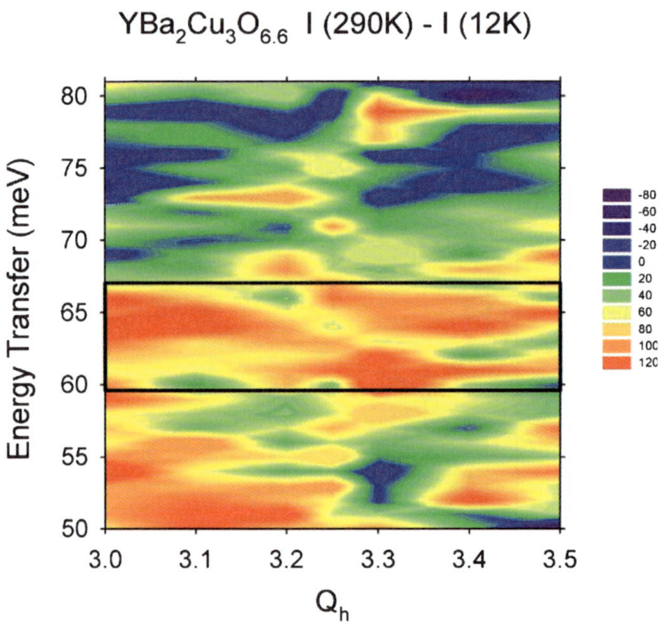

Fig. 13 Difference in $S(Q, E)$ between 295 K and 10 K for $YBa_2Cu_3O_{6.6}$ showing the presence of a localized mode at high temperatures [74]

sion. More importantly, the strength of this mode is only about 1/4 of the full mode. This indicates that there is an expansion of the unit cell, at least by a factor of 4. Thus each original mode splits into four, and only one of them shows the localization and anomalous temperature dependence. This inflation in complexity is consistent with the superlattice formation as discussed later.

Electronic Inhomogeneity in Space

Stripe Phase and STM/STS Results

As mentioned above initially the majority opinion in the field was that the HTSC occurs in the perfectly periodic square CuO_2 lattice through Cu spin fluctuations. The first challenge to this belief was made by the observation of the spin-charge stripe structure in $La_{1.475}Nd_{0.4}Sr_{0.125}CuO_4$ [18]. Built upon the idea that the AFM domain wall attracts charge [75], it was proposed that the stationary stripe structure of alternating AFM order and metallic stripe with the superperiodicity of $4a$ exists in this compound. In the original picture three rows of Cu chains in the AFM state and one metallic Cu row were thought to comprise the stripe structure [18]. The one-dimensional hole density is 0.5 per Cu ion. The recent soft x-ray anomalous scattering study [76], however, finds the holes to be centered on oxygen and spread over two rows of Cu chains (bond-centered for the Cu lattice). In this picture the AFM portion is also made of two rows of Cu chains, and resembles the spin-ladder structure.

The static stripe structure apparently competes against the superconductivity [77]. However, the periodicity of the incommensurate spin excitation spectrum observed by inelastic neutron scattering for superconducting compounds is proportional to the doped hole density (and to T_c) [78], and the periodicity of the stripe structure falls right in line, even though the stripe structure is static. This coincidence forms the basis of the conjecture that the dynamic stripe structure exists even in the superconducting cuprates. It is now widely believed, without hard evidences, that the dynamic stripe structure is the origin of the pseudogap state.

Several theories have been advanced using the stripe structure as the framework for the HTSC mechanism. The Kivelson-Fradkin-Emery theory assumes that the spin excitations in the AFM portion of the stripes pair the carriers in the metallic portion [27]. The Zaanen theory focuses on the topology of the stripe state, and argues that the quantum mechanical gliding of the dislocations in the tripe structure produces superfluidity [79]. Tranquada focuses on the spin-ladder of the AFM structure, and argue that the one-dimensional nature of the spin-ladder produces unconventional superconductivity [80]. Indeed the spin-ladder compound, $Sr_2Ca_{12}Cu_{24}O_{41}$, shows superconductivity under pressure [81]. However, T_c is quite low, and the symmetry could be *s*-type [82].

The next serious challenge on the idea of perfect periodicity came from the observation of extensive spatial inhomogeneity by the STM/STS studies [18–21]. While much of the inhomogeneity appears to be extrinsic and is related to the randomness introduced by the interstitial oxygen ions in BiSCCO-2212 [83], the fact that superconductivity survives such strong spatial inhomogeneity is surprising. Furthermore, the presence of a regular $4a \times 4a$ charge checkerboard structure [84, 85] as well as the stripe-like structure [86] suggests that the CuO_2 plane is far from homogeneous, and some local self-organization of spin and charge may be taking place. Furthermore the environment of Cu rarely has the four-fold symmetry necessary for the ZR state, and holes are more often localized on oxygen as in the Emery model [86]. This presents another challenge to the t–J model.

Intermediate Phase

It is useful to reconsider the nature of the stripe phase as an example of the intermediate phase between the Mott-Hubbard insulator and the Fermi-liquid metal. Doped holes in the AFM state are immobile, since they have to disturb the AFM spin ordering to move [4]. This immobility costs a high kinetic energy, and in addition magnetic dilution cost the magnetic energy. Thus the free energy of the doped AFM system increases quickly with doping. On the other hand the free energy of the Fermi-liquid phase is a weak function of the mobile hole density. Thus the theory of this phase transition from the doped AFM insulator to a Fermi-liquid metal [87] predicts high propensity of phase separation, even though phase separation into hole-rich and hole-poor regions will cost a high Coulombic energy. Such phase separation has been predicted by many, starting from Gor'kov [88].

The idea of electronic phase separation is consistent with the STM/STS observations that the surface of BiSCCO is made of nano-scale superconducting and pseudogap regions coexisting without an apparent order. It is reasonable to assume that the superconducting regions are metallic, and the pseudogap regions are dominated by the local AFM order. However, the random picture of the STM/STS may be a little misleading, since it is now known that the electronic inhomogeneity is related to interstitial oxygen impurities [83]. It is possible that without chemical disorder the AFM regions and metallic regions would self-organize, for instance in the stripe state, or as has already been seen by the STM/STS, the $4a \times 4a$ checkerboard state.

The stripe state, however, appears to be competing against the HTSC state [76]. It is also strongly one-dimensional, and appears to be inconsistent with the electronic state observed by the ARPES which is highly two-dimensional. Thus, while the stripe state could be one of the intermediate states, its universal relevance to the HTSC mechanism as proposed by Tranquada [80] remains to be proven. The checker-board state is seen more frequently in the underdoped state, and is associated with the pseudogap state, rather than the HTSC state [85]. It apparently is closely related to the nesting of the Fermi surface near the $[\pi, 0]$ point [21], and thus has the nature of charge density wave (CDW). This again will compete against the HTSC state, rather than help creating it.

On the other hand the recently observed superlattice may be more promising as an indicator of the intermediate state. In $YBa_2Cu_3O_{6.6}$ (YBCO-6.6) a diffuse superlattice diffraction peak of the type [0.25, 0.25, 0] was observed by neutron diffraction at $T = 8$ K, indicating the formation of the $2\sqrt{2}a \times 2\sqrt{2}b$ superlattice [89].

The [0.25, 0.25, 0] type superlattice order introduces eight sub-zones in the Brillouin zone, as shown in Fig. 14 [90]. Since the superlattice peaks are weak and the superlattice umklapp is weak, the extended zone scheme as in Fig. 14, rather than the folded zone scheme, is appropriate for discussion. It is important to note that the sub-zone 1 (gray colored) is far above the Fermi level, and the sub-zone 8 (hatched) is far below the Fermi level. On the other hand the Fermi surface cuts across other sub-zones. Therefore it is possible that the carriers in the sub-zones 1 and 8 are quite different in character from those in other sub-zones. It is also possible that the phonons in differ-

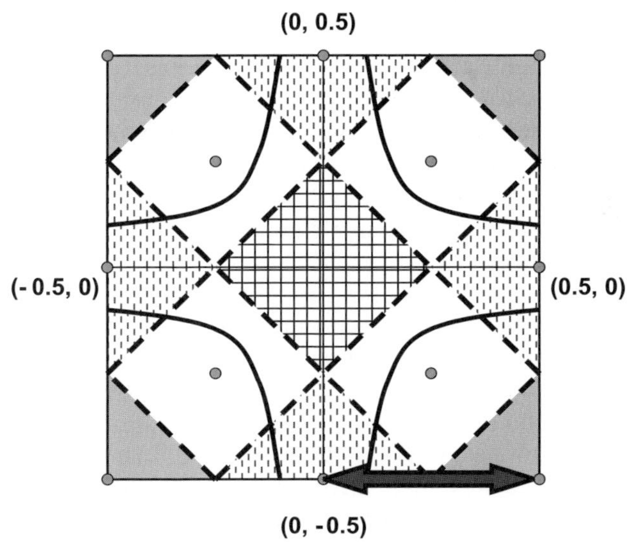

Fig. 14 Sub-Brillouin zones created by the $2\sqrt{2}a \times 2\sqrt{2}b$ superlattice [92]. The sub-zone 1 (*gray colored*) is far above the Fermi level, and the sub-zone 8 (*hatched*) is far below the Fermi level, while the Fermi surface cuts across other sub-zones. Also the anti-nodal particles near $[\pi, 0]$ are in sub-zones (*dashed*) different from those for the nodal particles along $[\pi, \pi]$. The half-breathing LO phonon ($\Leftrightarrow$) connects the sub-zone 1 and the anti-nodal particles

ent zones have different characters. The observed local mode with the 1/4 of the total intensity discussed above could reside in two of the eight subzones. Incidentally this superlattice creates a zone-boundary at [0.25, 0, 0], just as observed for the LO phonons in $La_{1.85}Sr_{0.15}CuO_4$ (Fig. 7) [60]. This discontinuity was later related to the stripe phase, but as pointed out above this explanation is inconsistent with the absence of a similar effect in the transverse mode. It is more natural to explain this in terms of the sub-Brillouin zone boundary.

It is interesting to note that the zone-boundary of the sub-zone 8 agrees well with the location of the "waterfall" in the electronic dispersion, as observed by the ARPES [25]. Around – 0.4 eV the dispersion jumps to about – 0.9 eV, creating the waterfall. The lower branch with the lower effective mass and wider band width agrees with the dispersion by the LDA calculation. The upper branch has a larger effective mass, probably because of heavy dressing by spins and phonons. Papers reporting this waterfall [91, 92] refer them as being in the ZR state, since the dispersion agrees with the *t–J* model prediction of the state dressed with spins. There is no question that the carriers in the upper branch must be heavily dressed by spins and phonons. The energy scale of the kink (0.4 eV $\sim 2J$) justifies the spin nature, but there is no direct evidence that they are described by the ZR model. It is possible that the waterfall is triggered by the sub-zone boundary.

Note that the intensity of the superlattice diffraction peaks is much weaker than those of the Bragg peaks. Such a weak superlattice cannot produce the gap in the dispersion as large as 0.5 eV observed for the waterfall, if this gap is created by potential scattering. On the other hand if this is caused by the many-body effect, and the gap in the single-particle excitation energy of 0.5 eV is compensated by other many-body interactions, even a weak superlattice may be able to produce such a change. This realization leads to a new speculation discussed below.

Incidentally, the explanation of the upper branch above the waterfall as the ZR or *t–J* state and the lower branch as the LDA state has a problem. As we mentioned above the *t–J* model basically assumes local atomic spin polarization, in a sense only $n_{i\uparrow}$ or $n_{i\downarrow}$ can be occupied since the upper Hubbard band is projected out, while the LDA state is spin degenerate, and both spin bands are occupied. Only by assuming the ZR singlet state the spinless *t–J* model is justified. If the ZR singlet state is broken, it is not possible to make a continuous change from the *t–J* state to the LDA state in the same Brillouin zone; it will cause the counting problem of the number of carriers. On the other hand if we introduce the Brillouin zone boundary, this problem can always be averted. It is more likely that the superlattice is playing that very role, and the waterfall is created as a result.

Role of Phonons in Pairing

Possibility of Bipolaron Formation

As discussed above the superconducting carriers in the cuprates have a strong polaronic, or vibronic, character. However, the polaronic character is not consistent with high T_c because of the heavy mass of the polarons. This conundrum can be nicely circumvented by the two-component model [87, 93–95]. If the system is made of two kinds of carriers, heavy bipolarons and light free carriers, then the pairing force is provided by the heavy bipolarons, while the mobility and high T_c is provided by the light free carriers.

In deed the strong anisotropy in the electronic states in the cuprates offers a natural ground for such a scenario [96, 97]. The carriers near the $[\pi, 0]$ point (anti-nodal carriers) are more localized, varies strongly in space [20], and is gapped at low temperatures, while those along the $[\pi, \pi]$ directions (nodal carriers) are nearly free and hardly see disorder [19]. At the same time Deutscher [98] pointed out that the superconducting gap and the pseudogap are different, with different composition dependence. Actually Fujimori et al. [99, 100] pointed out recently that the gap near the anti-nodal point, or the pseudogap, Δ_{PG}, and the superconducting gap near the nodal point, Δ_{SC}, appear to behave differently with doping in the underdoped state, and each of them is related to T_{PG} and T_c by the BCS-like relation,

$$\frac{2\Delta_{PG}}{kT_{PG}} \approx \frac{2\Delta_{SC}}{kT_c} \approx 4\,. \tag{6}$$

This explains why the gap ratio, $2\Delta/kT_c$, is as large as 8 for underdoped samples when Δ is determined at the anti-nodal point. It is, however, very strange to have different gaps in two different directions on the *same* Fermi surface. But the $2\sqrt{2}a \times 2\sqrt{2}b$ superlattice resolves this difficulty by creating a natural division between these two carries; the anti-nodal particles belong to the sub-zones 2/3 (dotted), while the nodal particles belong to different sub-zones (4–7) without mark.

The anti-nodal carriers can become gapped by the local AFM order, but they can also be helped by the phonons to become bipolarons or vibrons. It is most likely that the local dispersionless phonons discussed above (Fig. 13) are associated with this localized bipolarons. Here let us go back to the nature of the carriers in the sub-zones 8 in Fig. 14. In the sub-zone 8, the lower branch of the electron dispersion appears to be similar to the LDA results [91, 92]. This state is dominated by the oxygen p-orbital, and the fact that it is LDA-like suggests that this state does not heavily involve spin polarized Cu d-orbitals. Then what happens in the Cu d-orbitals in the sub-zones 1 and 8? They will naturally be in the Mott-Hubbard state. Now the half-breathing phonons of the $[\pi, 0]$ type induces charge transfer from Cu-d to O-p and vice versa, as discussed above. Thus they can connect the Cu d-orbitals in the sub-zones 1 and 8 with the $[\pi, 0]$ type anti-nodal carriers in the hybridized p–d orbitals in the sub-zones 2 and 3 (dotted regions), enhancing local spin polarization and producing the local spin-singlet bipolarons or vibrons (Fig. 14). The weight of the local phonon mode (1/4) in Fig. 13 suggests that the local singlet states are not static bipolarons, but they are vibronic, and these local phonons are associated with the vibrons.

The ARPES suggests that the $[\pi, 0]$ type anti-nodal carriers interact with the bosons with the energy of about 40 meV [101]. However, this conclusion assumes that the quasiparticle energy at the kink is equal to $\Delta + \Omega$, the sum of the gap energy and the boson energy. If the carriers form a bound bipolaron or vibronic state, since the photon energy to excite the electron is about 20 eV, much higher than the bipolaron binding energy, the Frank-Condon mechanism should be at work, and an electron will be ejected from the solid by a photon without dissolving the bipolaron. Thus the gap energy does not have to be counted and in the ARPES measurement the quasiparticle energy at the kink should be equal to the boson energy Ω. Indeed for the optimally doped BiSCCO-2212 the boson energy above T_c is nearly constant at 60 meV in moving from the nodal to anti-nodal direction, and remains little changed below T_c [102]. This result is consistent with the boson being the half-breathing phonon [59, 65], and the quasiparticle energy at the kink being equal to the boson energy due to the Frank-Condon mechanism. In the STS [22], however, since the energy scale of the experiment is the same as the quasiparticle energy the Frank-Condon mechanism does not apply, and the gap energy has to be counted.

If the carriers in the sub-zones 2 and 3 form local spin-singlet vibronic bipolarons, they and the free carriers in the sub-zones 4–7 can hybridize through the Coulomb term to produce two-component superconductivity [87, 93–95]. But this scenario appears to be inconsistent with the ARPES results, since the ARPES shows well-defined dispersion with apparently long life-time around the anti-nodal $[\pi, 0]$ saddle point for fully doped samples. However, we should note that the ARPES can pick up only the itinerant portion of the hybridized free particles, and does not see the localized state very well. In fact to explain the Hall effect data by Ando [103, 104], Gor'kov proposes that there are additional localized states around the $[\pi, 0]$ point [104]. It is possible that these hidden localized states are the vibronic bipolarons discussed above.

Role of the Superlattice

The principal role of the superlattice is to allow the coexistence of the Mott-Hubbard state and the Fermi-liquid metallic state in different sub-bands (sub-Brillouin zones), and to divide the Fermi surface into two regions, those relates to the pseudogap for the anti-nodal particles near $[\pi, 0]$ and those associated with the superconducting gap for the nodal particles.

Their coexistence naturally leads to two-component, or two-band, superconductivity.

While the Mott-Hubbard carriers and the Fermi liquid carriers belong to different sub-bands, they overlap in space, costing extra energy. But the situation is not very much different in the stripe model either. In the stripe model these two coexist in different locations in space, but they strongly overlap, resulting in an equally high interfacial energy. Thus while the benefit of the superlattice has to be carefully examined by further studies, it is at least as viable as the stripe state as the candidate for the intermediate phase. Although the presence of the superlattice has apparently been observed in YBCO, it has to be confirmed in other cuprates to confirm this idea to be universally relevant.

Conclusions

For a long time since the discovery of HTSC the lattice and phonons have been all but neglected by the majority in the field, in favor of the spin mechanism. However, developments in the last decade are rapidly changing the atmosphere, and now phonons are frequently discussed at HTSC conferences. There are expectations that we may be closing in to the new pathways to the HTSC phenomenon, which seriously takes the complexity into account. It is most likely that the lattice and phonon will play a major role in the new scenario, whatever their roles may be. In this chapter we discussed one of the likely scenarios involving the local bipolarons, or vibronic state, and the two-component superconductivity.

Acknowledgements It is my great pleasure and honor to thank K. Alex Müller for his amazing discovery, deep conviction in the role of the lattice, and continued encouragement to the community which shares his conviction. The history of the research in this field is by itself quite remarkable, full of surprises and unexpected developments. It is a great joy and privilege to work in this field, and the pleasure has been greatly enhanced by Alex. I would like to give him warm congratulation for his 80th birthday. I also thank my collaborators, in particular B. Fine, P. Piekarz, J.-H. Chung, F. Stercel, J. Karpinski, M. Yethiraj, H. Mook, M. Tachiki, P. Zschack and M. Arai, for their valuable contributions. I am very thankful to L. P. Gor'kov, A. Lanzara, J. C. Davis, E. Dagotto, A. Moreo, A. Bussmann-Holder, Y. Ando, A. Fujimori, A. Balatzky, A. Bianconi, J. Ranninger, C. Castellanni, A. R. Bishop, R. Zeyher, O. K. Andersen, P. Dai, N. Saini, H. Oyanagi, H. Keller, J. Zaanen, J. C. Phillips, D. Mihailovic and others for useful and stimulating discussions. This work was supported by the National Science Foundation through DMR04-04781.

References

1. Bednorz JG, Müller KA (1986) Z Phys B 64:189
2. Bardeen J, Cooper LN, Schrieffer JR (1957) Phys Rev 106:162; (1957) 108:1175

3. McMillan WL (1968) Phys Rev 167:331
4. Anderson PW (1987) Science 235:1196
5. Anderson PW (1997) The Theory of Superconductivity in the High-T_c Cuprates. Princeton University Press, Princeton
6. Bianconi A, Congiu Castellano A, De Santis M, Rudolf P, Lagarde P, Flank AM (1987) Solid St Commun 63:1009
7. Emery VJ (1988) Phys Rev Lett 58:2794
8. Zhang FC, Rice TM (1988) Phys Rev B 37:3759
9. Bulut N, Hone D, Scalapino DJ, Bickers NE (1990) Phys Rev B 41:1797; (1990) Phys Rev Lett 64:2723
10. Millis AJ, Monien H, Pines D (1990) Phys Rev B 42:167
11. Wallman DA, Van DJ Harlingen, Lee WC, Ginsberg DM, Leggett AJ (1993) Phys Rev Let 71:2134
12. Tsuei CC, Kirtley JR, Chi CC, Yu-Jahnes LS, Gupta A, Shaw T, Sun JZ, Ketchen MB (1994) Phys Rev Lett 73:593
13. Dagotto E (1994) Rev Mod Phys 66:763
14. Zaanen J, Chakravarty S, Senthil T, Anderson PW, Lee PK, Schmalian J, Imada M, Pines D, Randeria M, Varma C, Vojta M, Rice TM (2006) Nature Physics 2:138
15. Maier TA, Jarrell M, Schulthess TC, Kent PRC, White JB (2005) Phys Rev Lett 95:237001
16. Egami T, Billinge SJL (1996) Lattice Effects in High-T_c Superconductors. In: Ginsberg DM (ed) Physical Properties of High Temperature Superconductors V. World Scientific, p 265
17. Emery VJ, Reiter G (1988) Phys Rev B 38:4547
18. Tranquada JM, Sternlieb BJ, Axe JD, Nakamura Y, Uchida S (1995) Nature 375:561
19. Pan SH, J. P. O'Neal, Badzey RL, Chamon C, Ding H, Engelbrecht JR, Wang Z, Eisaki H, Uchida S, Gupta AK, Ng K-W, Hudson EW, Lang KM, Davis JC (2001) Nature (London) 413:282
20. Lang KM, Madhavan V, Hoffman JE, Hudson EW, Eisaki H, Uchida S, Davis JC (2002) Nature (London) 415:412
21. McElroy K, Simmonds RW, Hoffman JE, Lee DH, Orenstein J, Eisaki H, Uchida S, Davis JC (2003) Nature (London) 422:592
22. Jinho Lee, Fujita K, McElroy K, Slezak JA, Wang M, Aiura Y, Bando H, Ishikado M, Matsui T, Zhu J-X, Balatsky AV, Eisaki H, Uchida S, Davis JC (2006) Nature 442:546
23. Lanzara A, Bogdanov PV, Zhou XJ, Kellar SA, Feng DL, Lu ED, Yoshida T, Eisaki H, Fujimori A, Kishio K, Shimoyama J-I, Noda T, Uchida S, Hussain Z, Shen Z-X (2001) Nature (London) 412:510
24. Gweon G-H, Sasagawa T, Zhou SY, Graf J, Takagi H, Lee DH, Lanzara A (2004) Nature (London) 430:187
25. Graf J, Gweon G-H, Lanzara A, cond-mat/0610313
26. Castellani C, Di Castro C, Grilli M (1997) Z Phys 130:137
27. Kivelson SA, Fradkin E, Emery VJ (1998) Nature 393:550
28. Molegraaf HJA, Presura C, van der Marel D, Kes PH, Li M (2002) Science 295:2239
29. Boris AV, Kovaleva NN, Holden T, Lin CT, Keimer B, Bernhard C (2004) Science 304:708
30. Deutscher G, Santander-Syro AF, Bontemps N (2005) Phys Rev B 72:092504
31. Moriya T, Takahashi Y, Ueda K (1990) J Phys Soc Japan 59:2905
32. Monthoux P, Pines D (1992) Phys Rev Lett 69:961; (1993) Phys Rev B 47:6069
33. Abanov Ar, Chubukov AV (1999) Phys Rev Lett 83:1652

34. Bourges P (2000) In: Furrer A (ed) Neutron Scattering in Novel Materials. World Scientific, Singapore, p 252; cond-mat/0009373
35. Woo H, Dai P, Hayden SM, Mook HA, Dahm T, Scalapino DJ, Perring TG, Dogan F (2006) Nature Physics 2:600
36. Bednorz JG, Müller KA (1988) Rev Mod Phys 60:585
37. Frank JP (1994) Experimental Studies of the Isotope Effect in High Temperature Superconductors. In: Ginsberg DM (ed) Physical Properties of High Temperature Superconductors IV. World Scientific, p 189
38. Geballe TH, Matthias BT, Hull GW Jr, Corenzwit E (1961) Phys Rev Lett 6:275
39. Zhao G-M, Hunt MB, Keller H, Müller KA (1997) Nature 385:236
40. Khasanov R, Shengelaya A, Conder K, Morenzoni E, Savic IM, Keller H (2003) J Phys Cond Mat 15:L17
41. Lanzara A, Zhao GM, Saini NL, Bianconi A, Conder K, Keller H, Müller KA (1999) J Phys Cond Mat 11:L541
42. Rubino Temprano D, Mesot J, Janssen S, Conder K, Furrer A, Mutka A, Müller KA (2000) Phys Rev Lett 84:1990
43. Alexandrov A, Ranninger J (1981) Phys Rev B 24:1164
44. Alexandrov AS (1988) Phys Rev B 38:925
45. Alexandrov AS (2005) J Superconductiity: Incl Novel Magn 18:603
46. Schafloth MR (1955) Phys Rev 100:463
47. Bianconi A, Saini NL, Lanzara A, Missori M, Rosetti T, Oyanagi H, Yamaguchi H, Oka K, Itoh T (1996) Phys Rev Lett 76:3412
48. Saini NL, Oyanagi H, Itoh T, Scagnoli V, Filippi M, Agrestini S, Campi G, Oka K, Bianconi A (2003) Eur Phys J B 36:75
49. Toby BH, Egami T, Jorgensen JD, Subrmanian MA (1990) Phys Rev Lett 64:2414
50. Sendyka TR, Dmowski W, Egami T, Seiji N, Yamauchi H, Tanaka S (1995) Phys Rev B 51:6747
51. Mook HA, Mosteller M, Harvey JA, Hill N, Chakoumakos abd B Sales BC (1990) Phys Rev Lett 65:2712
52. Sharma RP, Rehn LE, Baldo PM, Liu JZ (1989) Phys Rev Lett 62:2869
53. Sharma RP, Ogale SB, Zhang ZH, Liu JR, Chu WK, Veal B, Paulikas A, Zheng H, Venkatesan T (2000) Nature 404:736
54. Egami T, Billinge SJL (2003) Underneath the Bragg Peaks: Structural Analysis of Complex Materials. Pergamon Press, Elsevier Science, Oxford
55. Egami T, Toby BH, Dmowski W, Billinge S, Davies PK, Jorgensen JD, Subramanian MA, Gopalakrishnan J, Sleight AW (1989) Physica C 93:162–164
56. Pickett W (1989) Rev Mod Phys 61:433
57. Bogdanov PV et al. (2000) Phys Rev Lett 85:2581
58. Johnson PD, Valla T, Fedorov AV, Yusof Z, Wells BO, Li Q, Moodenbaugh AR, Gu GD, Koshizuka N, Kendziora C, Sha Jian, Hinks DG (2001) Phys Rev Lett 87:177007
59. Pintchovius L, Reichardt W (1994) In: Ginsberg D (ed) Physical Properties of High Temperature Superconductors IV. Singapore, World Scientific, p 295
60. McQueeney RJ, Petrov Y, Egami T, Yethiraj M, Shirane G, Endoh Y (1999) Phys Rev Lett 82:628
61. Zaanen J, Sawatzky GA, Allen JW (1985) Phys Rev Lett 55:418
62. Egami T, Ishihara S, Tachiki M (1993) Science 261:1307
63. Ishihara S, Egami T, Tachiki M (1997) Phys Rev B 55:3163
64. Tachiki M, Machida M, Egami T (2003) Phys Rev B 67:174506
65. Chung J-H, Egami T, Mc Queeney RJ, Yethiraj M, Arai M, Yokoo T, Petrov Y, Mook HA, Endoh Y, Tajima S, Frost C, Dogan F (2003) Phys Rev B 67:014517

66. Piekarz P, Egami T (2005) Phys Rev B 72:054530
67. Su WP, Schrieffer JR, Heeger AJ (1979) Phys Rev Lett 42:1698
68. Basov DN, Singley EJ, Dordevic SV (2002) Phys Rev B 65:054516
69. Santander-Syro F, Lobo RPSM, Bontemps N, Konstantinovic Z, Li ZZ, Raffy H (2003) Eur Phys Lett 62:568
70. Resta R (1994) Rev Mod Phys 66:899
71. King-Smith RD, Vanderbilt D (1993) Phys Rev B 47:1651
72. Reznik D, Pintschovius L, Ito M, Ikubo S, Sato M, Goka H, Fujita M, Yamada K, Gu GD, Tranquada JM (2006) Nature 440:1170
73. Zaanen J, private communication
74. Egami T, Physica C, in press
75. Zaanen J, Gunnarsson O (1989) Phys Rev B 40:R7391
76. Abbamonte P, Rusydi A, Smadici S, Gu GD, Sawatzky GA, Feng DL (2005) Nature Physics 1:155
77. Tranquada JM, Axe JD, Ichikawa N, Nakamura Y, Uchida S, Nachumi B (1996) Phys Rev B 54:7489
78. Yamada K, Lee CH, Kurahashi K, Wada J, Wakimoto S, Ueki S, Kimura H, Endoh Y, Hosoya S, Shirane G, Birgeneau RJ, Greven M, Kastner MA, Kim YJ (1998) Phys Rev B 57:6165
79. Zaanen J, Osman Y, Kruis HV, Nissinov Z, Tworzydlo I (2001) Phil Mag B 81:1485
80. Tranquada JM, Woo H, Perring TG, Goka H, Gu GD, Xu G, Fujita M, Yamada K (2004) Nature 429:534
81. Uehara M, Nagata T, Akimitsu J, Takahashi H, Mori N, Kinoshita K (1996) J Phys Soc Japan 65:2764
82. Fujita N, Mori N, Uwatoko Y, Matsumoto T, Motoyama N, Uchida S (2003) Phys Rev Lett 90:137001
83. Mc Elroy K, Lee J, Slezak JA, Lee D-H, Eisaki H, Uchida S, Davis JC (2005) Science 309:1048
84. Vershinin M, Misra S, Ono S, Abe Y, Ando Y, Yazdani A (2004) Science 303:1995
85. Hanaguri T, Lupien C, Kohsaka Y, Lee D-H, Azuma M, Takano M, Takagi H, Davis JC (2004) Nature 430:1001
86. Davis JC, unpublished.
87. Dagotto E, Hotta T, Moreo A (2001) Physics Report 344:1
88. Gor LP'kov, Sokol AV (1987) JETP Lett 46:420
89. Chung J-H, Mook MA, Dogan F, Egami T, unpublished
90. Egami T (2006) J Superconductivity: Incl. Novel Magn 19:203
91. Graf J, Gweon G-H, Mc Elroy K, Zhou SY, Jozwiak C, Rotenberg E, cond-mat/0607319
92. Xie BP, Yang K, Shen DW, Zhao JF, Ou HW, Wei J, Gu SY, Arita M, Qiao S, Namatame H, Taniguchi M, Kaneko N, Eisaki H, Yang ZQ, Feng DL, cond-mat/0607450
93. Micnas R, Ranninger J, Robaszkiewicz S (1990) Rev Mod Phys 62:113
94. Cuoco M, Noce C, Ranninger J, Romano A (2003) Phys Rev B 67:224504
95. Bussmann-Holder A, Müller KA, Micnas R, Büttner H, Simon A, Bishop AR, Egami T (2001) J Phys: Cond Matt 13:L169
96. Egami T, Dmowski W, Mc Queeey RJ, Sendyka TR, Ishihara S, Tachiki M, Yamauchi H, Tanaka S, Hinatsu T, Uchida S (1995) In: Salje EKH et al. (eds) Polarons and Bipolarons in High-T_c Superconductors and Related Materials. Cambridge University Press, Cambridge, p 155
97. Geshkenbein VB, Ioffe LB, Larkin AI (1997) Phys Rev B 55:3173
98. Duetscher G (1999) Nature 397:410

99. Tanaka K, Lee WS, Lu DH, Fujimori A, Fujii T, Risdiana, Terasaki I, Scalapino JD, Devereaux TP, Hussain Z, Shen Z-X. Science, in press
100. Hashimoto M, Yoshida T, Tanaka K, Fujimori A, Okusawa M, Wakimoto S, Yamada K, Kakeshita T, Eisaki H, Uchida S. Phys Rev Lett, to be published
101. Cuk T, Baumberger F, Lu DH, Ingle N, Zhou XJ, Eisaki H, Kaneko N, Hussain Z, Devereaux TP, Nagaosa N, Shen Z-X (2004) Phys Rev Lett 93:117003
102. Lanzara A (2004) Physica C 412–414:46
103. Ando Y, Kurita Y, Komiya S, Ono S, Sagawa K (2004) Phys Rev Lett 92:197001
104. Ono S, Komiya S, Ando Y (2007) Phys Rev B 75:024515
105. Gor'kov LP, Teitl'baum GB, cond/mat-0607010.

Fossheim K (2007) In: Bussmann-Holder A, Keller H (eds)
High T_c Superconductors and Related Transition Metal Oxides.
Springer-Verlag Berlin Heidelberg, 131–134
DOI 10.1007/978-3-540-71023-3

Shared Fascinations

K. Fossheim

Department of Physics, Norwegian University of Science and Technology, Trondheim, Norway
Kristian.Fossheim@phys.ntnu.no

My first contact with Alex was established in 1968 through Jens Feder in Oslo, who had spent a postdoctoral period at the IBM Zurich lab in Rüschlikon. When I asked him about news from Zurich, he told me of the very interesting work by Alex and coworkers on $SrTiO_3$, about the discovery of a soft mode, and about the 105 K structural transition. Having just returned from University of Maryland where I studied electron- phonon interaction in superconductors, I had become much intrigued by the superconducting transition. In indium I had observed the transition at T_c to occur with 2/10000 degrees sharpness and had concluded that superconductors had extreme mean field character. Fluctuation phenomena apparently played no observable role. I had become aware of the generality of Landau theory through a paper on helium by Khalatnikov, and the beauty of the superfluid transition in helium through the specific heat measuremets by Buckingham and Fairbank where μK-scale resolution was reported. My fascination with superconductivity and superfluidity had therefore become linked to the fascinating field of critical phenomena and phase transitions in general.

I now could see how the *insulating* material $SrTiO_3$ with its structural aspects in a deeper sense was a clear parallel to the *superconducting* transitions I had studied, in spite of the huge differences in other physical properties, specially concerning electrical conductivity! In $SrTiO_3$ the Landau theory had just been worked out by Harry Thomas and Alex in a PRL. The soft mode was tied to the rotational motion of the oxygen octahedra whose average angle in turn represented the order parameter in the Landau approach. The order parameter temperature exponent determined in masterly EPR experiments of Alex and coworkers, was 0.5, the classical value, just like in the superconductors I had studied!

Yet, here I felt that in $SrTiO_3$ thermal fluctuations must play a role somehow, since the phonon population of the soft mode must increase more and more on approaching T_c. One should expect a peak in attenuation of sound at T_c, distinctly different from superconductors! It was time to contact Alex to obtain a crystal, which Jens Feder did on my behalf. This was still in 1968, and everything about soft phonon modes was new. Trying to examine the nature of structural phase transitions by means of phonon scattering seemed to me like a very interesting prospect. Alex immediately provided the crystal. I asked Bjørn Berre to join me in the experiments, and right away we could observe the expected effect: An attenuation peak at the 105 K cubic-to-tetragonal transition in $SrTiO_3$. After careful discussions on the telephone with Alex, and with the advice from Erling Pytte who was still at the Rüschlikon lab, we published a paper in Physical Review Letters: "Critical attenuation of sound by soft modes in $SrTiO_3$." The use of the word "critical" in the title was a subject of considerable discussions because this would be indicating that the scattering involved non-classical behaviour. So far the soft mode theory was purely based on mean field like arguments. A full theory of ultrasonic scattering had not yet been developed, so suggesting non-classical behaviour was a bit bold.

In Oslo we continued the work on $SrTiO_3$. I suggested suppressing the structural domain formation, a source of strong additional scattering of sound waves, by applying pressure along two of the three major axes of the cubic crystal. This worked beautifully and "cleaned away" this source of scattering. What remained was a nice symmetric peak in attenuation around the phase transition point T_c. What was more, we even observed a minimum in the elastic constant at T_c, while the typical Landau-like order parameter relaxation type of behaviour would only lead to a gradual shift of elastic constant.

In the summer of 1970 a very nice meeting on critical phenomena was organised in Varenna by M. S. Green. I met Alex for the first time there, and showed him our results on the elastic constant where there was a minimum at T_c. He urged us to write a PRL, but I explained that the data were not of sufficient quality. However, soon after Alex did much more careful experiments on the order parameter temperature dependence closer to T_c than before, and found a non-classical exponent, near 0.3 instead of 0.5. This was a major breakthrough, pointing in the direction of non-classical universality.

Later, Alex invited me to spend two summers, 1973 and 1975, at the IBM lab in Rüschlikon, away from university life in Trondheim where I had now moved. Afterwards I went to the IBM Yorktown Heights lab for the year 1975–76. The idea was that I would work with Norm Shiren on acoustic "echo" phenomena, not the type caused by reflections at a boundary, but in the style of spin echo phenomena in magnetic resonance, although due to propagating acoustic signals. "Phase conjugation" was later adopted as a more appropriate term. Here I got in contact with both Norm Shiren and Bob Melcher, and with Koji Kajimura visiting from Electrotechnical Laboratory, ETL, in Japan. One day Melcher showed us a letter from none other than Alex Müller after Alex had paid a visit to the Ioffe Institute in Leningrad. The letter pointed out that a group there had observed "phonon echoes" in powders. Little was understood about the mechanism, and it was decided we would go for it. The result was again a PRL, and later a very long paper in Physical Review, with very rich observational material, matched by rather convincing theoretical calculations.

After my return to Trondheim, my student Rune Holt and I developed the phase conjugation method in single crystals as a tool to "repair" acoustic waves which had been distorted on passing through an inhomogeneous medium, a new and useful tool, in particular near phase transitions.

During my stay at Yorktown Heights Alex wrote me about his intention to study the pressure dependence of T_c in $SrTiO_3$ in collaboration with Hatta in Japan. However, the experiment did not yield the expected results, as I recall. A few years later, in Trondheim, with Sigmund Stokka, we set up an experiment to measure heat capacity. We could follow the phase transition line in $SrTiO_3$ under pressure, and found a curved phase transition line. We got in contact with Amnon Aharony and one of his students, and everything fell into place: We could determine, for the first time, the crossover exponent of a structural phase transition. However, this time, the PRL referee told us he had seen enough critical exponents, and would have no more of them! We had to publish in PR Rapid Communications. Interestingly, this behaviour was matched in studies by Heini Rohrer and coworkers in certain magnetic phase diagrams. Universality of critical phenomena is indeed a fascinating aspect of matter.

In the summer of 1982 Alex was my guest in Trondheim for a couple of weeks, and the theme was multicritical phenomena in perovskites under pressure. The visit was too brief to give immediate results. But Alex followed up our work on the crossover exponent. In $LaAlO_3$ he found very similar results as we had found in $SrTiO_3$.

When, in 1979, I applied for the position of full professor in Trondheim I promised that if any major development in superconductivity would take place in a foreseeable future, I would take the responsibility to enter the field again to ensure the supply of competence to Norwegian industry and science. I had 77 K specifically in mind, but dared not say so, being afraid I would be laughed at by the committee. Little did I know that Alex and George Bednorz, would be the persons to provide the opportunity. A little later I might have come close to knowing, since Alex mentioned their project to me during my visit to the IBM Rüschlikon lab in 1983. Frankly, I was skeptical since I knew the low value of superconducting T_c obtained in $SrTiO_3$. It turned out that Alex, during a long sabbatical stay at the IBM Yorktown Heights lab had compensated for his previous lack of experience in superconductivity and become just as fascinated by this field as I had in the 60's.

The new development, i.e. the discovery of high-T_c superconductivity in 1986, was an obvious challenge for me to pick up: Superconductivity in perovskites! It was an almost unbelievable dream come true, combining two areas that had been at the forefront in Alex' work as well as in mine. When the high-T_c avalanche set off in earnest in 1987 a telephone call to Alex was all I needed to get started. I was again on the same road as Alex, sharing the same fascinations.

My first idea was to try to show that fluctuations of the order parameter would play a role in these materials. This would be a significant break with earlier experience in low-T_c materials. Indeed, the specific heat curve was peaked like in helium where spontaneously generated vortices dominated the behaviour near T_c, rather than having a sharp jump at T_c like in BCS type superconductors. This was a different world, and most people did not understand how radically different until Sudbø and coworkers' theoretical work ten years later, revealing the important role of phase fluctuations due to spontaneous vortex generation. Superconductivity had now joined the ranks of non-classical critical phenomena, a fascination already shared by Alex and myself since earlier times in perovskites.

Thank you Alex, for inspiration, collaboration, and shared fascinations!

Furrer A (2007) In: Bussmann-Holder A, Keller H (eds)
High T_c Superconductors and Related Transition Metal Oxides.
Springer-Verlag Berlin Heidelberg, 135–141
DOI 10.1007/978-3-540-71023-3

Admixture of an *s*-Wave Component to the *d*-Wave Gap Symmetry in High-Temperature Superconductors

A. Furrer

Laboratory for Neutron Scattering, ETH Zurich & PSI Villigen, CH-5232 Villigen PSI, Switzerland
albert.furrer@psi.ch

Abstract Neutron crystal-field spectroscopy experiments in optimally doped $La_{1.81}Sr_{0.15}Ho_{0.04}CuO_4$ and in slightly underdoped $HoBa_2Cu_4O_8$ high-temperature superconductors are presented. By this bulk-sensitive technique, information on the gap function is obtained from the relaxation behaviour of crystal-field transitions associated with the Ho^{3+} ions which sit as local probes close to the superconducting copper-oxide planes. The relaxation data exhibit distinct anomalies at the pseudogap temperature T^* where the pairing of the charge carriers occurs, as well as at the superconducting transition temperature T_c where long-range phase coherence sets in. Our results are compatible with an unusual temperature dependence of the gap function in the pseudogap region ($T_c \leq T \leq T^*$), i.e., a breakup of the Fermi surface into disconnected arcs. Moreover, the relaxation data can only be modelled satisfactorily if the gap function of predominantly *d*-wave symmetry includes an *s*-wave component of the order of 20–25%.

Ever since the discovery of high-temperature superconductivity by Bednorz and Müller [1] there has been a debate concerning the symmetry of the superconducting gap function. So far tunneling measurements are the only experiments that probe the sign of the order parameter [2]. Angle-resolved photoemission spectroscopy (ARPES) measurements, on the other hand, yield very direct information about the magnitude and the momentum dependence of the order parameter as well as its evolution with temperature [3, 4]. Both techniques, however, are only sensitive to the surface. For underdoped compounds, notably for $Bi_2Sr_2CaCu_2O_{8+x}$, it was found that a pseudogap with symmetry $d_{x^2-y^2}$ opens up in the normal state below the pseudogap temperature T^* and develops into the *d*-wave superconducting gap below the superconducting transition temperature $T_c < T^*$ [3, 4]. Other techniques [5] probing both the surface and the bulk, e.g., nuclear magnetic resonance (NMR) and muon-spin rotation (μSR) experiments, gave partially conflicting results by supporting the existence of a sizeable *s*-wave component. As pointed out by Müller [6] the gap function is different at the surface and in the bulk, thus bulk experiments are indispensable to unambiguously establish the nature of the gap function. Here we present bulk-sensitive neutron crystal-field spectroscopy experiments in La- and Y-type high-temperature superconductors, namely in optimally doped $La_{1.81}Sr_{0.15}Ho_{0.04}Cu^{18}O_4$ ($T_c \approx 32$ K, $T^* \approx 70$ K) and in slightly underdoped $HoBa_2Cu_4{}^{18}O_8$ ($T_c \approx 79$ K, $T^* \approx 220$ K), respectively, which give evidence for the coexistence of s- and d-wave components of the gap function. Moreover, the present analysis supports the unusual temperature dependence of the gap function in the pseudogap region ($T_c \leq T \leq T^*$) as evidenced by ARPES measurements in $Bi_2Sr_2CaCu_2O_{8+x}$ [7].

The principle of neutron crystal-field spectroscopic investigations of the crystal-field interaction in rare-earth based high-temperature superconductors was described in several review articles [8–10]. By this technique transitions between different crystal-field levels associated with the rare-earth ions can be directly measured. In the normal metallic state the excited crystal-field levels interact with phonons, spin fluctuations and charge carriers, which limit the lifetime of the excitation, thus the observed crystal-field transitions exhibit line broadening. The interaction with the charge carriers is by far the dominating relaxation mechanism [11]. The corresponding intrinsic linewidth $W_n(T)$ increases almost linearly with temperature according to the well-known Korringa law [12], i.e., $W_n(T) \sim T$. In the pseudogap as well as in the superconducting state, however, the pairing of the charge carriers creates an energy gap Δ_k in the single-electron spectral function ($\boldsymbol{k}$ is the wave vector), thus crystal-field excitations with energy $< 2\Delta_k$ do not have enough energy to span the gap. Consequently the interaction with the charge carriers is suppressed, and for an isotropic gap function $\Delta_k = \Delta$ the intrinsic linewidth is given by $W_s(T) = W_n(T) \cdot \exp[-\Delta/k_B T]$. This means that $W_s(T \ll T_c, T^*) \approx 0$, and line broadening sets in just below T_c (or T^*) where the superconducting gap (or the pseudogap) opens. For anisotropic gap functions the situation is more complicated, since certain relaxation channels exist even at the lowest temperatures [13].

Recently, detailed considerations concerning the analysis of neutron crystal-field relaxation data have been presented by Häfliger et al. [14] where several complications and modifications of the line-width behaviour outlined in the preceding paragraph have been identified. These include doping dependent effects on both the superconducting volume fraction V and the temperature exponent A of the gap function. In order to avoid uncertainties with respect to V and A, we consider in the following high-T_c compounds at or close to optimal doping where these parameters are undisputed, i.e., $V \approx 1$ and $A \approx 4$.

The temperature dependence of the intrinsic linewidth W(T) (HWHM) observed for $La_{1.81}Sr_{0.15}Ho_{0.04}Cu^{18}O_4$ [15] is shown in Fig. 1a. The linewidth is very small below $T_c \approx 32$ K, then it starts to raise up to 70 K; above 70 K it increases with reduced slope and almost linearly as expected for the normal state, i.e., we set $T^* \approx 70$ K. Figure 1b shows the temperature dependence of the instrinsic linewidth observed for the lowest crystal-field ground-state transition in $HoBa_2Cu_4{}^{18}O_8$ [16]. The linewidth is essentially zero at low temperatures. It starts to increase slightly below $T_c \approx 79$ K until a steplike enhancement occurs at 220 K, from whereon it increases linearly as expected

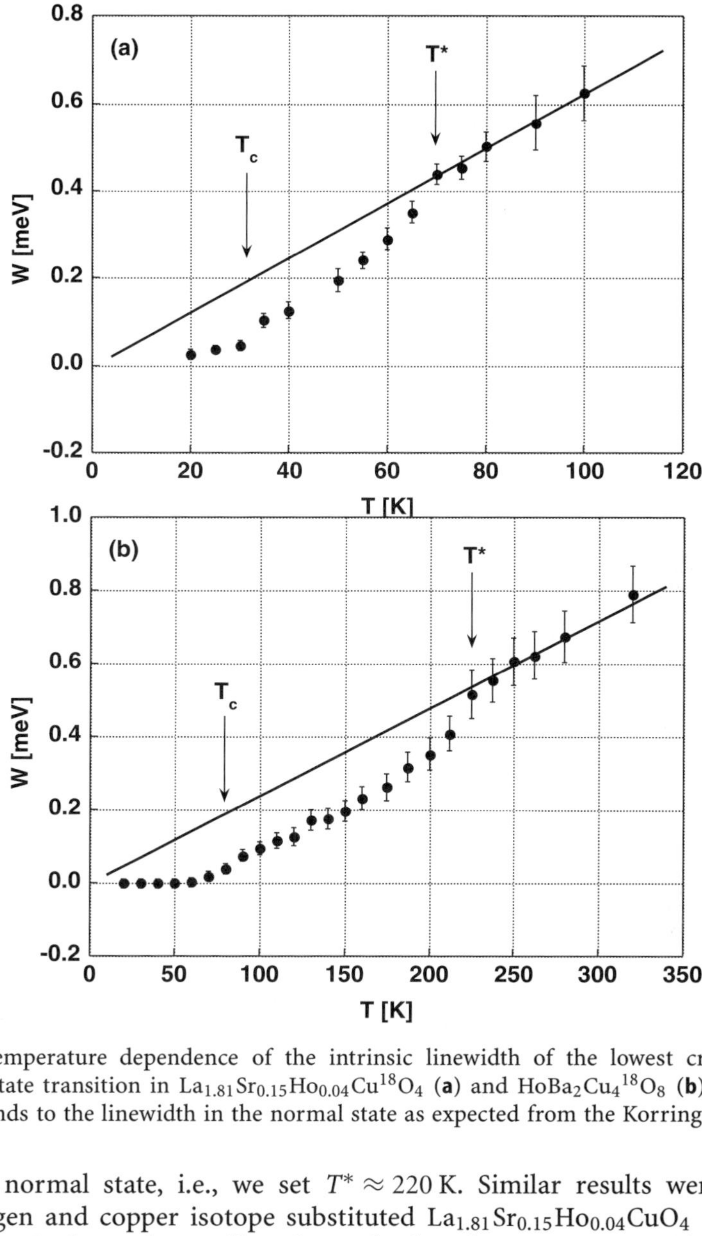

Fig. 1 Temperature dependence of the intrinsic linewidth of the lowest crystal-field ground-state transition in $La_{1.81}Sr_{0.15}Ho_{0.04}Cu^{18}O_4$ (**a**) and $HoBa_2Cu_4{}^{18}O_8$ (**b**). The *line* corresponds to the linewidth in the normal state as expected from the Korringa law

for the normal state, i.e., we set $T^* \approx 220\,K$. Similar results were found for oxygen and copper isotope substituted $La_{1.81}Sr_{0.15}Ho_{0.04}CuO_4$ [15] and $HoBa_2Cu_4O_8$ [16, 17] as well as for underdoped $HoBa_2Cu_3O_x$ with $x = 6.46$ and $x = 6.56$ [18], the only difference being the different sizes of the characteristic temperatures T_c and T^*. It is therefore useful to express the relaxation data in reduced units, i.e., the temperature T in units of T^* and the linewidth $W(T)$ in units of $W(T^*)$ as visualized for the relevant temperature range $0 \leq T \leq T^*$ in Fig. 2b–c.

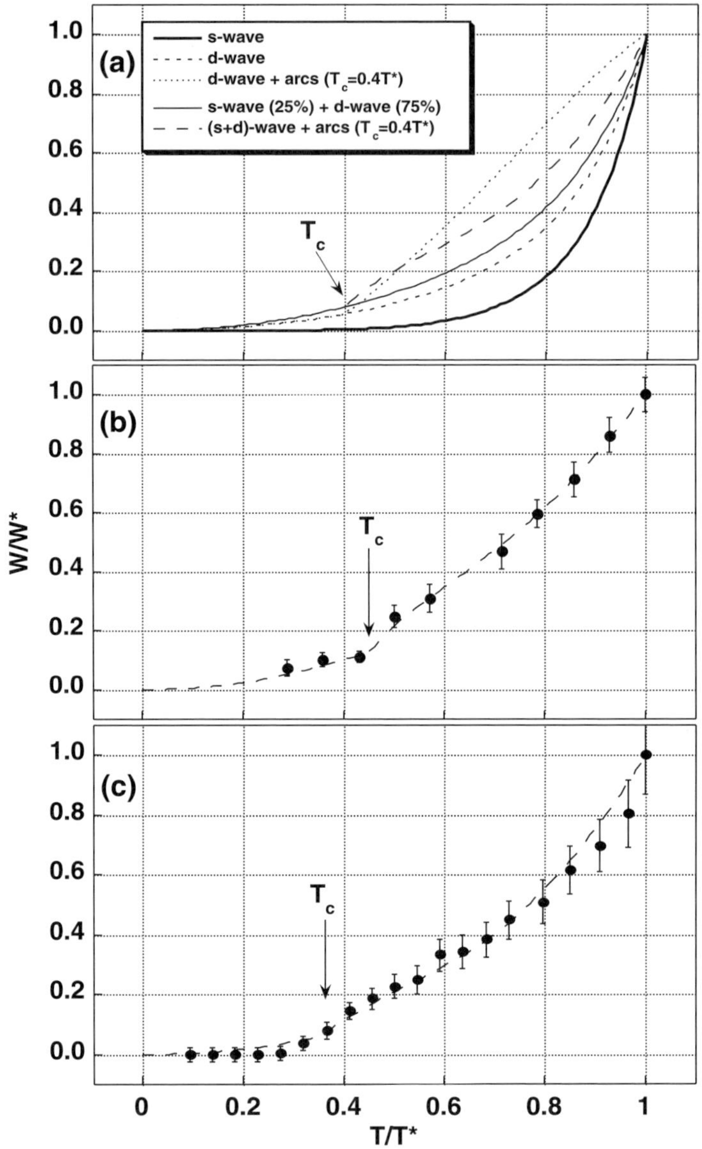

Fig. 2 Temperature dependence of the linewidth of low-energy crystal-field transitions in high-temperature superconductors in reduced units. **a** Model calculations for different types of gap functions (see text). **b** Relaxation data derived for $La_{1.81}Sr_{0.15}Ho_{0.04}Cu^{18}O_4$. **c** Relaxation data derived for $HoBa_2Cu_4{}^{18}O_8$. The *lines* in **b–c** denote least-squares fits on the basis of a mixed $(s + d)$-wave gap function including the occurrence of gapless arcs of the Fermi surface in the pseudogap region ($T_c < T < T^*$) as described in the text

We now proceed to analyse the relaxation data in terms of different gap functions. The model calculations were based on the procedure described in [14]. For the temperature dependence of the gap amplitude we used the expression $\Delta(T) = \Delta(0) \cdot [1 - (T/T^*)^A]$ with $A = 4$ [3], and the maximum gap amplitude was set to $\Delta_{max}/k_B = 2T^*$ which is a realistic value for the compounds under consideration. Figure 2a shows some calculations for an s-wave, a d-wave ($d_{x^2-y^2}$), and a mixed ($s + d$)-wave gap function. It can readily be seen that none of these models is able to explain the observed linewidth behaviour displayed in Fig. 2b–c. The s-wave model reproduces the low-temperature behaviour quite well, but it fails completely at high temperatures. The d-wave model is inadequate at both low and high temperatures. The mixed ($s + d$)-wave model has similar deficiencies. In particular, for the very detailed $HoBa_2Cu_4O_8$ data the behaviour around T_c where the slope of $W(T)/W(T^*)$ turns from being convex to concave cannot be reproduced by any of the above models. This is due to the neglect of the temperature dependence of the gap function in the pseudogap region. According to ARPES experiments performed for underdoped $Bi_2Sr_2CaCu_2O_{8+x}$ [7] the pseudogap opens up at different temperatures for different points in momentum space as sketched in Fig. 3. Below T^* there is a breakup of the Fermi surface into disconnected arcs, which then shrink with decreasing temperature before collapsing to the point nodes of the d-wave component below T_c. This unusual temperature evolution of the pseudogap opens additional relaxation channels in crystal-field linewidth studies above T_c. Indeed, model calculations including the presence of such gapless arcs produce relevant modifications of the linewidth for $T_c \leq T \leq T^*$ as visualised in Fig. 2a; in particular, the observed change of the linewidth from a convex to a concave temperature dependence around T_c can be nicely reproduced for a mixed ($s + d$)-wave gap function.

We have performed a least-squares fitting procedure to the relaxation data displayed in Fig. 2b–c on the basis of a mixed ($s + d$)-wave gap function including the gapless arc features of the Fermi surface in the pseudogap region

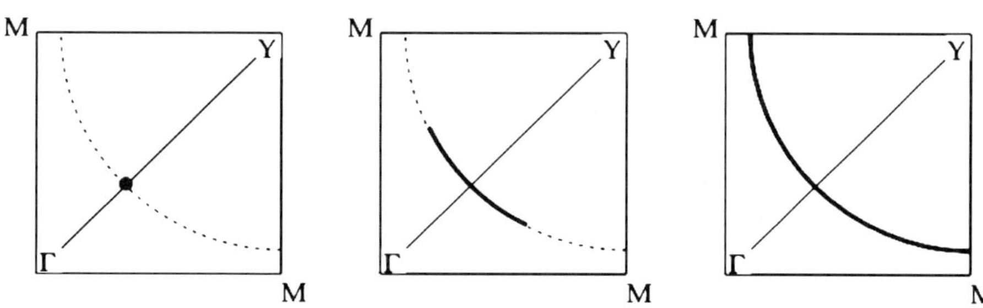

Fig. 3 Schematic illustration of the temperature evolution of the Fermi surface in the pseudogap region ($T_c \leq T \leq T^*$). The d-wave node below T_c (*left panel*) becomes a gapless arc above T_c (*middle panel*) which expands upon increasing the temperature to form the full Fermi surface at T^* (*right panel*)

Table 1 Gap parameters resulting from the least-squares fitting procedure applied to the crystal-field relaxation data displayed in Fig. 2b–c

	T_c [K]	T^* [K]	Δ_s [meV]	Δ_d [meV]	Δ_{max} [meV]
$La_{1.81}Sr_{0.15}Ho_{0.04}Cu^{18}O_4$	32	70	2.2 ± 0.7	8.7 ± 0.9	10.9 ± 1.6
$HoBa_2Cu_4{}^{18}O_8$	79	220	9.1 ± 1.2	28.0 ± 3.6	37.1 ± 4.8

($T_c \leq T \leq T^*$). Since the latter effect was only observed at a few selected temperatures [7], we assumed a linear evolution of the gapless arcs with temperature. The only fitting parameters were the amplitudes Δ_s and Δ_d of the *s*- and *d*-wave components, respectively. The temperature evolution of the linewidths resulting from the least-squares fitting procedure are in good agreement with the experimental data as shown in Fig. 2b–c. The corresponding model parameters are listed in Table 1. We recognise that the maximum gap amplitudes are consistent with the expectation $\Delta_{max}/k_B \approx 2T^*$.

In conclusion, our analyses of bulk-sensitive crystal-field relaxation experiments on high-temperature superconductors give two important results. Firstly, we gave evidence that the theoretically predicted [19, 20] and for underdoped $Bi_2Sr_2CaCu_2O_{8+x}$ observed [7] gapless arcs of the Fermi surface are also present in the Y- and La-based high-temperature superconductors, thereby being presumably a generic feature of all copper-oxide perovskites. Secondly, there is evidence for an admixture of an appreciable *s*-wave component to the predominant *d*-wave gap function in Y- and La-based high-T_c superconductors. This evidence is essentially based on the fact that the slope of the linewidth around T_c turning from convex to concave cannot be reproduced without an *s*-wave component. Recent angle-resolved electron tunneling experiments in $YBa_2Cu_3O_7$ report similar results, namely a 17% *s*-wave admixture to the *d*-wave gap function [21]. Moreover, the gap values listed in Table 1 are in excellent agreement with the results of recent μSR experiments performed by Khasanov et al. [22] who investigated the in-plane magnetic field penetration depth in $La_{1.83}Sr_{0.17}CuO_4$ which gave $\Delta_s = 1.57(8)$ meV and $\Delta_d = 8.2(1)$ meV. A subsequent theoretical study performed by Bussmann-Holder et al. [23] confirmed these findings by calculations based on a two-band model.

The author is very much indebted to K.A. Müller for stimulating discussions.

References

1. Bednorz JG, Müller KA (1986) Z Phys B 64:189
2. Tsuei CC, Kirtley JR (2000) Rev Mod Phys 72:969

3. Ding H, Yokoya T, Campuzano JC, Takahashi T, Randeria M, Norman MR, Mochiku T, Kadowaki K, Giapintzakis J (1996) Nature 382:51
4. Mesot J, Norman MR, Ding H, Randeria M, Campuzano JC, Paramekanti A, Fretwell HM, Kaminski A, Takeuchi T, Yokoya T, Sato T, Takahashi T, Mochiku T, Kadowaki K (1999) Phys Rev Lett 83:840
5. Müller KA, Keller H (1997) In: Kaldis E (ed) High-T_c Superconductivity: Ten Years after the Discovery. Kluwer, Dordrecht, pp 7–29
6. Müller KA (2002) Phil Mag Lett 82:279
7. Norman MR, Ding H, Randeria M, Campuzano JC, Yokoya T, Takeuchi T, Takahashi T, Mochiku T, Kadowaki K, Guptasarma P, Hinks DG (1998) Nature 392:157
8. Mesot J, Furrer A (1997) J Supercond 10:623
9. Mesot J, Furrer A (1998) In: Furrer A (ed) Neutron Scattering in Layered Copper-Oxide Superconductors. Kluwer, Dordrecht, pp 335–374
10. Furrer A (2005) Structure and Bonding 114:171
11. Lovesey SW, Staub U (2000) Phys Rev B 61:9130. These authors put forward the idea that the relaxation rate is dominated by phonon interactions. In their model calculations for $Ho_{0.1}Y_{0.9}Ba_2Cu_3O_7$ they use a truncated crystal-field level scheme, i.e., they neglect all but three out of the seventeen crystal-field states which leads to an unreasonably good agreement with the experimental data. However, as shown by A.T. Boothroyd in a comment published in Phys Rev B 64:066501 (2001), the inclusion of the complete set of crystal-field levels produces a drastically different temperature dependence of the linewidth, i.e., the phonon damping picture is no longer supported. Moreover, phonon relaxation exhibits a continuous temperature behaviour of the linewidths and cannot reproduce the step-like features observed in several high-T_c cuprates at or above T_c
12. Korringa J (1950) Physica (Utrecht) 16:601. For rare-earth based compounds the Korringa law is slightly modified due to crystal-field effects
13. Boothroyd AT, Mukherjee A, Murani AP (1996) Phys Rev Lett 77:1600
14. Häfliger PS, Podlesnyak A, Conder K, Pomjakushina E, Furrer A (2006) Phys Rev B 74:184520
15. Rubio Temprano D, Conder K, Furrer A, Mutka H, Trounov V, Müller KA (2002) Phys Rev B 66:184506
16. Rubio Temprano D, Mesot J, Janssen S, Conder K, Furrer A, Mutka H, Müller KA (2000) Phys Rev Lett 84:1990
17. Rubio Temprano D, Mesot J, Janssen S, Conder K, Furrer A, Sokolov A, Trounov V, Kazakov SM, Karpinski J, Müller KA (2001) Eur Phys J B (Rapid Note) 19:5
18. Rubio Temprano D, Mesot J, Conder K, Janssen S, Mutka H, Furrer A (2000) J Supercond 13:727
19. Wen XG, Lee PA (1996) Phys Rev Lett 76:503
20. Engelbrecht JR, Nazarenko A, Randeria M, Dagotto E (1998) Phys Rev B 57:13406
21. Smilde HJH, Golubov AA, Ariando G, Rijnders G, Dekkers JM, Harkema S, Blank DHA, Rogalla H, Hilgenkamp H (2005) Phys Rev Lett 95:257001
22. Khasanov R, Shengelaya A, Maiuradze A, La Mattina F, Bussmann-Holder A, Keller H, Müller KA (2007) Phys Rev Lett 98:057007
23. Bussmann-Holder A, Khasanov R, Shengelaya A, Maisuradze A, La Mattina F, Keller H, Müller KA, cond-mat/0610327

Bill A, Hizhnyakov V et al. (2007) In: Bussmann-Holder A, Keller H (eds)
High T_c Superconductors and Related Transition Metal Oxides.
Springer-Verlag Berlin Heidelberg, 143–156
DOI 10.1007/978-3-540-71023-3

Electronic Inhomogeneities and Pairing from Unscreened Interactions in High-T_c Superconductors

A. Bill[1] · V. Hizhnyakov[2,3] (✉) · G. Seibold[3] · E. Sigmund[3]

[1]Department of Physics & Astronomy, California State University Long Beach, Long Beach, CA 90840, USA

[2]Institute of Physics, University of Tartu, EE-2400 Tartu, Estonia
hizh@eeter.fi.tartu.ee

[3]Lehrstuhl für Theoretische Physik, BTU Cottbus, 03013 Cottbus, Germany

Abstract The presence of electronic inhomogeneities and layering strongly reduces the screening of the electron-ion interaction in high-temperature superconductors. This implies the existence of a non-totally screened long-range contribution to the electron-lattice coupling and opens an additional channel for the formation of Cooper pairs. We calculate the superconducting order parameter taking into account a) the long-range and the short-range parts of the electron-lattice interaction and b) the Coulomb repulsion between charge-carriers. We show that whereas the long-range electron-lattice coupling determines the anisotropy of the order parameter, the Coulomb repulsion and the short-range interactions determine its symmetry.

Introduction

Early on after the discovery of high-T_c cuprates by Bednorz and Müller [1] it was realized by both experimentalists and theoreticians that many of the unusual properties of these compounds may be understood by the assumption of an inhomogeneous distribution of charge carriers. The apparent occurence of phase separation (PS) in the high-T_c materials stimulated the discussion wether this phenomenon is induced by the dopant ions (chemically driven PS) or may be due to a vanishing of the compressibility of the electronic subsystem (electronically driven PS).

However, another point of view soon emerged [2–4] which pointed out the existence of microsocopic or nanoscale inhomogeneities (for a colletion of papers on this issue see [5–7]). While on a local scale this kind of charge segregation is now well established from NMR and NQR measurements for a large variety of cuprate materials, information on the spatial structure of the inhomogeneous charge distribution is up to now limited to the lanthanum compounds. Within elastic neutron scattering Tranquada and collaborators observed a splitting of both spin and charge order peaks in $La_{1.48}Nd_{0.4}Sr_{0.12}CuO_4$ [8] which bears resemblance to similar data in the nickelates, where both incommensurate antiferromagnetic (AF) order [9, 10] and the ordering of charges [10, 11] has been detected by neutron scattering and electron diffraction, respectively. In [10] it was shown that the magnetic ordering is revealed in the occurrence of first and third harmonic Bragg peaks whereas the charge ordering is associated with second harmonic peaks. From this it was concluded that the doped holes arrange themselves in quasi-onedimensional structures, called stripes, which simultaneously constitute antiphase domain walls for the AF order. Whereas the existence of static charge and spin order in the nickelates is well established by electron, neutron and X-ray diffraction [9–12] the interpretation of experimental results in the cuprate compounds [8–19] with respect to stripes is still controversial.

Direct evidence for static stripes in cuprates has been most clearly established in $La_{2-x}Sr_xCuO_4$ (LSCO) co-doped with Nd, Eu or Ba. The associated charge modulation in LBCO at doping $x = 1/8$ has been recently detected by resonant X-ray scattering experiments [20]. Note that in this regard the stripe concept also helped to clarify the origin of the dip in the T_c vs. x curve at doping $x \approx 0.12$ in $La_{2-x}Ba_xCuO_4$ [21].

The formation of stripe structures in LSCO has profound consequences on the excitation spectra of the system in both charge and magnetic channels. Within the framework of the unrestricted Gutzwiller approximation [22] extended to include fluctuations beyond the saddle-point solution [23–25] a large variety of experimental data has been shown to be consistent with the stripe scenario. This formalism applied to the three-band Hubbard model [26] yields metallic mean-field stripes consistent with the observed doping behavior of the magnetic incommensurability $\epsilon \equiv 1/(2d)$ [14–18] (d is the distance between charged stripes in units of the lattice constant). Also the chemical potential [27, 28], and transport experiments [29, 30] as a function of doping are explained in a parameter free way [26]. Within the stripe scenario the mid-infrared peak observed in the optical conductivity of cuprates [31, 32] can be attributed to a collective phason mode [33]. Therefore, this excitation strongly softens with increasing doping and merges with the Drude peak due to the increased lateral stripe fluctuations.

Very recent experiments were dedicated to explore the problem of universality in the magnetic excitations between different cuprate families [34–37]. In particular Tranquada and collaborators [34] reported INS measurements of the magnetic excitations in $La_{1.875}Ba_{0.125}CuO_4$ which shows static charge *and* spin order. Also in this case the stripe model provides a rather accurate describtion of the "hour-glass shape" of the spectrum [38, 39]. This includes the low energy spin-wave like excitations which disperse towards a saddle-point at the commensurate antiferromagnetic wave-vector which can be associated with the resonance peak. Above the saddle-point the spin fluctuations resemble again those of (high-energy) antferromagnetic fluctuations similar to those obtained by spin-leg ladder theories [40, 41].

The tendency of strongly correlated systems towards the formation of inhomogeneous structures can also be approached from alternative points of view. One of them is based on the concept of frustrated phase separation [42, 44, 45] where a phase separation instability is prevented by long-range Coulomb interactions. As a result the long-wavelength density fluctuations associated with phase separation are suppressed in favor of shorter-wavelength density fluctuations, giving rise either to dynamical slow density modes [42] or to incommensurate charge density waves [44].

Two of the authors have discussed early on the process by which electronic phase separation and the formation of stripes can arise (see, e.g. [2, 46–50]). The approach is especially useful to describe the low and intermediate doping regime in high-temperature superconductors. The basic concept can be summarized as follows. Doping the parent compound of a high-temperature superconductor results in the introduction of holes in the antiferromagnetically ordered CuO_2 planes. It has been shown that an additional hole induces a local distortion of the spin and lattice structure and breaks the symmetry of the system [46–56]. In other words, introducing a hole in the CuO_2 plane results in the simultaneous formation of a magnetic and lattice cluster or polaron. The size of the magnetic cluster has been evaluated to about five to ten neighbouring Cu atoms of the CuO_2 plane [46–49]. Furthermore, given the strength of the electron–phonon coupling and the structure of the lattice, the size of the lattice polaron has been estimated to be of the same order [55–57]. Thus, the quasiparticles can be characterized as polarons of *intermediate* size.

Two time scales are associated with the motion of the magnetic cluster. The local distortion of the antiferromagnetic background is hardly mobile ($\tau_1 \gg 10^{-12}$ s) since it is associated with a high spin-flip energy, whereas the (undressed) hole inside the cluster is highly mobile ($\tau_2 \gg 10^{-15}$ s) [46–49]. The fact that the dressed hole is also an intermediate sized lattice polaron reduces further the (already low) mobility of the cluster.

The existence of these two time-scales has an important consequence on the electronic structure when doping further the CuO_2 planes. Indeed, adding holes to the system results in the overlap of some of the clusters. These interconnected clusters form a conducting subsystem in a background dominated by strong antiferromagnetic correlations. The magnetic interaction between the clusters is weakly attractive [54]. However, due to the competition between potential energy (including Coulomb and magnetic contributions) and kinetic energy, the clusters do not form a homogeneous sea in an antiferromagnetic background. Instead, the clusters (or magnetic-lattice polarons) form a percolative network in which the holes can move easily. If one assumes that the holes are mainly confined in the CuO_2 planes, the dimension D of this network is lower than two ($1 \leq D \leq 2$). We stress the fact that the hole subsystem (percolation network) is not static but is dynamic with characteristic time scale τ_1.

The concept of the magnetic-lattice polaron and the percolation network has several other important features. First, the network is ordered over intermediate distances (≤ 100 Å). The hole-clusters form stripes aligned in the (11) and ($1\bar{1}$) directions. One of the reasons for this ordering is that the interaction between clusters is attractive along these directions, but repulsive along (10) and (01) [51–54]. As a result, holes can easily move in in (11) and ($1\bar{1}$) directions whereas their mobility is strongly reduced along (10) and (01). This leads to the strongly anisotropic energy spectrum of holes with large dispersion in M-direction and small dispersion in X- and Y-directions. In this way the percolative model naturally explains the experimentally observed strongly anisotropic Fermi surface in high-T_c superconductors with extended saddle points (Van Hove singularities) in X- and Y-directions (see also [58]). It is important to note that doping changes the characteristic size of the ordered network, but has little effect of the band structure itself. Note also that because of the weak attraction between clusters and the strong electron-lattice interaction, some holes will not be part of the network, but remain as quasi-localized entities. The occurrence of pinning is important in this context since it contributes to determine the shape of the cluster network and the concentration of isolated clusters.

Another interesting property of the percolative network of clusters (magnetic-lattice polarons) is that the structure of the lattice is also inhomogeneous on a local scale. This implies that local and global (crystal) symmetries differ. This idea is supported by numerous experimental observations of local distortions (see, e.g. [59–64]). These lattice distortions contribute to the stability of the network and suggest that the picture of local electronic inhomogeneities and stripes extends to the intermediate-doping regime of high–T_c superconductors.

The inhomogeneous electronic structure resulting from the competition between Coulomb and magnetic interactions is not only supported by the presence of strong electron–phonon interactions, but also affects the latter in an important way. It leads to an enhancement of the long-wavelength optical phonon contribution to the interaction between charge carriers. As is well known, the electron–phonon interaction with optical phonons is long-range and strong in insulators but screened out by the high-frequency plasmons in normal metals. In high-T_c superconductors the presence of AF fluctuations results in the local reduction of charge carrier concentration as best visualized in the cluster model. Therefore, in AF-clusters and in their vicinity the electric field associated with the long-wave optical phonon remains poorly screened, and the total screening of the interaction is reduced. This effect of inhomogeneities on the screening bears some resemblance with the reduction of screening caused by insulating granulas in granular superconductors. It was shown in this latter case that the reduction of screening implied an enhancement of the critical temperature in these systems [65].

When discussing the reduction of screening it is important to remember that cuprate superconductors have a layered structure and anisotropic transport properties. This implies the occurrence of low-energy electronic collective modes (plasmons) and a dynamic screening of the Coulomb interaction [66–69]. These low-energy collective modes have been shown to contribute constructively to the pairing mechanism [67–69]. Furthermore, the anistropic transport properties reduce the screening of the Coulomb and the electron–phonon interactions along the c-direction (perpendicular to the layers). This property, together with the electronic inhomogeneities in the CuO_2 planes, results in the appearance of poorly screened three-dimensional electron–phonon interaction with long-wave optical phonons. Thus, the superconducting order due to this interaction is also three-dimensional. This aspect of the problem is important since one cannot have a finite T_c in purely 2D systems, and it is known that high-T_c superconductivity is indeed a 3D phenomenon [70]. Note also that the pairing interaction caused by spin fluctuations is very weak in c-direction and this in itself is a serious concern when describing the observed 3D superconducting state in terms of magnetic fluctuations.

One essential and seemingly counter-intuitive implication of long-range electron–phonon interaction for the formation of the superconducting state is the well-defined $\boldsymbol{k}$-dependence of the order parameter (gap) Δ despite the locally inhomogeneous structure: the long-wave interaction averages over these inhomogeneities. The formation of anomalously correlated singlet pairs occurs despite the spin-polarized stripe-structure of the CuO_2 planes on the microscopic scale. In this case it is rather irrelevant that charge carriers with opposite spins are located in different stripes: the long-range anomalous correlations of the charge carriers forming the singlet pairs depend only little on the microscopic-scale structure of the system.

Another important consequence of long-range interactions is that the resulting pairing interaction mixes states with close wave vectors and leads therefore to the essential dependence of the order parameter (gap) Δ on the direction in $\boldsymbol{k}$-space: the resulting anisotropy of $|\Delta|$ is of the same type as the density of states on the Fermi surface [71]. This is just the situation found in high-T_c superconductors. Moreover, as will be shown in the next sections (see also [73]), the actual symmetry is due to the contribution of the Coulomb repulsion between charge carriers and is therefore d-wave, in agreement with experiment. Note that the sign-alternating behavior of Δ is often considered as a support of pairing mechanisms caused by spin fluctuations. However, we show here that a mechanism involving long-range electron–phonon interactions in the presence of a strong local Coulomb repulsion provides the sign-alternating anisotropic order parameter observed in experiment.

In the following we study the impact of electronic inhomogeneities on the superconducting pairing mechanism in high-T_c materials. We focus our attention on the anisotropy and symmetry of the order parameter. A large body of experimental evidences exist that reveals the anisotropy of the superconducting order parameter in high-temperature superconductors [75–82].

However, the discussion of the underlying pairing mechanism remains controversial. We show below (see also [73]) that: 1) the *anisotropy* of the order parameter in high-T_c materials is easily explained by the presence of long-range interactions, and 2) the *symmetry* of the order parameter requires a precise knowledge of the strengths of the various competing interactions. We stress the fact that the microscopic pairing mechanism presented here and based on [71, 73] allows an understanding both of the anisotropy and the symmetry of the order parameter.

As mentioned above, there are alternative descriptions of the electronic phase separation and its relation to superconductivity [42, 44]. However, the following discussion deals with rather general implications of the presence of inhomogeneities and does, therefore, not rely on the particular model used for their description.

In Sect. II we discuss in more detail the influence of electronic phase separation on the electron-lattice interaction. We discuss the pairing potential induced by long-range and short-range electron-lattice interactions and take into account the Coulomb repulsion between charge-carriers. In Sect. III we determine numerically the anisotropy and symmetry of the superconduction order parameter for the pairing potentials obtained in Sect. II.

Electronic Inhomogeneities and the Electron-Lattice Interaction

In this section we focus on the component of the interactions along the CuO_2 plane. Regardless of the specific structure of the electronic inhomogeneities, its very existence has dramatic consequences for the electron-lattice interaction. Indeed, electronic phase separation implies a reduced screening of the electron-ion (attractive) coupling. In particular, according to the picture involving two time scales as presented in the introduction, the interaction of charge carriers with ions of the antiferromagnetic, insulating regions is barely screened, resulting in an enhancement of the *long-range* part of the electron-lattice interaction. Thus, the electron-lattice interaction can be divided into two parts [55–57]

$$H_{\mathrm{el}} = \sum_{\boldsymbol{k},\sigma} \sum_{\boldsymbol{q},\nu} \left[g_{\mathrm{S}}(\boldsymbol{k}, \boldsymbol{q}\nu) + g_{\mathrm{L}}(\boldsymbol{k}, \boldsymbol{q}\nu) \right] c^{+}_{\boldsymbol{k},\sigma} c_{\boldsymbol{k}+\boldsymbol{q},\sigma} b^{+}_{\boldsymbol{q},\nu} + \mathrm{h.c.}\,, \tag{1}$$

where $c^{+}_{\boldsymbol{k},\sigma}$ ($b^{+}_{\boldsymbol{q},\nu}$) is the creation operator of a hole (phonon) with wave-vector $\boldsymbol{k}(\boldsymbol{q})$ and spin σ (branch ν). The functions g_{S} and g_{L} describe the short- and long-range interactions, and are functions with a maximum at long and short wavelengths, respectively. The first term in Eq. 1 is the dominant coupling in metals. The appearence of the second part g_{L} is specific to systems displaying electronic inhomogeneities as high-temperature "ionic" superconductors and is due to the ineffective screening of the electron–ion interaction.

We now show that the anisotropy of the order parameter in high-T_c superconductors is explained in a natural way by the long-range part of the electron-lattice interaction. We also demonstrate that the Coulomb repulsion or the short-range part of the electron-lattice interaction determines the symmetry of the order parameter. Without Coulomb interaction, both *s*- and *d*-wave types of function satisfy the gap equation, but the long-range part leads to an *s*-wave stable solution (lower free energy). However, adding the Coulomb repulsion between charge-carriers and/or the short-range part of the electron-lattice interaction leads to an enhancement of the free energy for the *s*-wave solution without affecting the *d*-wave solution. Thus, depending on the values of the different interaction strengths, either the *s*-wave or the *d*-wave solution is stable. It appears that the Coulomb repulsion is strong enough in high-T_c superconductors to privilege the *d*-wave solution.

The main part of the pairing potential comes from the long-range in-plane interaction of charge-carriers (holes) with totally symmetric (A_{1g}) and non-totally symmetric (B_{1g}) optical phonons (we remind that we do not consider here the contribution to the pairing arising from the out-of-plane coupling and the dynamic screened Coulomb interaction [67–69]). The in-plane interaction mixes hole states with close $|\boldsymbol{k}|$. Therefore, the pairing potential given by this interaction directly depends on the density of states on the Fermi surface ρ_F [71]. In cuprate superconductors, ρ_F is strongly anisotropic having much larger values in *a* and *b* directions than in between. Thus, the long-range part of the pairing potential is strongly anisotropic too. As explained in Sect. I, this anisotropy is a direct consequence of the presence of electronic inhomogeneities. Another important property of the pairing interaction discussed here is the weak correlation of Cooper pairs with different $|\boldsymbol{k}|$ (they are asymptotically independent). Hence, the actual phase difference of Δ in *a* and *b* directions depends essentially on other (here short-range) interactions that only weakly influence $\Delta_{\max}$ and the critical temperature T_c. We show below that even a relatively small Coulomb repulsion (which is totally symmetric) in the plane has a remarkable effect on the free energy of the *s*-wave type superconducting state (it is enhanced) and has no effect on the free energy of the *d*-wave state. As a consequence, rather moderate variations of the short-range part of the Coulomb repulsion or the short-range interaction with phonons or other sign-alternating interactions may change *s*-wave to *d*-wave type superconductivity without remarkable change of $\Delta_{\max}$ (and T_c).

Although a strong-coupling consideration is needed to describe correctly high-T_c superconductors the effects of different pairing interactions can be elucidated within the framework of standard BCS theory. A detailed study of the solutions obtained from the long-range and short-range interactions was given in [73]. Here we present the main results of our numerical study of the BCS gap equation at zero temperature with account of the different pairing potentials mentioned above.

The BCS gap equation requires the knowledge of the electronic dispersion relation and the pairing potential. The in-plane parametrization of the dispersion for the conduction holes, $\varepsilon_{\vec{k}}$, is well-established [71, 73, 74]. For reasons given earlier we neglect at this stage the dispersion perpendicular to the CuO_2 planes. The corresponding density of states were given in [71, 73]. We note that for such band $\varepsilon_{\vec{k}}$ the density of states $\rho_F = v_F^{-1}$ has strong maxima along $\varphi = n\pi/2$ ($n = 0; 1; 2; 3$) in the $(k_x; k_y)$-plane.

In the following we discuss three contributions to the pairing interaction: 1) the coupling to the totally symmetric optical mode A_{1g}, 2) the coupling to the non-totally symmetric modes B_{1g} and B_{2g}, and 3) the coupling to short-range interactions.

We first consider the attraction of charge carriers mediated by totally symmetric optical phonons and the Coulomb repulsion. Here we take into account the fact that in the limit $(\vec{k}, k_z) = (k_x, k_y, k_z) \to 0$ these interactions compensate each other (due to the charge neutrality) and should therefore be considered simultaneously. The corresponding interaction potential caused by an optical A_{1g}-mode with frequency ω_ν in a layered crystal equals [71, 83]

$$V_0\left(\vec{k};\mathbf{q}\right) = -\frac{4\pi e^2}{q^2\epsilon\left(\mathbf{q}\right)}\frac{\omega_{k;q}^2}{\omega_{k;q}^2 - \left(\omega_\nu^2 + i\gamma_\nu\right)^2}, \tag{2}$$

where $\omega_{k;q} = (\varepsilon_{\vec{k}+\mathbf{q}} - \varepsilon_{\vec{k}})/\hbar$, γ_ν is the damping rate of phonon ($\gamma_\nu \ll \omega_{k;q}$), and $\epsilon(\boldsymbol{q})$ is the static dielectric function (see [71, 73]). In the limit $\boldsymbol{q} \to 0$, $\omega_{k;q}$ and the interactions vanish. However, taking into account the finite bandwidth along the c-direction the main contribution to the gap equation comes from terms with $q_\perp$ of the order of the reciprocal superconducting correlation length in c direction ($\sim \pi/d_c$). For such $q_\perp$, the interaction (Eq. 1) is finite and given by [71]

$$V_0\left(\mathbf{q}_\parallel\right) \simeq \frac{U_0\kappa^2}{\kappa^2 + q_x^2 + q_y^2}. \tag{3}$$

$\mathbf{q} = \vec{k} - \vec{k}'$ and $\vec{k} = (k_x, k_y)$ is the wave-vector component *parallel* to the CuO_2 plane. U_0 and κ were estimated in [73] and are equal to $U_0 = 0.1$ eV and $\kappa \sim 0.3$. These values are in agreement with the values of the corresponding parameters used in [84–86] to describe the experimentally observed phonon renormalizations caused by the superconducting transition.

The second contribution to the pairing is the attractive potential mediated by *non-totally symmetric* B_{1g} and B_{2g} phonons. Using the results derived in [87] for the coupling function to B_{1g} phonons, the long-range interaction has the following form

$$V_{B_{1g}}\left(\vec{k};\mathbf{q}\right) = U_{B_{1g}}\left(\cos k_x - \cos k_y\right)^2. \tag{4}$$

The pairing potential due to B_{2g}-phonons is analogous, but with k_x, k_y replaced by k'_x, $k'_y = k_x \pm k_y/\sqrt{2}$. Note that for small angles $|\phi - n\pi/2|$, $n = 0; 1; 2; 3$ the factor $(\cos k'_x - \cos k'_y)^2$ is small, while the factor $(\cos k_x - \cos k_y)^2$ is of the order of 1 (the average value being 3/2). Therefore, the interaction $V_{B_{2g}}$ does practically not contribute to $\Delta_{\max}$ (and T_c) whereas the contribution of $V_0 + V_{B_{1g}}$ is decisive. From the analysis of phonon renormalization data given [84–86] follows that $U_{B_{1g}} \sim U_0 \sim 0.1$ eV.

The third and last contribution to the pairing is the potential caused by the *short-range* part of the electron–phonon interaction and by the Coulomb repulsion between charge-carriers. These interactions are dominant in usual metals. For totally symmetric phonons and taking into account the Coulomb repulsion one obtains:

$$\bar{V}_0\left(\vec{k};\mathbf{q}\right) \simeq \frac{\bar{U}_0\omega_{k;q}^2}{\omega_{k;q}^2 - \left(\omega_\upsilon + i\gamma_\upsilon\right)^2} \,. \tag{5}$$

This potential is attractive for small $|\omega_{k;q}|$ and repulsive for large $|\omega_{k;q}|$; in the limit $|\omega_{k;q}| \gg \omega_\upsilon$ it reduces to the short-range part ($U_c =- \bar{U}_0$) of the Coulomb repulsion. On the other hand, the short-range part of the pairing potential mediated by B_{1g} phonons has the form:

$$\bar{V}_{B_{1g}}\left(\vec{k};\mathbf{q}\right) \simeq \frac{\bar{U}_{B_{1g}}\omega_\upsilon^2}{\omega_{\mathbf{k;q}}^2 - \left(\omega_\upsilon + i\gamma_\upsilon\right)^2}\left(\cos k_x - \cos k_y\right)^2 \,. \tag{6}$$

This potential is also attractive for small $|\omega_{k;q}|$ and repulsive for large $|\omega_{k;q}|$ but unlike $\bar{V}_0(\vec{k};\mathbf{q})$ it vanishes in the limit $|\omega_{k;q}| \gg \omega_\upsilon$; this property reflects the absence of non-totally symmetric short-range repulsion.

Note that the change of sign of the potential $\bar{V}_{B_{1g}}$ for $|\omega_{k;q}| > \omega_\upsilon$ is the result of retardation effects related to the electron–phonon coupling. The short-range part of the potential caused by B_{2g} phonons is analogous to Eq. 6, but with k_x, k_y replaced by k'_x, $k'_y = (k_x \pm k_y)/\sqrt{2}$. $V_{B_{2g}}$ is effective in the region of $\mathbf{q}$-space where $|k_x|$ and $|k_y|$ essentially differ from zero and where ρ is small. As a consequence, the contribution of $\bar{V}_{B_{2g}}$ to $\Delta(\vec{k})$ is small and can be neglected (this conclusion has also been verified numerically).

Numerical Results and Discussion

The numerical solutions of the BCS gap-equation with the pairing potentials given by Eqs. 3–6 are presented in Table 1 and exemplified in Figs. 1 and 2.

$F_{s;d}$ describes the lowering (gain) of the free energy due to the superconducting transition. As can be seen in Table 1 (first row), the unscreened interaction of charge-carriers with long-wave optical phonons A_{1g} and B_{1g} leads to an anisotropic order parameter with *s*-wave symmetry. We re-

Table 1 Comparative values of F_s, F_d, Δ_{min}, Δ_{max} for different pairing potentials (in eV, $U_l = 0.03$ eV, $\kappa = 0.3$)

Type of interaction	F_s	F_d	Δ^d_{max}	Δ^s_{max}	Δ^s_{min}
Long-range A_{1g} and B_{2g} phonons	0.6448	0.2485	0.0237	0.0319	0.0052
Long-range A_{1g} and B_{2g} phonons + short-range Coulomb repulsion $U_c = -0.0025$	0.3529	0.2484	0.0236	0.0255	0.0012
Long-range A_{1g} and B_{2g} phonons + short-range Coulomb repulsion $U_c = -0.005$	0.1674	0.2484	0.0235	0.0199	0.0001
Long-range A_{1g} and B_{2g} phonons + short-range Coulomb repulsion $U_0 = 0.05$ + short-range A_{1g} and B_{1g} phonons $U_{B_{1g}} = 0.004$	0.1812	0.2705	0.0248	0.0295	0.0060

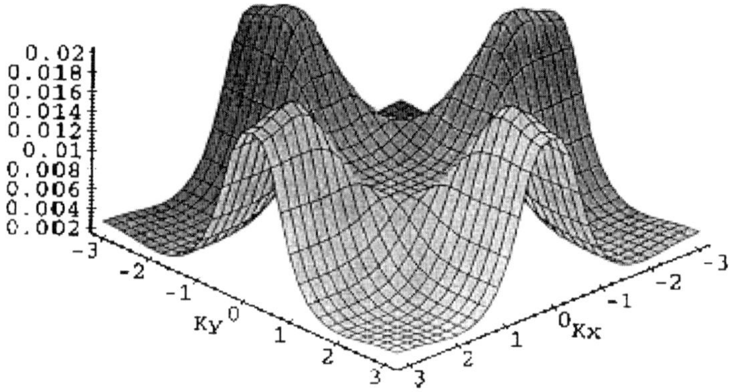

Fig. 1 *s*-wave solution of the BCS gap-equation. The pairing interaction contains contributions from both the long-wave and short-wave interactions with A_{1g} and B_{1g} phonons as well as the Coulomb repulsion between charge-carriers. The stability of the solution depends on the values of the different coupling strengths (see text and Table 1).

mind that this pairing is specific to systems containing electronic inhomogeneities and results from the ineffective screening of the electron–ion interaction.

Introducing the important (short-range, totally symmetric) Coulomb repulsion between charge-carriers into the pairing potential (second and third row of Table 1) leads to the enhancement of the free energy for the *s*-wave solution of the gap equation, whereas the free energy for the *d*-wave solution remains almost constant. A rather moderate strength of the Coulomb repulsion can favor the stability of a *d*-wave order parameter with respect to the

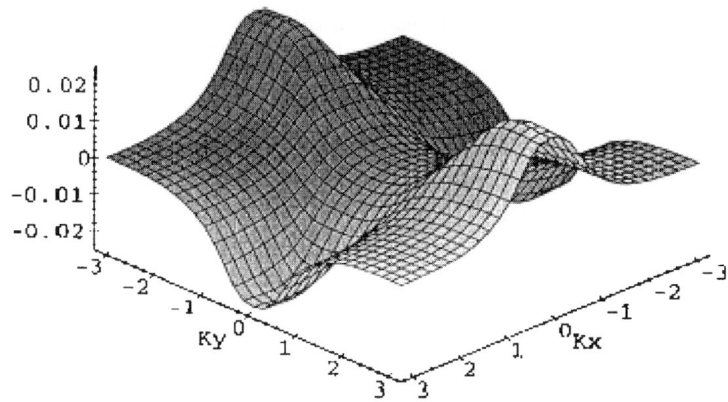

Fig. 2 *d*-wave solution of the gap-equation for the same interactions as in Fig. 1. The stability of the solution depends on the value of the different coupling strengths (see text and Table 1)

s-wave solution. Therefore, the short-range repulsive part of the interaction can stabilize the *d*-wave solution.

The last row of Table 1 shows the solutions of the gap-equation obtained if one takes into account all pairing interactions (Eqs. 3–6) (including the short-range part of the electron-lattice coupling). Figures 1 and 2 display these solutions. One notes that for the parameters used Δ_{max} (and T_c) is mainly determined by the long-range attractive pairing potential. On the other hand, the symmetry of the most stable solution is essentially determined by the short-range interactions of the holes with totally symmetric (A_{1g}) phonons and the Coulomb repulsion between holes.

We also performed calculations with repulsive interactions having four peaks $(\pm\pi, \pm\pi)$ in $\vec{k}$-space and which are usually associated with antiferromagnetic fluctuations (see, e.g. [88, 89]). Such interaction turns out to significantly increase F_{d} without changing F_{s} and, therefore, help stabilizing the *d*-wave type solutions of the gap equation. We remind the reader, however, of the problem related to the three-dimensional character of superconductivity that spin-fluctuations cannot reproduce.

Conclusion

The picture resulting from our work on high-temperature superconductors ([2–5, 22–26, 33, 38, 39, 46–57, 67–69, 71, 73, 84, 85]) can be summarized as follows: the presence of strong competing Coulomb and electron-lattice interactions result in the existence of dynamic local electronic inhomogeneities. These inhomogeneities order over a few hundreds nanometers into stripes. This, together with the layered structure of the material results in the absence of perfect screening and the appearance of long-range interactions, both of the Coulomb and the electron-lattice interactions. The long-range part of the interaction is absent in conventional metallic systems because of the "perfect" screening properties of these materials. The long-range interactions affect various properties of high-T_c materials among which the superconducting pairing interaction and order parameter.

We studied numerically the influence of the long-range electron-lattice coupling, together with short-range (electron-lattice and Coulomb) interactions, on the superconducting pairing state. We found that the long-range part of the interaction leads to a strongly anisotropic order parameter. This interaction determines the value of Δ_{max} and T_c and would lead to an *s*-wave superconducting state if it were the only interaction present in the system. However, a sufficiently strong short-range component of the interactions (Coulomb repulsion and/or short-range part of the electron-lattice coupling) stabilizes the solution with *d*-wave symmetry. Therefore, the symmetry of the superconducting order parameter depends on the relative coupling strengths of the different contributions to the pairing potential presented above.

Acknowledgements A.B. acknowledges support through the SCAC at CSU Long Beach. V.H. acknowledges support under the Mercator programme of the Deutsche Forschungsgemeinschaft and under the Estonian Science Foundation, Grant 6534. Finally, we are deeply grateful to K. A. Müller for his important participation in our research and who over all the years stimulated our work with interesting discussions and who gave and still gives major contributions to our present understanding of the subject.

References

1. Bednorz JG, Müller KA (1986) Z Physik B 64:189
2. Hizhnyakov V, Sigmund E (1988) Physica C 156:655
3. Kremer R, Sigmund E, Hizhnyakov V, Hentsch F, Simon A, Müller KA, Mehring M (1992) Physik Z B 86:319
4. Kremer R, Hizhnyakov V, Sigmund E, Simon A, Müller KA (1993) Z Physik B 94:17
5. Müller KA, Benedek G (eds) (1993) Proceedings of the First workshop on phase separation in cuprate superconductors. World Scientific, Singapure
6. Sigmund E, Müller KA (eds) (1994) Proceedings of the Second workshop on phase separation in cuprate superconductors. Springer, Berlin, Heidelberg
7. Di Castro C, Sigmund E (eds) (1996) Proceedings of the Third workshop on phase separation, electronic inhomogeneitites and related mechanisms in high-T_c superconductors. J Supercond, p 9
8. Tranquada JM, Sternlieb BJ, Axe JD, Nakamura Y, Uchida S (1995) Nature 375:56
9. Hayden SM, Lander GH, Zarestky J, Brown PJ, Stassis C, Metcalf P, Honig JM (1992) Phys Rev Lett 68:1061
10. Tranquada JM, Buttrey DJ, Sachan V, Lorenzo JE (1994) Phys Rev Lett 73:1003
11. Chen CH, S-W. Cheong, Cooper AS (1993) Phys Rev Lett 71:2461
12. Vigliante A, von Zimmermann M, Schneider JR, Frello T, Andersen NH, Madsen J, Buttrey DJ, Doon Gibbs, Tranquada JM (1997) Phys Rev B 56:8248

13. Tranquada JM, Axe JD, Ichikawa N, Nakamura Y, Uchida S (1996) Nachumi B Phys Rev B 54:7489
14. Tranquada JM, Axe JD, Ichikawa N, Moodenbaugh AR, Nakamura Y, Uchida S (1997) Phys Rev Lett 78:338
15. Yamada K, Lee CH, Kurahashi K, Wada J, Wakimoto S, Ueki S, Kimura H, Endoh Y, Hosoya S, Shirane G, Birgeneau RJ, Greven M, Kastner MA, Kim YJ (1998) Phys Rev B 57:6165
16. Niemöller T et al. (1999) Eur Phys J B 12:509
17. Arai M, Nishijima T, Endoh Y, Egami T, Tajima S, Tomimoto K, Shiohara Y, Takahashi M, Garrett A, Bennington SM (1999) Phys Rev Lett 83:608
18. Arai M, Endoh Y, Tajima S, Bennington SM (2000) Int J Mod Phys B 14:3312
19. Fujita M, Goka H, Yamada K, Tranquada JM, Regnault LP (2004) Phys Rev B 70:104517
20. Abbamonte P et al. (2005) Nature Physics 1:166
21. Moodenbaugh AR, Xu Y, Suenaga M, Folkerts TJ, Shelton RN (1988) Phys Rev B 38:4596
22. Seibold G, Sigmund E, Hizhnyakov V (1998) Phys Rev B 57:6937
23. Seibold G, Lorenzana J (2001) Phys Rev Lett 86:2605
24. Seibold G, Becca F, Lorenzana J (2003) Phys Rev B 67:085108
25. Seibold G, Becca F, Rubin P, Lorenzana J (2004) Phys Rev B 69:155113
26. Lorenzana J, Seibold G (2002) Phys Rev Lett 89:136401
27. Ino A, Mizokawa T, Fujimori A, Tamasaku K, Eisaki H, Uchida S, Kimura T, Sasagawa T, Kishio K (1997) Phys Rev Lett 79:2101
28. Harima N, Matsuno J, Fujimori A, Onose Y, Taguchi Y, Tokura Y (2001) Phys Rev B 64:R220507
29. Noda T, Eisaki H, Uchida S (1999) Science 286:265
30. Yayu Wang, Ong NP (2001) Proc Natl Acad Sci USA 98:11091
31. Uchida S, Ido T, Takagi H, Arima T, Tokura Y, Tajima S (1991) Phys Rev B 43:7942
32. Suzuki M (1989) Phys Rev B 39:2312
33. Lorenzana J, Seibold G (2003) Phys Rev Lett 90:066404
34. Tranquada JM, Woo H, Perring TG, Goka H, Gu GD, Xu G, Fujita M, Yamada K (2004) Nature 429:534
35. Reznik D, Bourges P, Pintschovius L, Endoh Y, Sidis Y, Matsui T, Tajima S, cond-mat/0307591
36. Christensen NB, McMorrow DF, Ronnow HM, Lake B, Hayden SM, Aeppli G, Perring TG, Mangkorntong M, Nohara M, Takagi H (2004) Phys Rev Lett 93:147002
37. Pailhes S, Sidis Y, Bourges P, Hinkov V, Ivanov A, Ulrich C, Regnault LP, Keimer B (2004) Phys Rev Lett 93:167001
38. Seibold G, Lorenzana J (2005) Phys Rev Lett 94:107006
39. Seibold G, Lorenzana J (2006) Phys Rev B 73:144515
40. Vojta M, Ulbricht T (2004) Phys Rev Lett 93:127002
41. Uhrig GS, Schmidt KP, Grüninger M (2004) Phys Rev Lett 93:267003
42. Emery VJ, Kivelson SA (1993) Physica C 209:597
43. Löw U et al. (1994) Phys Rev Lett 72:1918
44. Castellani C, Di Castro C, Grilli M (1995) Phys Rev Lett 75:4650
45. Seibold G, Castellani C, C. Di Castro, Grilli M (1998) Phys Rev B 58:13506
46. Hizhnyakov V, Sigmund E (1988) Physica C
47. Hizhnyakov V, Kristoffel N, Sigmund E (1989) Physica C 161:435
48. Hizhnyakov V, Kristoffel N, Sigmund E (1989) ibid 160:119
49. Hizhnyakov V, Sigmund E, Schneider M (1991) Phys Rev B 44:795
50. Hizhnyakov V, Sigmund E, Seibold G, in [2], p 46

51. Seibold G, Sigmund E, Hizhnyakov V (1993) Phys Rev B 48:7537
52. Seibold G, Sigmund E, Hizhnyakov V (1996) J Supercond 9:407
53. Seibold G (1995) PhD thesis, Stuttgart, Shaker
54. Essl W, Sigmund E (1996) J Supercond 9:467
55. Bill A, Hizhnyakov V, Sigmund E in [3, 4], p 290
56. Bill A (1995) PhD thesis, Stuttgart, Shaker
57. Sigmund E, Bill A, Hizhnyakov V (1995) In: Mihailovi'c D, Ruani G, Kaldis E, Müller KA (eds) Proceeding of the workshop on Anharmonic properties of high-T_c cuprates. Singapore World Scientific, p 187
58. Salkola MI, Emery VJ, Kivelson SA (1996) J Supercond 9:401
59. Hammel PC, in [3, 4], p 185
60. Egami T, Billinge S (1996) In: Ginsberg DM (ed) Physical Properties of High-Temperature Supercondutors. Singapore World Scientific
61. Sharma RP et al. (1997) Phys Rev Lett 77:4624
62. Bianconi A et al. (1995) Europhys Lett 31:411
63. Bianconi A et al. (1996) Phys Rev Lett 76:3412
64. Mihailovic D, Müller KA (1997) Proceedings of the NATO conference on Material Aspects of High-Temperature Superconductivity: 10 Years after the Discovery, Kluwer
65. Deutscher G, Fenichel H, Gershenson M, Grunbaum E, Ovadyahu Z (1973) J Low Temp Phys 10:213
66. Kresin VZ, Morawitz H (1988) Phys Rev B 37:7854
67. Bill A, Morawitz H, Kresin VZ (1999) J Low Temp Phys 117:283
68. Bill A, Morawitz H, Kresin VZ (2002) Phys Rev B 66:100501
69. Bill A, Morawitz H, Kresin VZ (2003) Phys Rev B 68:144519
70. Schneider T (2002) Europhys Lett 60:141
71. Hizhnyakov V, Sigmund E (1996) Phys Rev B 53:5163
72. Hizhnyakov V, Sigmund E (1996) J Supercond 9:335
73. Sigmund E, Hizhnyakov V, Nevedrov D, Bill A (1997) J Supercond 10:441
74. Norman MR (2000) Phys Rev B 61:14751
75. Shen Z-X, Spicer WE, King DM, Dessau DS, Wells BO (1995) Science 267:343
76. Ding et al. (1996) Phys Rev B 54:9678
77. Wollman DA, Van Harlingen DJ, Lee WC, Ginsberg DM, Leggett AJ (1993) Phys Rev Lett 71:2134
78. Tsuei CC et al. (1997) Nature 387:481
79. Mathai A et al. (1995) Phys Rev Lett 74:4523
80. Chaudhari P, Shawn-Yu Lin (1994) Phys Rev Lett 72:1084
81. Sun AG et al. (1994) Phys Rev Lett 72:2267
82. Deutscher G, in [3, 4]
83. Pines D, Nozi'eres P (1989) The Theory of Quantum Liquids, Normal Fermi Liquids. Addison-Wesley New York
84. Bill A, Hizhnyakov V, Sigmund E (1995) Phys Rev B 52:7637
85. Bill A, Hizhnyakov V, Sigmund E (1996) J Supercond 10:493
86. Maly J, Lin DZ, Levin K (1996) Phys Rev B 53:6786
87. Devereaux TP, Virosztek A, Zawadowski A (1995) Phys Rev B 51:505
88. Millis AJ, Monien H, Pines P (1990) Phys Rev B 42:167
89. Monthoux P, Pines D (1993) Phys Rev B 47:6069

Kamimura H, Ushio H et al. (2007) In: Bussmann-Holder A, Keller H (eds)
High T_c Superconductors and Related Transition Metal Oxides.
Springer-Verlag Berlin Heidelberg, 157–165
DOI 10.1007/978-3-540-71023-3

Interplay of Jahn-Teller Physics and Mott Physics in Cuprates

H. Kamimura[1] (✉) · H. Ushio[2] · S. Matsuno[3]

[1]Department of Applied Physics, Tokyo University of Science, 1-3 Kagurazaka, Shinjuku-ku, 162-8601 Tokyo, Japan
kamimura@rs.kagu.tus.ac.jp

[2]Tokyo National College of Technology, 1220-2 Kunugidai-chou, 193-0997 Hachioji, Japan

[3]General Education Program Center, Tokai University, Shimizu Campus 3-20-1, Shimizu-Orido, 424-8610 Shizuoka, Japan

Introduction

In January of 1986 George Bednorz and Alex Müller discovered high temperature superconductivity in copper oxides with motivation that higher superconducting transition temperatures T_c could be achieved by combining Jahn-Teller active Cu ions with the structural complexity of layer-type perovskite oxides [1]. Their discovery, honored by the Nobel physics prize a year later, marked an historic milestone in the fields of not only superconductivity but also condensed matter science. In particular, the idea of Jahn-Teller poralon developed by Alex Müller has brought the remarkable development of the Jahn-Teller physics.

In celebrating his 80th birthday we would like to show that the interplay between his idea of Jahn-Teller physic and Mott physics based on the electron-electron interaction developed by Sir Nevill Mott plays an essential role in determining the electronic structures and properties of cuprates in their normal and superconducting phases.

Let us start with Jahn-Teller physics advocated by Müller. Paying an attention to the CuO_6 octahedrons elongated along the *c*-axis by the Jahn-Teller interaction, most of theoretical models have considered that the doped holes itinerate through an orbital extended over a CuO_2 plane. We call these models "a *single-component scenario theory*". Since undoped La_2CuO_4 is an antiferromagnetic insulator, some of the single-component scenario theories pointed out that the electron-correlation is very important in cuprates and that the superconductivity occurs in doped Mott insulators near the metal-insulator transition. Those models, however, have met a serious difficulty that, in the presence of local antiferromagnetic (AF) order the hole carriers cannot move smoothly because of different spin directions at the neighboring sites. In this context the idea of an assembly of spin-singlet pairs called

"RVB model" [2] and the t–J model of a spin-singlet quasiparticle [3] were proposed for doped-Mott insulators.

Anti-Jahn-Teller Effect

However, when Sr^{2+} ions are substituted for La^{3+} ions in LSCO, one may think intuitively that apical oxygen in the CuO_6 octahedrons tend to approach toward central Cu^{2+} ions in order to gain the attractive electrostatic energy. Theoretically it was shown by the first-principles variational calculations of the spin-density-functional approach [4, 5] that the optimized distance between apical O and Cu in LSCO which minimizes the total energy of LSCO decreases with Sr concentration. As a result the elongated CuO_6 octahedrons by the Jahn-Teller (JT) interactions shrink by doping holes. We have called this shrinking effect against the Jahn-Teller distortion "*anti-Jahn-Teller effect*" [6].

The Important Role of Two Kinds of Multiplets

By this anti-Jahn-Teller effect, the energy separation between the two kinds of orbital states, the a^*_{1g} anti-bonding orbital state constructed by Cu dz^2 orbital and six surrounding oxygen p orbitals and the b_{1g} bonding orbital state consisting of four in-plane p_σ orbitals with a small Cu dx^2–y^2 component, becomes smaller when spin-triplet Hund's coupling interaction and spin-singlet exchange interaction with a localized spin in the antibonding b^*_{1g} orbital, respectively, are taken into account for the a^*_{1g} and b_{1g} states. Thus, as a many-electron states both the Hund's coupling spin-triplet and the Zhang-Rice spin-singlet appear at nearly the same energy, as shown in Fig. 1.

Figure 1 shows the energy-level landscape starting from the e_g and t_{2g} orbitals of a Cu^{2+} ion in a regular CuO_6 octahedron with octahedral symmetry shown at the left column in the case of optimum doping ($x = 0.15$) in $La_{2-x}Sr_xCuO_4$. By the JT effect the Cu e_g orbital state splits into a_{1g} and b_{1g} orbital states, which form antibonding and bonding molecular orbitals of A_{1g} and B_{1g} symmetry with oxygen p_σ orbitals in an elongated CuO_6 cluster, respectively, as shown at the middle column, which are denoted by a^*_{1g}, a_{1g}, and b^*_{1g} and b_{1g}. In an undoped case, 9 electrons occupy these orbitals, so that the highest occupied b^*_{1g} state is half filled, resulting in an $S = 1/2$ state, where a b^*_{1g} orbital has Cu dx^2–y^2 character. Following Anderson [2], we introduced the Hubbard U interaction ($U = 10$ eV) as an electron-correlation effect. Then the b^*_{1g} state splits into the lower and upper Hubbard bands, L.H. and U.H., and this gives rise to a localized spin around a Cu site. These localized spins form the antiferromagnetic (AF) order by the superexchange interaction via an intervening O^{2-} ion in undoped La_2CuO_4.

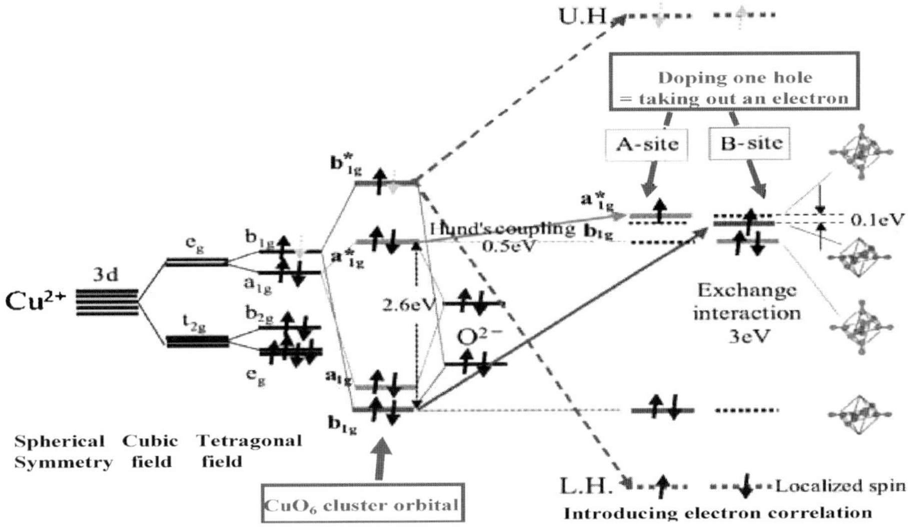

Fig. 1 Energy diagram including electron correlation and anti-JT effects (Results of the first-principles calculations)

Now we dope one hole into this CuO_6 cluster embedded in LSCO material. This means that we take out an electron from this system. In this case, there appear two multiplets, as shown at the right column in Fig. 1. One case is that a dopant hole occupies an antibonding a^*_{1g} orbital, and its spin becomes parallel by Hund's coupling of 0.5 eV with a localized spin in the b^*_{1g} orbital. This spin-triplet multiplet is called the "Hund's coupling triplet" denoted by ${}^3B_{1g}$. The other case is that a dopant hole occupies a bonding b_{1g} orbital, and its spin becomes antiparallel to the localized spin in the b^*_{1g} orbital. This spin-singlet multiplet corresponds to the "Zahn-Rice singlet" in the t–J model [3], and is denoted by ${}^1A_{1g}$. Sketch of a^*_{1g} and b_{1g} orbitals is shown at the extremely right column in Fig. 1.

By the first-principles cluster calculations which takes into account the Madelung potential due to all the ions surrounding a CuO_6 cluster, Kamimura and Eto showed that the lowest-state energies of these two multiplets are nearly equal when the anti-Jahn-Teller effect is taken into account [7–9] As seen in the right column of Fig. 1, the energy difference between the highest occupied orbital states a^*_{1g} in ${}^3B_{1g}$ multiplet and b_{1g} in ${}^1A_{1g}$ multiplet is only 0.1 eV for $x = 0.15$ in LSCO. Thus, when the two CuO_6 clusters with localized up and down spins are nearest neighbors, these states are easily mixed by the transfer interaction between a^*_{1g} and b_{1g} orbitals, which is about 0.3 eV for LSCO.

An Extended Two-Story House Model (The K–S Model)

By using the calculated results of Kamimura and Eto [7, 8] and assuming that the localized spins form an AF order in a spin-correlated region, Kamimura and Suwa [6, 9] constructed a metallic state of LSCO for its underdoped regime. We call this theoretical result "an *extended two-story house model*", which is now called "the K–S model". The physics of the extended two-story house model is sketched in Fig. 2.

In this figure the first story of a Cu house is occupied by the Cu localized spins, which form the AF order in the spin-correlated region by the superexchange interaction. The second story in a Cu house consists of two parts, lower a^*_{1g} floor and upper b_{1g} floor. These second stories are connected by Oxygen bridges in between Cu houses. In the second story a hole-carrier with up spin enters into the a^*_{1g} floor at the left-hand Cu house due to Hund's coupling with Cu localized spin in the first story (Hund's coupling triplet), as shown in the figure. By the transfer interaction marked by a long arrow in the figure, the hole is transferred into the b_{1g} floor at the neighboring Cu house (the second from the left) through the Oxygen bridge, where the hole with up spin forms a spin-singlet state with a localized down spin at the second Cu house from the left (Zhang-Rice singlet). The key feature of the K–S model is that the hole-carriers in the underdoped regime of LSCO form a metallic state, by taking the Hund's coupling triplet and the Zhang-Rice singlet alternately in the presence of the local AF order without destroying the AF order, as indicated in the figure. In the extended two-story house model, the second story consists of the two floors of different symmetry. Thus this model has been considered as a prototype of a "*two-component scenario theory*". In the case of the two-component scenario theory, the AF order is preserved when a hole-carrier itinerates. This feature in the two-component theory is different from that of the single-component theory.

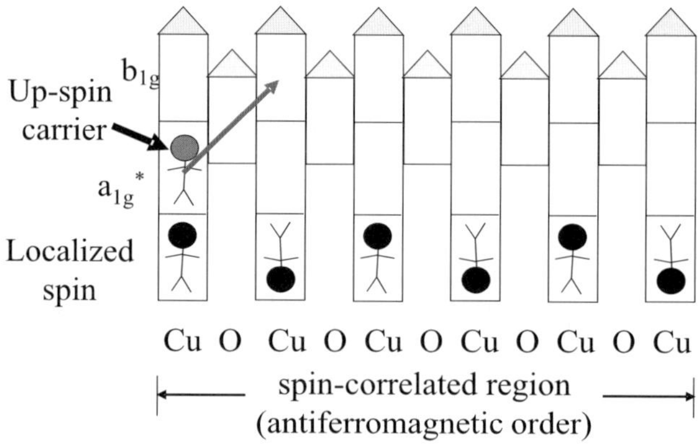

Fig. 2 An extended two-story house model (K–S model)

By adopting the mean-field approximation for the exchange interaction between localized and carrier spins in the K–S model, Ushio and Kamimura [6, 10, 11] calculated the many-body-effect-included energy band, Fermi surfaces, density of states, thermal, transport and optical properties in a similar way to the case of a single-electron-type band structure, although the many-body-effect-included energy band is completely different from the energy bands obtained from an ordinary band structure calculation of the local spin-density approximation. The appearance of small Fermi surfaces calculated for LSCO is supported by ARPES experiments of Stanford group for LSCO, where only outer sections of the FS called "Fermi arcs" in the AF Brillouin zone appear due to the finite size of a metallic region ([12] and Yoshida T et al., (2005) unpublished results). Further the heavy mass of a hole-carrier due to the polaronic nature leads to the appearance of anomalous electronic entropy, which is consistent with experimental results by Loram et al. [13]. Actually the K–S model has showed that a hole-carrier is an anti-Jahn-Teller polaron, when a dynamical nature of polaronic behavior is taken account. Theoretically the LDA + U calculations by Anisimov, Ezhov, and Rice supported the K–S model [14].

In this way one can understand that a serious difficulty which the single-component scenario theories have met with regard to a coherent motion of a carrier has been solved by the two-components scenario theories and that the anti-Jahn-Teller effect plays an essential role in creating the metallic and superconducting state of cuprates. Thus the K–S model has shown that the interplay of the Jahn-Teller physics and the Mott physics is essentially important in determining the characteristic properties of a normal state of metallic cuprates. In this context the K–S model has proved that the prediction of Jahn-Teller materials for cuprates by Bednorz and Müller is correct.

Mechanism of High Temperature Superconductivity

On the basis of the K–S model we have proposed the interplay between electron-phonon-mechanism and the AF local order as the mechanism of superconductivity in LSCO [6, 15]. As seen in Fig. 1, the wavefunctions of a hole-carrier with up and down spins have the phase relation of

$$\Psi_{k\downarrow}(\boldsymbol{r}) = \exp(\mathrm{i}\boldsymbol{k}\cdot\boldsymbol{a})\Psi_{k\uparrow}(\boldsymbol{r})\,. \tag{1}$$

From this relation (Eq. 1), the electron-phonon interaction matrix elements from states $\boldsymbol{k}$ to $\boldsymbol{k}'$ with down spin scattered by phonon with wave vector $\boldsymbol{q}$, $V_\downarrow(\boldsymbol{k},\boldsymbol{k}')$, has the following spin-dependent property:

$$V_\uparrow(\boldsymbol{k},\boldsymbol{k}') = \exp(\mathrm{i}\boldsymbol{K}\cdot\boldsymbol{a})V_\downarrow(\boldsymbol{k},\boldsymbol{k}')\,, \tag{2}$$

where $\boldsymbol{K} = \boldsymbol{k} - \boldsymbol{k}' - \boldsymbol{q}$ is a reciprocal lattice vector in the AF Brillouin zone and a is a Cu – O – Cu distance.

From the relation (Eq. 2) and $K = (n\pi/a, m\pi/a, 0)$ with $n + m$ = even, the effective interactions of a pair of holes from $(k\uparrow, -k\downarrow)$ to $(k'\uparrow, -k'\downarrow)$ is expressed as

$$V_{\uparrow}(k, k')V_{\downarrow}(-k, -k') = \exp(\mathrm{i}K \cdot a)|V_{\uparrow}(k, k')|^2 . \quad (3)$$

Since $\exp(\mathrm{i}K \cdot a) = +1$ for n = even and $\exp(\mathrm{i}K \cdot a) = -1$ for n = odd, the effective interaction for forming a Cooper pair becomes attractive for n = even while repulsive for n = odd. This remarkable result in the K–S model leads to the superconducting gap of dx^2–y^2 symmetry.

Kamimura et al. calculated the electron-phonon spectral function $\alpha^2F(\Omega)$ for LSCO with tetragonal symmetry, where $F(\Omega)$ is the density of phonon states in energy and α^2 is the square of the electron-phonon coupling constant [6, 15, 16], and their calculated result of d-wave component of the spectral function for LSCO with tetragonal symmetry, $\alpha^2F^{(2)}_{\uparrow\downarrow}(\Omega)$, is shown in Fig. 3 as a function of phonon frequency Ω. The phonon modes of LSCO with tetragonal symmetry are classified into in-plane modes and out-of-plane modes with regard to a CuO_2 plane. Figure 3 shows that the out-of-plane modes yield positive spectral function so that they contribute to the formation of Cooper pairs by acting as an attractive force while the in-plane modes

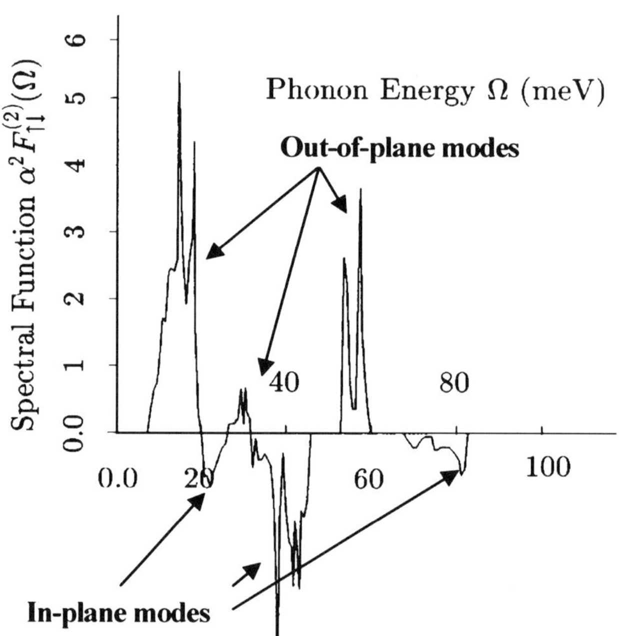

Fig. 3 The d-wave component of $\alpha^2F(\Omega)$ calculated for tetragonal LSCO with an optimum doping

yield negative spectral function so that these do not contribute to the formation of Cooper pairs by acting as a repulsive force.

As mentioned above, we have shown that, even though the electron-phonon interaction is involved in the mechanism of superconductivity, the interplay of the AF order and phonon mechanism in the K–S model creates superconductivity of "*d-wave symmetry*", when a system is infinite and homogeneous.

Now we will present the calculated results of T_c and the isotope effect for LSCO with tetragonal symmetry. The calculated results of the hole-concentration dependence of T_c and the isotope effects for LSCO by Kamimura et al. [6, 15, 16] are shown in Fig. 4, and the isotope effect is compared with experimental results by Crawford et al. [17, 18] and Ronay [19]. Here the isotope effect α is defined by

$$\alpha = - \, d\ln T_c / d\ln M \,, \tag{4}$$

where M denotes the mass of constituent atoms. In their calculation a well-known relation between the Debye frequency and a mass, $\Omega_D \sim M^{-0.5}$, is used. Further, in their calculation, the masses of all constituent atoms have been changed by the mass ratio of ^{18}O to ^{16}O. Thus their calculated results may be overestimated.

Here we should make a remark on the calculated results of T_c. On the K–S model the size of a metallic state is finite, because a two-story house model (the K–S model) holds in the spin-correlated region of the AF local order, as shown in Fig. 2. In order to obtain the observed value of 40 K for the highest T_c of LSCO at the optimum doping ($x = 0.15$), Kamimura et al. [6, 16] have chosen the size of a metallic region to be 300 Å. The T_c curve shown in Fig. 4 is the result calculated for the case of 300 Å.

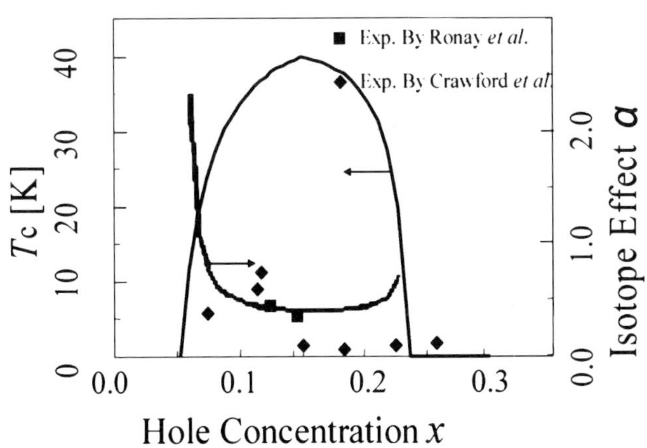

Fig. 4 The calculated isotope effect α and T_c curve (*thick line*) for tetragonal LSCO as a function of x

From the calculated results of the isotope effect shown in Fig. 4, the following conclusions emerge: The isotope effect on T_c in LSCO depends on the hole concentration critically, and the isotope constant α is remarkably large near the onset concentration of superconductivity while it is small around the optimum concentration. Recently Bishop et al. published a review article on recent and earlier results on isotope effects in cuprates, pointing out that (1) the isotope effect on T_c vanishes at optimum doping but increases with decreasing doping level to be substantially larger than the BCS value ($\alpha = 0.5$) at the border to the AF state and that (2), from the experiment of site selective isotope effect on T_c, the observed isotope effects are found to stem to nearly 100% from oxygen ions in the CuO_2 planes [20, 21]. Among these experimental results (1) and (2), the first one (1) seems consistent with the theoretical result obtained from the K–S model shown in Fig. 4.

Conclusion

(1) By showing that the anti-Jahn-Teller effect plays an essential role in creating the metallic and superconducting state of cuprates, we have proved that the prediction of Jahn-Teller materials for cuprates by Bednorz and Müller is correct. (2) Then we have shown that the existence of local AF spin order as well as the local distortion of CuO_6 octahedrons due to the anti-Jahn-Teller effect induce the alternating appearance of the Zhang-Rice singlet and the Hund's coupling triplet, leading to a two-component scenario theory (the K–S model). (3) We have showed that the electronic structures, and electronic and superconducting properties of underdoped cuprates can be explained well by the the K–S model. (4) In particular, we have emphasized that the interplay of the Jahn-Teller physics and the Mott physics is essentially important in understanding the characteristic properties in the normal and superconducting phases of underdoped cuprates. (5) Finally we have shown that the interplay of the AF order and phonon mechanism in the K–S model creates *d-wave* superconductivity, when a system is infinite and homogeneous. However, in an inhomogeneous case with a finite size of a metallic state such as 300 Å in real cuprates, the symmetry of a superconducting gap may be a mixture of *s*- and *d*-wave symmetries, which may be consistent with argument by Müller [22].

Acknowledgements It is a great pleasure for one of the authors (H.K.) to dedicate this article to Alex Müller to commemorate his eightieth birthday and a friendship of many years since H.K. met him at the Adriatico Research Conference on High Temperature Superconductors held in the International Centre for Theoretical Physics at Trieste on July 5 to 8, 1987.

References

1. Bednorz JG, Müller KA (1986) Z Phys B 64:189
2. Anderson PW (1987) Science 235:1196
3. Zhang FC, Rice TM (1988) Phys Rev B37:3759
4. Shima N, Shiraishi K, Nakayama T, Oshiyama A, Kamimura H (1989) In: Sugano T et al. (eds) Proc JSAP-MRS Int'l Conf on Electronic Materials. Materials Research Society, p 51
5. Oshiyama A, Shima N, Nakayama T, Shiraishi K, Kamimura H (1989) In: Kamimura H, Oshiyama A (eds) Springer Series in Materials Sciecne Vol. 11 Mechanism of High Temperature Superconducitivity, Springer, p 111
6. Kamimura H, Ushio H, Matsuno S, Hamada T (2005) Theory of Copper Oxide Supercondcutors. Springer, Heidelberg
7. Kamimura H, Eto M (1990) J Phys Soc Jpn 59:3053
8. Eto M, Kamimura H (1991) J Phys Soc Jpn 60:2311
9. Kamimura H, Suwa Y (1993) J Phys Soc Jpn 62:3368–3371
10. Kamimura H, Ushio H (1954) Solid State Commun 91:97
11. Usho H, Kamimura H (1995) J Phys Soc Jpn 64:2585
12. Yoshida T, Zhou XJ, Nakamura M, Keller SA, v. Bogdanov P, Lu ED, Lanzara A, Hussain Z, Ino A, Fujimori A, Eisaki H, Shen Z-X, Kakeshita T, Uchida S (2003) Phys Rev Lett 91:027001
13. Loram JW, Mirza KA, Cooper JR, Tallon JL (1998) J Phys Chem Solids 59:2091
14. Anisimov VL, Yu Ezhov S, Rice TM (1997) Phys Rev B55:12829
15. Kamimura H, Matsuno S, Suwa Y, Ushio H (1996) Phys Rev Lett 77:723
16. Kamimura H, Hamada T, Matsuno S, Ushio H (2002) J Supercond 15:379
17. Crawford MK, Farneth WE, McCarron EM III, Harlow RI, Moudden AH (1990) Science 250:1390
18. Crawford MK, Kunchur MN, Farneth WE, McCarron EM III, Poon SJ (1990) Phys Rev B41:282
19. Ronay M, Frisch MA, McGuire TR (1992) Phys Rev B45:355
20. Bishop AR, Bussmann-Holder A, Cardona M, Dolgov OV, Furrer A, Kamimura H, Keller H, Khasanov R, Kremer RK, Manske D, Müller KA, Simon A (2007) to be published in J Supercond (cond-mat/0610036)
21. Zech D, Keller H, Conder K, Kaldis E, Liarokapis E, Poulakis N, Müller KA (1994) Nature (London) 371:681
22. Müller KA (2002) Phil Mag Lett 82:279

Karpinski J (2007) In: Bussmann-Holder A, Keller H (eds)
High T_c Superconductors and Related Transition Metal Oxides.
Springer-Verlag Berlin Heidelberg, 167–175
DOI 10.1007/978-3-540-71023-3

20 years High Pressure Materials Synthesis Group Activity After Discovery of High-T_c Superconductors

J. Karpinski

Laboratory for Solid State Physics, ETH 8093 Zürich, Switzerland
karpinski@solid.phys.ethz.ch

Abstract I present a short review of selected activities of the High Pressure Materials Synthesis Group at the Laboratory for Solid State Physics ETH Zürich during 20 years from 1987 until 2006, initiated by the discovery of Dr. George Bednorz and Prof. Alex Müller.

On the occasion of the 80th birthday of Prof. Alex Müller I decided, following a suggestion of Hugo Keller, not to write one more scientific paper about recent results, but a review of our 20 years work initiated by the discovery of high-T_c superconductivity.

In 1986 Prof. Alex Müller gave a lecture at the Physik-Institut der Universität Zürich about his and George Bednorz's new discovery [1]. I learned for the first time about high temperature superconductivity in the $La_{2-x}Ba_xCuO_4$ compound. We felt immediately that something extremely important happened, that will change our scientific activity. Short time later at the beginning 1987 we have learned about the new compound, $YBa_2Cu_3O_{7-x}$ with T_c above 90 K [2, 3].

At that time I had a postdoc position at the Laboratory for Solid State Physics ETH in the solid state chemistry group of E. Kaldis. I was working on the high pressure synthesis of hydrides and nitrides. We had just finished the construction of the high pressure system for active gas pressure up to 2.5 kbar at high temperature up to 1300 °C, which was a unique device not available in any other laboratory. For the synthesis we used a double chamber system with internal chamber being a metallic crucible made of tungsten or molybdenum. High nitrogen or hydrogen gas pressure inside the crucible was supported with the same pressure of Ar from outside of the crucible. This was important because separation of the furnace atmosphere from the sample atmosphere allowed high purity work with a reactive gas in the sample space.

It has been reported that instead of substitution of Ba or Sr in La_2CuO_4 for hole doping an excess of oxygen, introduced by a high oxygen pressure treatment, make this compound superconducting [4, 5]. This told us that high pressure may be very useful for synthesis of superconductors.

We decided to search for new high T_c compounds at high oxygen pressure. But nobody worked at that time with very high gaseous oxygen pressure at high temperature. One possible reason was that metallic crucibles cannot work in oxygen atmosphere at high temperature. Another problem could be self-ignition of metal parts in the pressure manifold in contact with oxygen at high pressure leading to an explosion or at least to a big fire. No data which could avoid those difficulties were available. We decided to use a ceramic crucible as internal pressure vessels for oxygen pressure. After several trials we solved technical difficulties and the system worked at oxygen pressure up to 2.5 kbar and temperature up to 1300 °C. Due to very small amount of gaseous oxygen in the sample space the experiments were safe.

As a first step we have chosen to investigate the Y – Ba – Cu – O system. Our high pressure gas system allowed studies of the P–T–x phase diagram of oxides. We investigated the decomposition pressure above $YBa_2Cu_3O_{7-x}$ (Y123) and soon realized that some reaction leading to another compound took place and the decomposition pressure above a new compound is in the kbar range. In a short time we were able to define the structures of the new compounds $YBa_2Cu_4O_8$ (Y124) and $Y_2Ba_4Cu_7O_{15-x}$ (Y247). Three papers in Nature [6–8] and a patent for Y124 [9] followed these results. Especially, Y124 turned out to be a very interesting compound, because it is the only stoichiometric cuprate, as it contains a fixed amount of oxygen in the structure. Y124 contains stable double CuO_2 chains instead of single chains as in Y123. Therefore it does not undergo an orthorhombic to tetragonal phase transition with heating or cooling, which causes twinning in Y123 crystals. Y247 has alternating double and single chains. Figure 1a shows the P–T phase diagram with stability ranges of three superconducting compounds in the Y – Ba – Cu – O system for a composition of the sample equal to Y124 [10]. The stability field of the Y124 phase in the high temperature range of the phase diagram is limited by the following decomposition reactions:

1. $2YBa_2Cu_4O_8 = Y_2Ba_4Cu_7O_{15-x} + CuO + {}^{x}/_{2}O_2$
2. $YBa_2Cu_4O_8 = YBa_2Cu_3O_{7-x} + CuO + {}^{x}/_{2}O_2$.

As the Gibbs free energies of these reactions are not equal, these two equilibrium lines cross. The crossing point is around $P_{O2} = 1$ bar, $T = 890$ °C. Reactions (1) and (2) take place at $P_{O2} \geq 1$ bar and $P_{O2} \leq 1$ bar, respectively. We did not observed any low temperature limitation of the stability range of Y124. Therefore, we concluded that Y124 is the only superconducting phase in Y – Ba – Cu – O system, which is thermodynamically stable at room temperature. Several attempts to form new phases with higher CuO content such as Y249 or Y125 at high oxygen pressures up to 3000 bar were not successful. We have grown Y124 and Y247 single crystals in our high pressure autoclave from the high temperature CuO – $BaCuO_2$ flux at $T = 1100$ °C under oxygen pressure of 600–1000 bar.

Y124 is an underdoped superconductor, therefore the pure compound has a superconducting critical temperature $T_c = 81$ K, lower than $T_c = 93$ K for Y123. Substitution of Ca^{2+} for Y^{3+} increases T_c up to 90 K while Zn substitution decreases T_c very rapidly. Although Zn is non-magnetic, it substi-

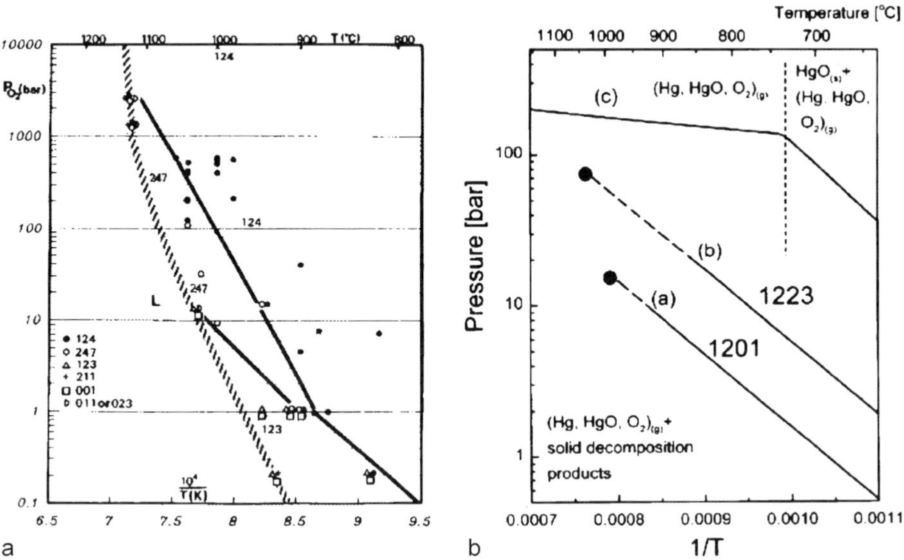

Fig. 1 **a** P–T phase diagram of the Y – Ba – Cu – O system for a composition Y124. It shows stability ranges of three superconducting phases Y124, Y123 and Y247. **b** P–T phase diagram of Hg1201 and Hg1223. It shows the total equilibrium pressure of volatile components Hg, HgO and O_2 for Hg1201 (**a**) and Hg1223 (**b**) as a function of temperature. Adapted from [17]

tutes Cu in the CuO_2 planes causing lack of spin on this site. So it works as a "pseudo-magnetic" scattering center and suppresses T_c (Fig. 2). Y124 shows also another interesting property: a very high pressure effect on T_c, $dT_c/dp = 5$ K/GPa. Substitution of smaller isovalent Sr ion for larger Ba ion

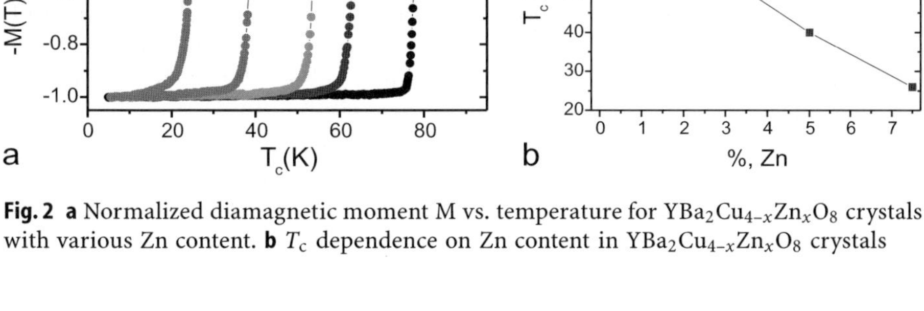

Fig. 2 **a** Normalized diamagnetic moment M vs. temperature for $YBa_2Cu_{4-x}Zn_xO_8$ crystals with various Zn content. **b** T_c dependence on Zn content in $YBa_2Cu_{4-x}Zn_xO_8$ crystals

in Y124 generates internal chemical pressure and causes redistribution of the charge in the CuO_2 planes [11]. Recently, Y124 is again investigated as the only example of stoichiometric cuprate superconductor. Properties of Y247 have been published in [12].

In several years after the first discovery, tens of new high T_c compounds have been synthesized, many of them by using high pressure methods. The high pressure anvil technique appeared to be very successful in the discovery of several homologous series of superconductors with similar structural elements: blocks of CuO_2 planes with Ca atoms between planes and charge reservoir blocks, which dope superconducting CuO_2 planes with holes.

On the beginning of the 90ties the research team of E. Antipov at the Chemical Department of the Moscow State University worked on Hg based superconducting compounds, which led to successful synthesis of the first member ($n = 1$ with one CuO_2 plane) of HgBaCaCuO series with astonishingly high critical temperature $T_c = 94$ K [13]. This led to the expectation, that the members with $n = 2$ or 3 will have very high T_c. Based on this result Andreas Schilling from our Institute with collaborators started to work on the synthesis of polycrystalline samples of mercury based superconductors for n>1 using ampoule method. Shortly later they obtained a mixed phase sample containing $HgBa_2Ca_2Cu_3O_{8+\delta}$ (Hg1223) compound with T_c onset at 133 K, which remained until now a world record (it can be increased by hydrostatic pressure up to 165 K) [14].

Our approach was to grow single crystals of mercury based superconductors, what required higher pressure, than available in our oxygen system. For these experiments we constructed a new high pressure gas system with working pressure up to 15 kbar of Ar or N_2 at high temperatures up to 1500 °C. The reason of application of high Ar pressure was that for the growth of Hg based superconducting single crystals we had to reach a melting temperature of the compound, which is above 1000 °C. At this temperature the total pressure of the volatile components Hg, HgO and O_2 of $HgBa_2Ca_2Cu_3O_{8+\delta}$ (Hg1223) compound is about 100 bar. Figure 1b shows the total pressure as a function of temperature. High pressure of the inert gas acts as a liquid encapsulation of the sample, suppresses evaporation of the volatile components, and prevents diffusion of components out of the melt. After application of $P = 10$ kbar Ar pressure we succeeded in growing the first single crystals of $HgBa_2Ca_{n-1}Cu_nO_{2n+2+\delta}$ (Hg12($n-1$)n) compounds with 1 to 7 CuO_2 planes ($n = 1$–7) and infinite layer ($n = \infty$) [15, 16]. Figure 3 shows T_c as a function of n for single crystals. For each n T_c varies as a function of oxygen content and substitutions. Figure 4 shows main blocks of the Hg12($n-1$)n structure with arrows indicating structural changes with increasing number of planes n.

For the next exciting discovery of new superconducting compounds with high T_c we were waiting several years and in 2001 J. Nagamatsu and J. Akimitsu of the Tokyo University [18] investigated the well known compound MgB_2, probably the only di-boride which was not yet studied at low tem-

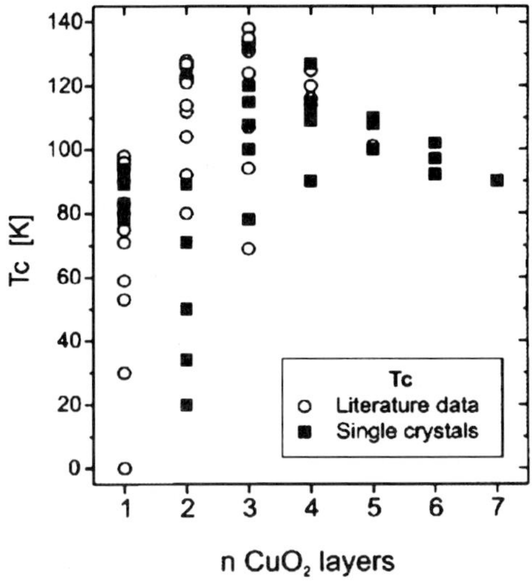

Fig. 3 T_c values of $HgBa_2Ca_{n-1}Cu_nO_{2n+2+\delta}$ single crystals for various n and δ in comparison with the literature data for ceramic samples. Adapted from [17]

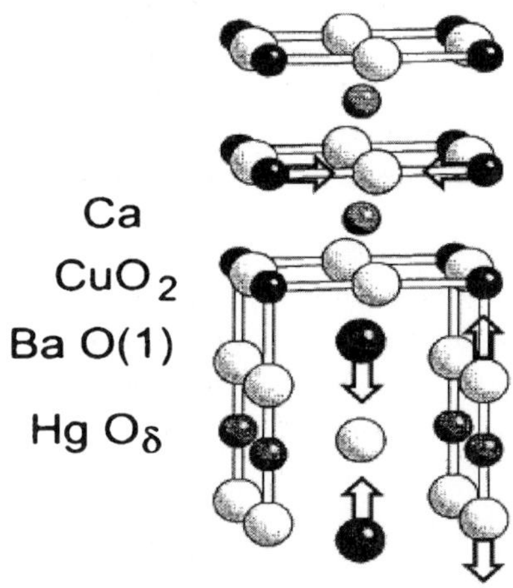

Fig. 4 Atom shifts in $HgBa_2Ca_{n-1}Cu_nO_{2n+2+\delta}$ with increasing n. Adapted from [17]

perature, and they found a superconducting transition with an astonishingly high critical temperature of 39 K. This triggered the avalanche of papers. Due to Internet and cond-mat page every new result was announced immediately and few days later confirmed (or denied) by another laboratory. MgB_2 is a two-band, two-gap superconductor with several anomalous properties originating from the existence of two separate sheets of the Fermi surface, one quasi 2D (σ band) and second quasi 3D (π band). Such electronic structure leads to the high critical temperature T_c of 39 K due to the large energy gap in one of the bands. Due to the strongly anisotropic character of MgB_2, single crystal studies of its anisotropic properties are very important. Superconducting and normal state properties of pure MgB_2 are now well evidenced by experiments and explained by theory [19, 20]. However, modification of properties through chemical substitutions is still not well understood. The critical temperature and other superconducting properties of a two-band superconductor depend on the doping level and on the interband and intraband scattering. Therefore, they can be modified by chemical substitutions. Substitutions change the electronic structure, inter- and intra-band scattering, gaps width and defect structure, and thus electronic properties such as T_c, upper critical fields and its anisotropy, etc.

It was a great challenge for us again to find a method of growing single crystals of MgB_2. Although MgB_2 is very easy to synthesize as a polycrystalline sample it turns out very difficult to grow crystals. Conventional methods of crystal growth did not work for MgB_2. High temperature solution growth in metals (Mg, Al, Cu, etc.) at normal pressure used for other borides were not applicable due to very low solubility of MgB_2 in these metals or formation of other compounds. The solubility of MgB_2 in Mg is extremely low up to the boiling temperature of Mg (1107 °C) at ambient pressure. At higher temperature solubility increases, but partial pressure of Mg vapor above molten Mg increases and at temperature of crystal growth (1800 °C) is of the order of 50 bar. This means that for the crystal growth of MgB_2 from solution in Mg at higher temperature, high pressure technique should be applied. Our phase diagram studies show, that for the stabilization of the MgB_2

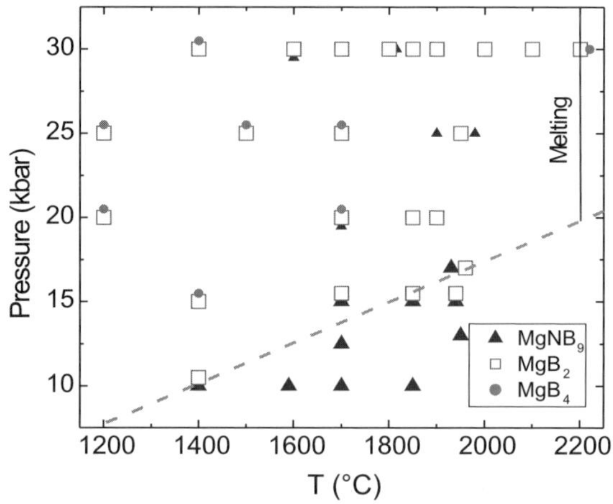

Fig. 5 *P–T* phase diagram of the Mg – B – N system. The source of nitrogen is the BN crucible. Symbols show phases observed in the samples. Additionally, BN and Mg were also present in all samples (not shown on the diagram). MgB_2 single crystals have been grown above 1800 °C

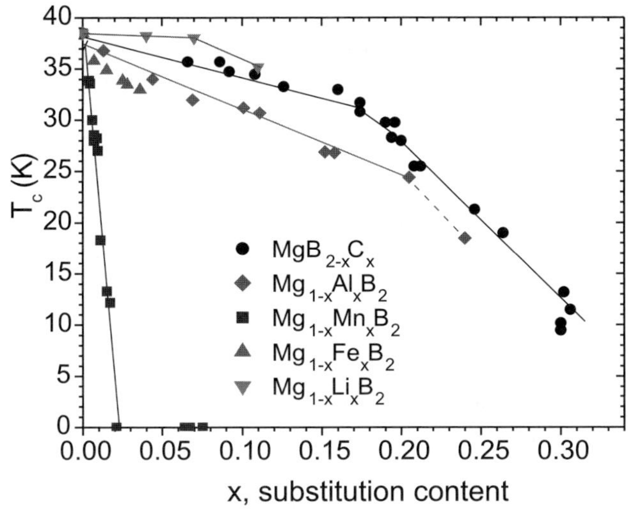

Fig. 6 Critical temperature dependence on substitutions in MgB_2 single crystals

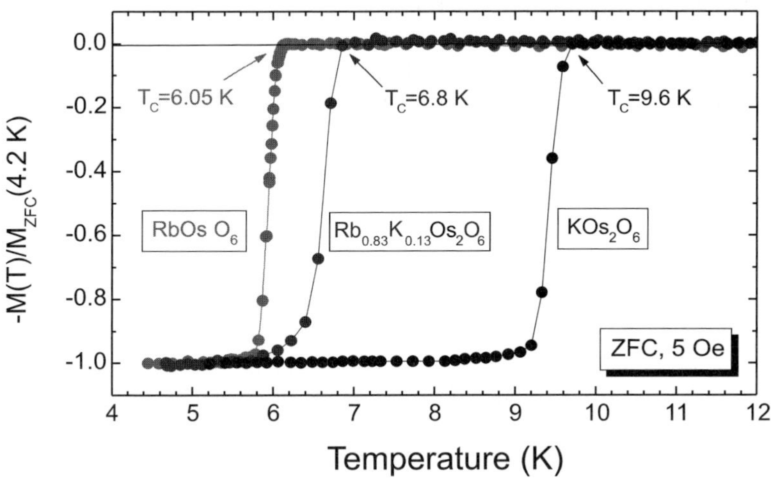

Fig. 7 Normalizad magnetic moment as a function of temperature for $Rb_{1-x}K_xOs_2O_6$ crystals

compound at high temperature not only high partial pressure of Mg vapor but also high hydrostatic pressure is necessary. At high pressure the gas phase disappears and only solid and liquid phases remain in the system in equilibrium. This allows high temperature crystal growth from solution. Figure 5 shows the p–T phase diagram with conditions used for single crystal growth of MgB_2. Using cubic anvil technique we have grown pure unsubstituted and substituted MgB_2 single crystals [21–23]. Figure 6 summarizes the results of our investigations of the influence of substitutions in MgB_2 single crystals on their T_c. Availability of sizeable MgB_2 crystals allowed investigations of superconducting properties [24–26] and results in about 250 papers published on measurements performed on our crystals.

Our next challenge for the single crystal growth of superconducting compounds is the recently discovered family of pyrochlore oxides AOs_2O_6 where A = Cs, Rb and K with T_c up to 9.5 K [27, 28]. In the pyrochlore structure corner sharing OsO_6 octahedra form a three-dimensional network of tetrahedrally arranged Os atoms, most likely leading to a geometrical frustration of the ground state. We succeeded in growing single crystals of KOs_2O_6, $RbOs_2O_6$ and mixed $K_xRb_{1-x}Os_2O_6$ compounds [29, 30]. Figure 7 shows the magnetic moment as a function of temperature for $Rb_{1-x}K_xOs_2O_6$ crystals. The mechanism of superconductivity in these oxides has not been established yet. Some experimental evidence indicates the possibility of unconventional superconductivity in KOs_2O_6.

In this short review I tried to show milestones of our work during last 20 years. On the end I would like to underline that the establishment of our laboratory and the whole activity described in this paper was initiated by the discovery of high T_c superconductivity.

Acknowledgements I would like to thank Prof. Alex Müller for his very important support. I also would like to thank all my collaborators who contributed to this work.

References

1. Bednorz JG, Müller KA (1986) Z Phys B 64:198
2. Wu MK, Ashburn JR, Torng CJ, Hor PH, Meng RL, Gao L, Huang ZJ, Wang YQ, Chu CW (1987) Phys Rev Lett 58:908
3. Cava RJ, Batlogg B, van Dover RB, Murphy DW, Sunshine S, Siegrist T, Remeika JP, Rietman EA, Zahurak SM, Espinosa GP (1987) Phys Rev Lett 58:1676
4. Schirber JE, Morosin B, Merrill RM, Hlava PF, Venturini EL, Kwak JF, Nigrey PJ, Baughman RJ, Ginley DS (1988) Physica C 152:121
5. Jorgensen JD, Dabrowski B, Pei S, Hinks DG, Soderholm L, Morosin B, Schirber JE, Venturini EL, Ginley DS (1988) Phys Rev B 38:11337
6. Karpinski J, Kaldis E (1988) Nature 331:242
7. Bordet P, Chaillout C, Chenavas J, Hodeau JL, Marezio M, Karpinski J, Kaldis E (1988) Nature 334:596
8. Karpinski J, Kaldis E, Jilek E, Rusiecki S, Bucher B (1988) Nature 336:660
9. Kaldis E, Karpinski J (1991) Ceramic high temperature superconductor in bulk form, and method of manufacturing the same. US Patent 5 032 569
10. Karpinski J, Rusiecki S, Kaldis E, Bucher B, Jilek E (1989) Physica C 160:449
11. Karpinski J, Kazakov SM, Angst M, Mironov A, Mali M, Ross J (2001) Phys Rev B 64:094518
12. Karpinski J, Conder K, Schwer H, Krüger Ch, Kaldis E, Maciejewski M, Rossel C, Mali M, Brinkmann D (1994) Physica C 227:68–76
13. Putilin SN, Antipov EV, Chmaissem O, Marezio M (1993) Nature 362:226

14. Schilling A, Cantoni M, Guo JD, Ott HR (1993) Nature 363:56
15. Karpinski J, Schwer H, Mangelschots I, Conder K, Morawski A, Lada T, Paszewin A (1994) Nature 371:661
16. Karpinski J, Schwer H, Mangelschots I, Conder K, Morawski A, Lada T, Paszewin A (1994) Physica C 234:10–18
17. Karpinski J, Meijer GI, Schwer H, Molinski R, Kopnin E, Conder K, Angst M, Jun J, Kazakov S, Wisniewski A, Puzniak R, Hofer J, Alyoshin V, Sin A (1999) Supercond Sci Technol 12:R153–R181
18. Nagamatsu J, Nakagawa N, Muranaka T, Zenitani Y, Akimitsu J (2001) Nature 410:63
19. Choi HJ, Roundy D, Sun H, Cohen ML, Louie SG (2002) Nature 418:758
20. Kortus J, Dolgov OV, Kremer RK, Golubov AA (2005) Phys Rev Lett 94:027002
21. Angst M, Puzniak R, Wisniewski A, Jun J, Kazakov SM, Karpinski J, Roos J, Keller H (2002) Phys Rev Lett 88:167004
22. Karpinski J, Angst M, Jun J, Kazakov SM, Puzniak R, Wisniewski A, Roos J, Keller H, Perucchi A, Degiorgi L, Eskildsen MR, Bordet P, Vinnikov L, Mironov A (2003) Supercond Sci Technol 16:221
23. Karpinski J, Kazakov SM, Jun J, Angst M, Puzniak R, Wisniewski A, Bordet P (2003) Physica C 385:42
24. Kazakov SM, Puzniak R, Rogacki K, Mironov AV, Zhigadlo ND, Jun J, Soltmann Ch, Batlogg B, Karpinski J (2005) Phys Rev B 71:024533
25. Karpinski J, Zhigadlo ND, Schuck G, Kazakov SM, Batlogg B, Rogacki K, Puzniak R, Jun J, Müller E, Wägli P, Gonnelli R, Daghero D, Ummarino GA, Stepanov VA (2005) Phys Rev B 71:174506
26. Rogacki K, Batlogg B, Karpinski J, Zhigadlo ND, Schuck G, Kazakov SM, Wägli P, Puźniak R, Wiśniewski A, Carbone F, Brinkman A, van der Marel D (2006) Phys Rev B 73:174520
27. Yonezawa S, Muraoka Y, Matsushita Y, Hiroi Z (2004) J Phys Condens Matter 16:L9
28. Yonezawa S, Muraoka Y, Matsushita Y, Hiroi Z (2004) J Phys Soc Jpn 73:819
29. Schuck G, Kazakov SM, Rogacki K, Zhigadlo ND, Karpinski J (2006) Phys Rev B 73:144506
30. Brühwiler M, Kazakov SM, Karpinski J, Batlogg B (2006) Phys Rev B 73:094518

Khasanov R, Shengelaya A et al. (2007) In: Bussmann-Holder A, Keller H (eds)
High T_c Superconductors and Related Transition Metal Oxides.
Springer-Verlag Berlin Heidelberg, 177–190
DOI 10.1007/978-3-540-71023-3

Two-Gap Superconductivity in the Cuprate Superconductor $La_{1.83}Sr_{0.17}CuO_4$

R. Khasanov[1] (✉) · A. Shengelaya[2] · A. Bussmann-Holder[3] · H. Keller[1]

[1]Physik-Institut der Universität Zürich, Winterthurerstrasse 190, CH-8057 Zürich, Switzerland
khasanov@physik.unizh.ch

[2]Physics Institute of Tbilisi State University, Chavchavadze 3, GE-0128 Tbilisi, Georgia

[3]Max-Planck-Institut für Festkörperforschung, Heisenbergstrasse 1D, 70569 Stuttgart, Germany

Abstract Based on the idea of K.A. Müller [Nature (London) **377**, 133 (1995)] that two superconducting condensates with *s*-wave and *d*-wave symmetries are present in high-temperature cuprate superconductors, we performed measurements of the in-plane magnetic field penetration depth (λ_{ab}) in single-crystal $La_{1.83}Sr_{0.17}CuO_4$ by means of the muon-spin rotation technique. The temperature dependence of λ_{ab}^{-2} has an inflection point around 10–15 K, suggesting the opening of two superconducting gaps: a large gap (Δ_1^d) with a *d*-wave and a small gap (Δ_2^s) with an *s*-wave symmetry. The zero-temperature values of the gaps at $\mu_0 H = 0.02T$, obtained from the global fit of muon data, were found to be $\Delta_1^d(0) = 8.2(1)$ meV and $\Delta_2^s(0) = 1.57(8)$ meV. With increasing magnetic field the contribution of Δ_2^s decreases substantially, in contrast to an almost constant contribution of Δ_1^d. The magnetic field dependence of the transition temperature T_c and the distribution of the local magnetic fields $P(B)$ in $La_{1.83}Sr_{0.17}CuO_4$ superconductor in the mixed state were also found to be consistent with the presence of two superconducting gaps.

Introduction

More than 20 years after the discovery of high temperature cuprate superconductors (HTS) their pairing mechanism is still not completely understood. For a long time, the role of phonons in this mechanism was dismissed, despite evidence of their relevance, for example by isotope experiments (see, e.g. [1, 2] and references therein). One of the reasons for this was the prevailing belief that phonons simply could not account for such high transition temperatures. The recent discovery of MgB_2, where phonons are clearly the pairing glue leading to an unexpectedly high transition temperature $T_c \simeq 40$ K, clearly falsified this belief. It is now widely accepted that the key element leading to the high transition temperature of MgB_2 is the presence of *two* superconducting gaps. The realization of the two-gap scenario in HTS was proposed already more then 10 years ago by Kresin and Wolf [3] theoretically and, independently, by Müller [4–6] based on the analysis of tunnelling data. The idea of Müller was generated partly because two gaps were observed in n-type $SrTiO_3$ [7], the first oxide in which superconductivity was detected [8]. Later on, the presence of two superconducting condensates were detected for various HTS by Andreev reflection [9], tunnelling [10], Raman [11, 12], nuclear magnetic resonance [13], inelastic neutron scattering [14], and muon [15] experiments.

The two-gap scenario can be probed by the magnetic field penetration depth (λ) experiments. In particular, $\lambda(T)$, which reflects the quasiparticle density of states available for thermal excitations, admits to probe the superconducting gap structure. Measurements of the field dependence of λ allow to study the anisotropy of the superconducting energy gap [16] and, in the case of two-gap superconductors, to obtain details on the relative contribution of each particular gap as a function of magnetic field [17].

In this paper we concentrate on a study of the in-plane magnetic penetration depth (λ_{ab}) in slightly overdoped single-crystal $La_{1.83}Sr_{0.17}CuO_4$ by means of the muon-spin-rotation (μSR) technique. It was found that at low magnetic fields ($\mu_0 H \leq 0.3$ T) $\lambda_{ab}^{-2}(T)$ exhibits an inflection point at $T \simeq$ 10–15 K. We interpret this feature as a consequence of the presence of two superconducting gaps, analogous to double-gap MgB_2 [18]. It is suggested that the large gap ($\Delta_1^d(0) = 8.2(1)$meV) has a d-wave and the small gap ($\Delta_2^s(0)$ = 1.57(8)meV) an s-wave symmetry. With increasing magnetic field the contribution of Δ_2^s decreases substantially, in contrast to an almost constant contribution of Δ_1^d. Both the temperature and the field dependences of λ_{ab}^{-2} were found to be similar to what was observed in double-gap MgB_2 [17, 18]. In addition, the shape of the local magnetic field distribution $P(B)$ probed by μSR was also found to be consistent with the presence of two superconducting gaps.

Experimental Details

The $La_{1.83}Sr_{0.17}CuO_4$ single-crystal was grown by the travelling solvent floating zone technique [19]. The transition temperature T_c and the width of the superconducting transition at $\mu_0 H \simeq 0$ T, obtained by means of AC-susceptibility [20], were found to be 36.2 K and 1.5 K, respectively.

The transverse-field muon-spin rotation (TF-μSR) experiments were performed at the πM3 beam line at the Paul Scherrer Institute (Villigen, Switzerland). Typical counting statistics were $\sim$ 16–18 million muon detections over three detectors. The sample was field cooled from above T_c to 1.6 K in a series of fields ranging from 0.02 T to 0.64 T. The sample was aligned such that the c-axis was parallel (within 1 degree, as measured by Laue X-ray diffraction) to the external magnetic field.

The discussion of the TF-μSR technique is given in [21], where details of the application of the technique to the determination of λ can be found. In the analysis presented below we used the well-known fact that for an extreme type-II superconductor in the mixed state λ^{-4} is proportional to the second moment of the local magnetic field distribution $P(B)$ probed by μSR [22]. The second moment of $P(B)$ was calculated by using a procedure similar to the one described in [23]. It includes fit of the μSR time-spectra with a three component Gaussian expression:

$$P(t) = \sum_{i=1}^{3} A_i \exp\left(-\sigma_i^2 t^2/2\right) \cos\left(\gamma_\mu B_i t + \phi\right) , \tag{1}$$

where A_i, σ_i, and B_i are the asymmetry, the relaxation rate, and the first moment of the i-th component, and ϕ is the initial phase of the muon-spin ensemble. The muon time-spectrum for $La_{1.83}Sr_{0.17}CuO_4$ at $T = 1.7$ K, $\mu_0 H = 0.02$ T together with the corresponding three component Gaussian fit by means of Eq. 1 is shown in Fig. 1. The first and the second moments of $P(B)$

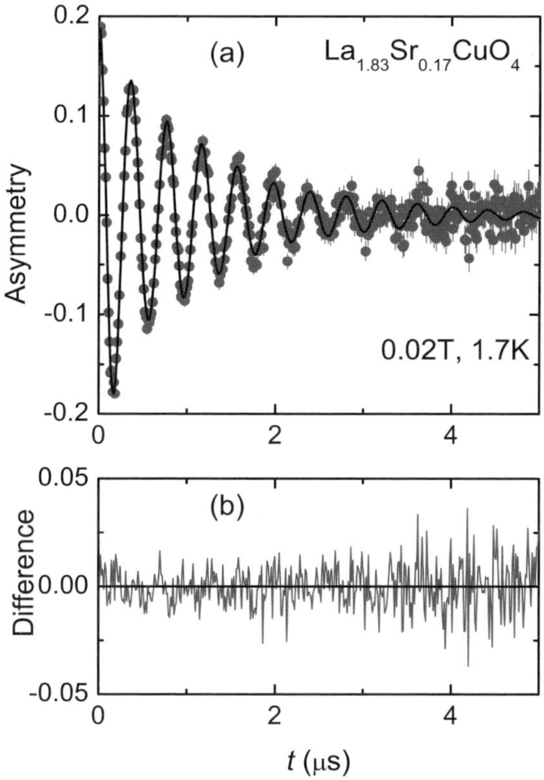

Fig. 1 **a** A muon time-spectrum for $La_{1.83}Sr_{0.17}CuO_4$ at $T = 1.7$ K, $\mu_0 H = 0.02$ T. The *solid line* represents a fit of the muon spectrum by means of Eq. 1. **b** Difference between the experimental data and the fit. See text for details

were calculated in accordance with the following expressions [23]:

$$\langle B\rangle = \sum_{i=1}^{3} \frac{A_i B_i}{A_1 + A_2 + A_3} \tag{2}$$

and

$$\left\langle \Delta B^2 \right\rangle = \frac{\sigma^2}{\gamma_\mu^2} = \sum_{i=1}^{3} \frac{A_i}{A_1 + A_2 + A_3} \left[(\sigma_i/\gamma_\mu)^2 + [B_i - \langle B\rangle]^2 \right] . \tag{3}$$

Here $\gamma_\mu = 2\pi \times 135.5342$ MHz/T is the muon gyromagnetic ratio. The superconducting part of the square root of the second moment ($\sigma_{sc} \propto \lambda_{ab}^{-2}$) was then obtained by subtracting the nuclear moment contribution (σ_{nm}) measured at $T > T_c$ according to $\sigma_{sc}^2 = \sigma^2 - \sigma_{nm}^2$ [21].

To ensure that the increase of σ below T_c is attributed entirely to the vortex lattice, zero-field μSR experiments were performed. The experiments show no evidence for static magnetism in $La_{1.83}Sr_{0.17}CuO_4$ down to 1.7 K.

Experimental Results

Dependence of λ_{ab} on Temperature

In Fig. 2 we plot the temperature dependence of $\sigma_{sc} \propto \lambda_{ab}^{-2}$ for $\mu_0 H = 0.02$ T. It is seen that λ_{ab}^{-2} increases rather monotonically with decreasing temperature. Most importantly, around 10–15 K an inflection point appears. In [24, 25] it was shown that an inflection point in $\lambda^{-2}(T)$ may appear in superconductors with two weakly coupled superconducting bands. Also, in MgB_2, where the σ- and π-bands are almost decoupled, an upward curvature of $\lambda^{-2}(T)$, similar to the one presented in Fig. 3, was detected (see e.g. [18]). In analogy to MgB_2, it was assumed that the total $\sigma_{sc}(T,H)$ is a sum of two components [26, 27]:

$$\sigma_{sc}(T,H) = \sigma_1(T,H) + \sigma_2(T,H) . \tag{4}$$

In order to separate the field and the temperature dependent contributions we used the following procedure. First, Eq. 4 was divided by the zero-temperature value $\sigma_{sc}(0,H)$. Then it was assumed that the temperature dependence of each component on the right hand side of Eq. 4 is determined by $\delta\sigma(\Delta(0),T)$ so that:

$$\frac{\sigma_{sc}(T,H)}{\sigma_{sc}(0,H)} = \frac{\sigma_1(0,H)}{\sigma_{sc}(0,H)} \delta\sigma(\Delta_1(0),T) + \frac{\sigma_2(0,H)}{\sigma_{sc}(0,H)} \delta\sigma(\Delta_2(0),T) . \tag{5}$$

In a last step, we define the weight factor $w(H) = \sigma_1(0,H)/\sigma_{sc}(0,H)$, $1 - w(H) = \sigma_2(0,H)/\sigma_{sc}(0,H)$, yielding:

$$\sigma_{sc}(T,H)/\sigma_{sc}(0,H) = w(H) \cdot \delta\sigma(\Delta_1(0),T) + (1 - w(H)) \cdot \delta\sigma(\Delta_2(0),T) . \tag{6}$$

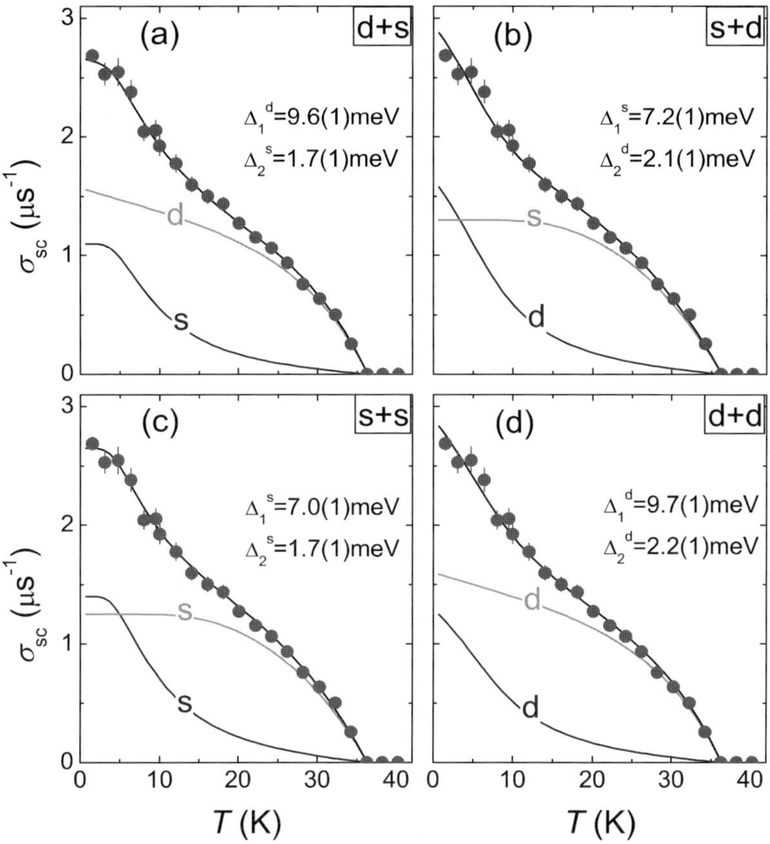

Fig. 2 Temperature dependence of $\sigma_{sc} \propto \lambda_{ab}^{-2}$ of single-crystal $La_{1.83}Sr_{0.17}CuO_4$ measured at 0.02 T (field-cooled). *Lines* represent results of the fit by means of Eq. 6. The data were analyzed assuming that the large and the small gaps entering Eq. 6 have the following symmetries: $d + s$ (**a**), $s + d$ (**b**), $s + s$ (**c**), and $d + d$ (**d**). The gap values obtained from the fits are shown in the corresponding figures. See text for details

As is seen, the field [$w(H)$ and $(1 - w(H)$] and the temperature [$\delta\sigma(\Delta_{1,2}(0), T)$] dependent contributions are now separated.

The $\delta\sigma(\Delta_{1,2}(0), T)$ dependences were assumed to be described by the BCS model extended for the gap having a certain angular symmetry [15]:

$$\delta\sigma(\Delta(0), T) = 1 + \frac{1}{\pi} \int_0^{2\pi} \int_{\Delta(T,\varphi)}^{\infty} \left(\frac{\partial f}{\partial E}\right) \frac{E\,\mathrm{d}E\,\mathrm{d}\varphi}{\sqrt{E^2 - \Delta(T,\varphi)^2}} \,. \tag{7}$$

Here, $f = [1 + \exp(E/k_B T)]^{-1}$ is the Fermi function, $\Delta(0)$ is the maximum value of the gap, and $\Delta(T, \varphi) = \Delta(0)\tilde{\Delta}(T/T_c)g(\varphi)$. For the normalized gap

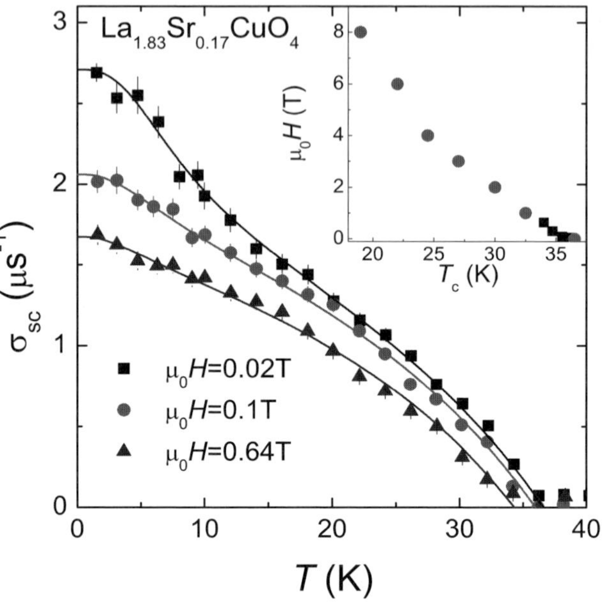

Fig. 3 Temperature dependence of $\sigma_{sc} \propto \lambda_{ab}^{-2}$ of single-crystal $La_{1.83}Sr_{0.17}CuO_4$ measured at 0.02 T, 0.1 T, and 0.64 T (field-cooled). *Lines* in the main figure represent the fit with the two gap model (Eq. 6). The *inset* shows the dependence of the transition temperature T_c on H obtained in the present study (*solid squares*) and from the AC-magnetization curves measured for the same crystal as used in this work (*solid circles*). See text for details

$\tilde{\Delta}(T/T_c)$ values tabulated in [28] were used. The function $g(\varphi)$ represents the angular dependence of the superconducting gap.

The data in Fig. 2 were analyzed by means of Eq. 6 assuming that the large and the small gaps have the following symmetries: $d + s$ (a), $s + d$ (b), $s + s$ (c), and $d + d$ (d). For the d-wave gap we used the well-known expression $g^d(\varphi) = |\cos(2\varphi)|$ [9] and for the s-wave one $g^s(\varphi) = 1$. The results of the fits are summarized in Fig. 2 and Table 1.

Comparison of the gap values with $\Delta(0) \simeq 10$ meV obtained on a similar sample by tunnelling experiments [29], suggests that the large gap has d-wave symmetry. Another argument in favor of a "large" d-wave gap comes from the observation of a square vortex lattice in the same crystal as used in this work in fields higher than 0.4 T [30, 31], where as shown below, the contribution from the large gap to σ_{sc} is dominant. A square vortex lattice is typical for d-wave superconductors [30]. Concerning the symmetry of the second gap (d vs. s), however, the situation is unclear. Based on the observation of a substantial s-wave contribution to the superconducting order parameter by Andreev reflection [9], tunnelling [10], Raman [12], nuclear magnetic resonance [13] and inelastic neutron scattering [14] experiments as well as by the analysis of tunnelling data [4–6], we assume that the second gap has isotropic

Table 1 Summary of the two-gap analysis for single-crystal $La_{1.83}Sr_{0.17}CuO_4$ at $\mu_0 H = 0.02$ T. The meaning of the parameters is – $\mu_0 H$: external magnetic field, T_c: superconducting transition temperature, $\sigma_{sc}(0)$: zero-temperature μSR relaxation rate, w: relative weighting factor, $\Delta_1(0)$ and $\Delta_2(0)$: the maximum values of the large Δ_1 and the small Δ_2 gap at zero temperature

Gaps	T_c (K)	$\sigma_{sc}(0)$ (μs^{-1})	w	$\Delta_1(0)$ (meV)	$\Delta_2(0)$ (meV)	$\frac{2\Delta_1(0)}{k_B T_c}$	$\frac{2\Delta_2(0)}{k_B T_c}$
$d+s$	36.3(1)	2.67(8)	0.59(3)	9.6(1)	1.7(1)	6.14(6)	1.09(6)
$s+d$	36.3(1)	2.92(9)	0.44(2)	7.2(1)	2.1(1)	4.60(6)	1.34(6)
$s+s$	36.3(1)	2.65(8)	0.47(2)	7.0(1)	1.7(1)	4.48(6)	1.09(6)
$d+d$	36.3(1)	2.90(8)	0.55(3)	9.7(1)	2.2(1)	6.20(6)	1.41(6)

s-wave symmetry. This implies that in $La_{1.83}Sr_{0.17}CuO_4$ the "$d+s$" scenario presented in Fig. 2a is realized. This is the most obvious scenario, even though other gap dependences cannot be fully ruled out.

Dependence of T_c and λ_{ab} on Magnetic Field

In Fig. 3 we plot the temperature dependences of $\sigma_{sc} \propto \lambda_{ab}^{-2}$ for $\mu_0 H = 0.02$ T, 0.1 T, and 0.64 T (for clarity, data for 0.05 T and 0.3 T are not shown). The solid lines represent the global fit of Eq. 6 assuming that the large and the small gaps have a d-wave (Δ_1^d) and an s-wave (Δ_2^s) symmetry (see Sect. 3.1). In the analysis all the $\sigma_{sc}(T)$ curves (0.02, 0.05, 0.1, 0.3, 0.64 T) were fitted simultaneously with $\sigma_{sc}(0)$, T_c, and w as individual parameters for each particular data set. $\Delta_1^d(0)$ and $\Delta_2^s(0)$ were assumed to scale linearly with T_c according to the relation $2\Delta(0)/k_B T_c = \text{const}$. The results of the global fit are summarized in Table 2.

A quick glance at Table 2 reveals that the transition temperature T_c depends rather strongly on the magnetic field. In the inset of Fig. 3 we plot T_c vs. H as obtained in the present study and from the AC-magnetization curves measured for the same crystal as used in this work [20]. According to the AC data [20], with increasing magnetic field from 0 T to 1 T the transition temperature T_c decreases from 36.3 K to 32.6 K in good agreement with the change of T_c observed from the fit (Table 2). It is also seen that the slope $|dT_c/dH|$ decreases with increasing magnetic field. Note that a similar T_c vs. H behavior was observed in MgB_2 [32], $Na_{0.35}CoO_2 \cdot yH_2O$ [33], YNi_2B_2C and $LuNi_2B_2C$ [34], and was explained within the framework of a two-band model [35, 36].

The dependence of $\sigma_{sc}(0)$ and w on magnetic field (Table 2) is shown in Fig. 4a. The decrease of $\sigma_{sc}(0)$ is associated with an increase of the contribution of the large gap to λ^{-2}. Similar field dependences of w and σ_{sc} were observed in MgB_2 by small angle neutron scattering [37], point-contact spec-

Table 2 Summary of the "$d+s$" two-gap analysis for single-crystal $La_{1.83}Sr_{0.17}CuO_4$

$\mu_0 H$ (T)	T_c (K)	$\sigma_{sc}(0)$ (μs^{-1})	w	$\Delta_1^d(0)$ (meV)	$\Delta_2^s(0)$ (meV)	$\frac{2\Delta_1^d(0)}{k_B T_c}$	$\frac{2\Delta_2^s(0)}{k_B T_c}$
0.02	36.3(1)	2.71(8)	0.68(3)	8.2(1)	1.57(8)		
0.05	36.1(1)	2.20(7)	0.78(2)	8.2(1)	1.56(8)		
0.1	35.5(1)	2.07(7)	0.88(2)	8.0(1)	1.54(8)	5.24(7)[1]	1.00(5)[1]
0.3	34.7(1)	1.82(6)	0.92(2)	7.8(1)	1.50(7)		
0.64	34.0(1)	1.71(5)	0.94(2)	7.7(1)	1.47(7)		

[1] Common fit parameter for all fields

troscopy [38], and μSR [17] experiments. This was explained by the fact that superconductivity within the weaker π-band is suppressed at much lower fields than that within the stronger σ-band [38]. As shown in Fig. 4b this is also the case for $La_{1.83}Sr_{0.17}CuO_4$. Indeed, while the contribution from the large gap $[\sigma_1(0) = w \cdot \sigma_{sc}(0)]$ changes only slightly, the contribution from the

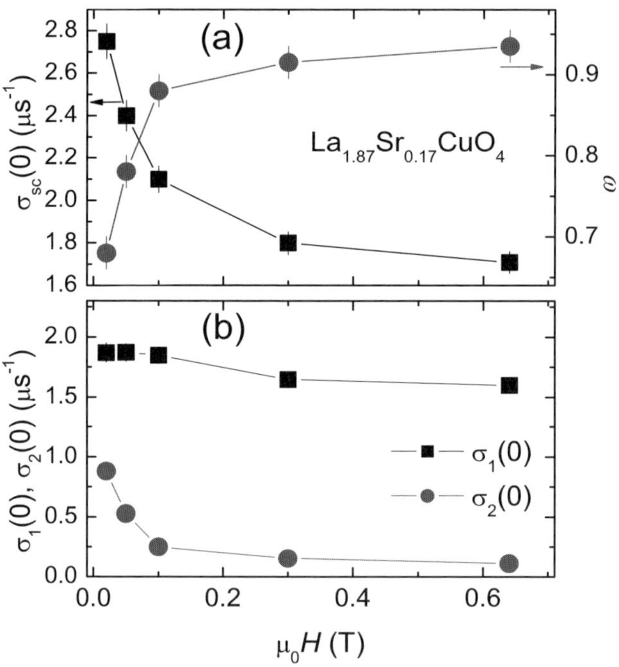

Fig. 4 **a** – Field dependences of $\sigma_{sc}(0)$ and w for single-crystal $La_{1.83}Sr_{0.17}CuO_4$ obtained from the fit of Eq. 6 to the data (Table 2). **b** – Contribution from the large $[\sigma_1(0)]$ and the small $[\sigma_2(0)]$ superconducting gap to the total $\sigma_{sc}(0)$

small gap $[\sigma_2(0) = (1 - w) \cdot \sigma_{sc}(0)]$ decreases by almost an order of magnitude with the field rising from 0.02 T to 0.64 T (Fig. 4b). This implies that the field dependences of λ_{ab}^{-2} in $La_{1.83}Sr_{0.17}CuO_4$ are similar to MgB_2, and consequently demonstrate the existence of two gaps.

Analysis of the Line Shape

It is important to emphasize that the observation of an inflection point in $\lambda^{-2}(T)$ is not restricted to MgB_2 and the particular $La_{1.83}Sr_{0.17}CuO_4$ sample studied in this work. Indication of an inflection point in $\lambda^{-2}(T)$ was also observed in hole-doped $YBa_2Cu_3O_{7-\delta}$ [16, 39], $YBa_2Cu_4O_8$ [40, 41], and $La_{1.85}Sr_{0.15}CuO_4$ [42], as well as in electron-doped $Pr_{1.855}Ce_{0.145}CuO_{4y}$ [43]. In [39] the increase of the second moment of $P(B)$ observed in $YBa_2Cu_3O_{7-\delta}$ at low temperatures was attributed to pinning effects. In order to investigate the role of pinning in our sample we compare the $P(B)$ distributions for 0.05 T and 0.64 T, obtained from the measured muon-time spectra by means of the maximum entropy Fourier transform technique (Fig. 5a), with the theoretical $P(B)$ curves. A standard way to account for pinning is to convolute the theoretical $P(B)$ for an ideal vortex lattice (black line in the inset of Fig. 5a) with a Gaussian distribution of fields [44]:

$$P(B) = \frac{1}{\sqrt{2\pi}\sigma_B} \int \exp\left[-\frac{1}{2}\left(\frac{B - B'}{\sigma_B}\right)^2 \right] P_{id}(B')\,\mathrm{d}B' . \tag{8}$$

σ_B is the width of the Gaussian distribution and $P_{id}(B)$ is the field distribution for an ideal vortex lattice

$$P_{id}(B) = \frac{\int \delta(B - B')\,\mathrm{d}A(B')}{\int \mathrm{d}A(B')} , \tag{9}$$

with the local field intensity at point $\mathbf{r}$ [39]

$$B(\mathbf{r}, \lambda, \xi) = B \sum_{\mathbf{g}} \frac{e^{-\xi^2 g^2/2}}{(1 + \lambda^2 g^2)} e^{i\mathbf{g}\cdot\mathbf{r}} . \tag{10}$$

Here $\mathrm{d}A(B')$ is an elemental piece of the vortex lattice unit cell for which the field is equal to B', $\mathbf{g}$ is a reciprocal lattice vector, ξ is a core cutoff parameter which in the ideal case should be equal to the coherence length, and B is an average field inside the superconductor. The integration in Eq. 9 is performed within the unit cell. The theoretical $P(B)$ profiles obtained by means of Eq. 8 for $\lambda = 220$ nm, $\xi = 2$ nm and $\sigma_B = 0$, 0.3, 0.6, and 1.0 μs^{-1} are shown in the inset of Fig. 5a. The direct comparison of the $P(B)$ data for $\mu_0 H = 0.05$ T and 0.64 T with theoretical $P(B)$ profiles clearly demonstrates that pinning is not the main source of the observed increase of the second moment of $P(B)$ at low

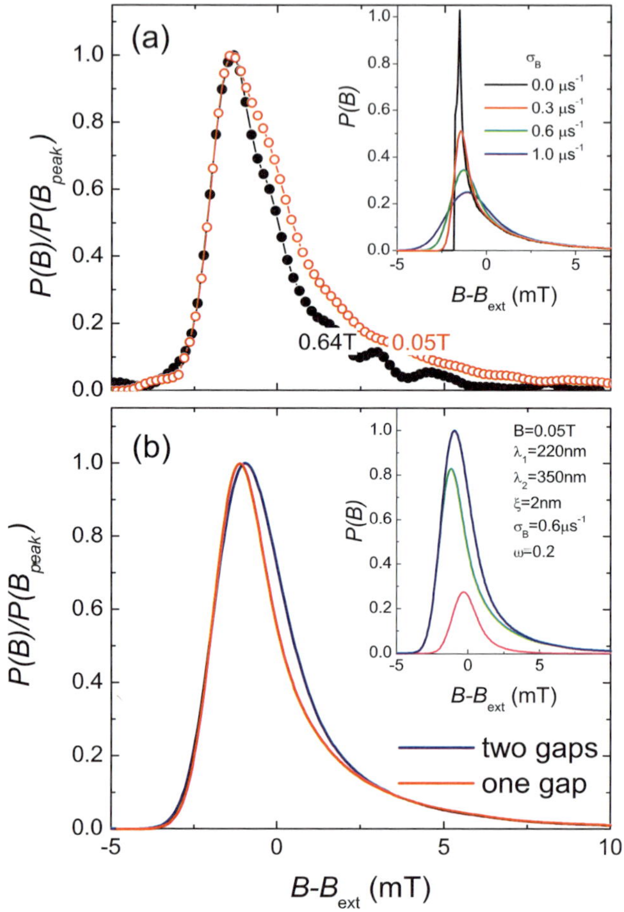

Fig. 5 **a** Local magnetic field distribution $P(B)$ in the mixed state of single-crystal $La_{1.83}Sr_{0.17}CuO_4$ ($T = 1.7$ K, field-cooled) normalized to their maximum value at $B = B_{peak}$ for 0.05 T and 0.64 T. The *inset* shows theoretical $P(B)$ distributions ($\lambda = 220$ nm, $\xi = 2$ nm, and $\mu_0 H = 0.05$ T) for different values of the smearing parameter $\sigma_B = 0$, 0.3, 0.6, and 1.0 μs^{-1}. **b** $P(B)$ profiles obtained within the one-gap and the two-gap models. In the *inset* the contributions to $P(B)$ arising from the first and the second terms of Eq. 11 are shown separately. See text for details

temperatures. Indeed, pinning leads to an almost symmetric (around B_{peak}) broadening of $P(B)$ (see inset of Fig. 5a), while the experimental $P(B)$ profiles very well coincide at low fields ($B < B_{peak}$). Deviations only occur in the high-field tail of $P(B)$ ($B > B_{peak}$).

Qualitatively, an additional broadening of $P(B)$ at $B > B_{peak}$ can again be explained within the framework of the two-gap model. In Fig. 5b we show the $P(B)$ profiles obtained within the one- and the two-gap models. In a later case the local field intensity at point **r** was assumed to be described within the model suggested by Serventi et al. [17,45]:

$$B(\mathbf{r},\lambda_1,\lambda_2,\xi) = w \cdot B(\mathbf{r},\lambda_1,\xi) + (1-w) \cdot B(\mathbf{r},\lambda_2,\xi)\,. \tag{11}$$

It is seen that the presence of the second superconducting gap leads to appearance of an additional weight in $P(B)$ just above B_{peak} in qualitative agreement with the experimental $P(B)$ profiles presented in Fig. 5a.

Discussions

As has been shown in the previous Section, the temperature and the magnetic field dependence of the in-plane magnetic penetration depth λ_{ab}, the magnetic field dependence of the superconducting transition temperature T_c, and the shape of $P(B)$ distribution, are consistent with the presence of two superconducting gaps in $La_{1.83}Sr_{0.17}CuO_4$ cuprate superconductor. The obvious question which now arises is where to locate the second superconducting gap in $La_{2-x}Sr_xCuO_4$? The phase diagram of cuprates is usually interpreted in terms of holes doped into the planar Cu $d_{x^2-y^2} - \text{O}\ p_\alpha$ ($\alpha = x, y$) antibonding band. In $La_{2-x}Sr_xCuO_4$ it is assumed that one hole per Sr atom enters this band. However, recent ab-initio calculations yielded additional features appearing on doping of $La_{2-x}Sr_xCuO_4$ [46]. According to these calculations part of the holes occupy the Cu $d_{3z^2-r^2} - \text{O}\ p_z$ orbitals. These results are further supported by neutron diffraction data [47], showing that the doped holes indeed appear in both the planar and the out-of-plane bands. In contrast to this finding, in angle-resolved photoemission (ARPES) experiments on HTS only the planar band was observed, suggesting a quasi-two-dimensional electronic structure with negligible *intercell* coupling of CuO_2-layers (see e.g. [48]). This is, however, inconsistent with in-plane and out-of-plane λ measurements [49], optical conductivity [50], and anisotropy parameter studies [51]. All these experiments demonstrate that with increasing doping cuprates become more and more three-dimensional. Recently a 3D Fermi surface was observed in overdoped $TlBa_2CuO_{6+\delta}$ [52]. In addition, a careful analysis of ARPES data reveals that the finite dispersion of the energy bands along the z-direction of the Brillouin zone (k_z dispersion) naturally induces an irreducible linewidth of the ARPES peaks which is unrelated to any scattering mechanism [53–55]. This implies that a 2D single band model is insufficient and out-of-plane hybridized bands have to be incorporated.

Conclusions

Systematic μSR studies of the in-plane magnetic penetration depth λ_{ab} in single-crystal $La_{1.83}Sr_{0.17}CuO_4$ were performed. It was found that at low magnetic fields ($\mu_0 H \leq 0.3$ T) $\lambda_{ab}^{-2}(T)$ exhibits an inflection point at $T \simeq 10{-}15$ K. We interpret the appearance of this point as being a consequence of the presence of two superconducting gaps, analogous to double-gap MgB_2 [18]. From the comparison of the gap values obtained from the fit with the literature data it was suggested that the large gap (Δ_1^d) has a d-wave and the small gap (Δ_2^s) an s-wave symmetry. The zero temperature values of the gaps at $\mu_0 H = 0.02$ T, obtained from the global fit of μSR data, were found to be $\Delta_1^d(0) = 8.2(2)$ meV and $\Delta_2^s(0) = 1.57(8)$ meV. With increasing magnetic field the contribution of Δ_2^s to the superfluid density decreases substantially (from 32% at $\mu_0 H = 0.02$ T to 6% at $\mu_0 H = 0.64$ T), in contrast to an almost constant contribution of Δ_1^d. Both the temperature and the field dependences of λ_{ab}^{-2} were found to be similar to what was observed in double-gap MgB_2 [17, 18]. The magnetic field dependence of the transition temperature T_c and the shape of the local magnetic field distribution $P(B)$ probed by μSR in $La_{1.83}Sr_{0.17}CuO_4$ in a mixed state were also found to be consistent with the presence of two superconducting gaps. The present results, added to the previously reported coexistence of two (s and d) superconducting condensates in various HTS [10–15], suggest that the two-gap behavior appears to be generic for high-temperature cuprate superconductors.

Acknowledgements This work was partly performed at the Swiss Muon Source (SμS), Paul Scherrer Institute (PSI, Switzerland). We gratefully acknowledge Karl Alex Müller for initiating this work and for many fruitful discussions. We are also grateful to N. Momono, M. Oda, M. Ido and J. Mesot for providing us the $La_{1.83}Sr_{0.17}CuO_4$ single crystal, A. Maisuradze for help in the data analysis, and A. Maisuradze, F. La Mattina, C.J. Juul, A. Amato, and D. Herlach for assistance during the μSR measurements. This work was supported by the Swiss National Science Foundation, the K. Alex Müller Foundation, and in part by the SCOPES grant No. IB7420-110784 and the EU Project CoMePhS.

References

1. Keller H (2005) In: Müller KA, Bussmann-Holder A (eds) Struct Bond 114:114–143
2. Khasanov R, Shengelaya A, Conder K, Morenzoni E, Savic IM, Karpinski J, Keller H (2006) Phys Rev B 74:064504
3. Kresin VZ, Wolf SA (1992) Phys Rev B 46:6458
4. Müller KA (1995) Nature (London) 377:133
5. Müller KA (1996) J Phys Soc Jap 65:3090
6. Müller KA, Keller H (1996) In: High-T_c Superconductivity: Ten years after discovery, 1997 Kluwer Academic Publishers pp 7–29
7. Binnig G, Baratoff A, Hoenig HE, Bednorz JG (1980) Phys Rev Lett 45:1352
8. Schooley JF, Hosler WR, Cohen ML (1964) Phys Rev Lett 12:474
9. Deutscher G (2005) Rev Mod Phys 77:109
10. Yeh N-C, Chen C-T, Hammerl G, Mannhart J, Schmehl A, Schneider CW, Schulz RR, Tajima S, Yoshida K, Garrigus D, Strasik M (2001) Phys Rev Lett 87:087003
11. Heyen ET, Cardona M, Karpinski J, Kaldis E, Rusiecki S (1991) Phys Rev B 43:12958

12. Masui T, Limonov M, Uchiyama H, Lee S, Tajima S, Yamanaka A (2003) Phys Rev B 68:060506(R)
13. Stern R, Mali M, Roos J, Brinkmann D (1995) Phys Rev B 51:15478
14. Furrer A (2005) In: Superconductivity in complex sysytems. Springer Berlin, Heidelberg, pp 171–204
15. Khasanov R, Shengelaya A, Maisuradze A, La Mattina F, Bussmann-Holder A, Keller H, Müller KA (2007) to be published in Phys Rev Lett 98:057007
16. Sonier JE, Brewer JH, Kiefl RF (2000) Rev Mod Phys 72:769
17. Serventi S, Allodi G, De Renzi R, Guidi G, Romanò L, Manfrinetti P, Palenzona A, Niedermayer C, Amato A, Baines C (2004) Phys Rev Lett 93:217003
18. Carrington A, Manzano F (2003) Physica C 385:205
19. Nakano T, Momono N, Oda M, Ido M (1998) J Phys Soc Jap 67:2622
20. Gilardi R (2004) PhD thesis, ETH Zurich
21. Zimmermann P, Keller H, Lee SL, Savic IM, Warden M, Zech D, Cubitt R, Forgan EM, Kaldis E, Karpinski J, Krüger C (1995) Phys Rev B 52:541
22. Brandt EH (1988) Phys Rev B 37:R2349
23. Khasanov R, Eshchenko DG, Di Castro D, Shengelaya A, La Mattina F, Maisuradze A, Baines C, Luetkens H, Karpinski J, Kazakov SM, Keller H (2005) Phys Rev B 72:104504
24. Xiang T, Wheatley JM (1996) Phys Rev Lett 76:134
25. Bussmann-Holder A, Khasanov R, Shengelaya A, Maisuradze A, La Mattina F, Keller H, Müller KA (2007) to be published in Europhys Lett 77:27002
26. Niedermayer C, Bernhard C, Holden T, Kremer RK, Ahn K (2002) Phys Rev B 65:094512
27. Kim M-S, Skinta JA, Lemberger TR, Kang WN, Kim H-J, Choi E-M, Lee S-I (2002) Phys Rev B 66:064511
28. Mühlschlegel B (1959) Z Phys 155:313
29. Oda M, Momono N, Ido M (2000) Supercond Sci Technol 13:R139
30. Gilardi R, Mesot J, Drew A, Divakar U, Lee SL, Forgan EM, Zaharko O, Conder K, Aswal VK, Dewhurst CD, Cubitt R, Momono N, Oda M (2002) Phys Rev Lett 88:217003
31. Drew AJ, Heron DOG, Divakar UK, Lee SL, Gilardi R, Mesot J, Ogrin FY, Charalambous D, Momono N, Oda M, Baines C (2006) Physica B 203:374–375
32. Suderow H, Tissen VG, Brison JP, Martinez JL, Vieira S, Lejay P, Lee S, Tajima S (2004) Phys Rev B 70:134518
33. Yang HD, Lin J-Y, Sun CP, Kang YC, Huang CL, Takada K, Sasaki T, Sakurai H, Takayama-Muromachi E (2005) Phys Rev B 71:020504(R)
34. Shulga SV, Drechsler S-L, Fuchs G, Müller K-H, Winzer K, Heinecke M, Krug K (1988) Phys Rev Lett 80:1730
35. Prohammer M, Schachinger E (1987) Phys Rev B 36:8353
36. Kao J-T, Lin J-Y, Mou C-Y (2007) Phys Rev B 75:012503
37. Cubitt R, Eskildsen MR, Dewhurst CD, Jun J, Kazakov SM, Karpinski J (2003) Phys Rev Lett 91:047002
38. Gonnelli RS, Daghero D, Ummarino GA, Stepanov VA, Jun J, Kazakov SM, Karpinski J (2002) Phys Rev Lett 89:247004
39. Harshman DR, Kossler WJ, Wan X, Fiory AT, Greer AJ, Noakes DR, Stronach CE, Koster E, Dow JD (2004) Phys Rev B 69:174505
40. Panagopoulos C, Tallon JL, Xiang T (1999) Phys Rev B 59:R6635
41. Khasanov R, Karpinski J, Keller H (2005) J Phys: Condens Matter 17:2453
42. Luke GM, Fudamoto Y, Kojima K, Larkin M, Merrin J, Nachumi B, Uemura YJ, Sonier JE, Ito T, Oka K, de Andrad M, Maple MB, Uchida S (1997) Physica C 1465:282–287

43. Skinta JA, Lemberger TR, Greibe T, Naito M (2002) Phys Rev Lett 88:207003
44. Brandt EH (1988) J Low Temp Phys 73:355
45. It was assumed that apperance of two superfluids leads to presence of two penetration depths rather than two coherence lenghths as suggested by Serventi et al. in [17]. This scenario is inderectly confirmed by calculations of Amin [Amin MHS, PhD thesis, cond-mat/0011455] showing that an *s*-wave admixture naturally leads to appearance of two peaks on $P_{\rm id}(B)$.
46. Perry JK, Tahir-Kheli J, Goddard WA (2002) Phys Rev B 65:144501
47. Božin ES, Billinge SJL (2005) Phys Rev B 72:174427
48. Damascelli A, Hussain Z, Shen Z-X (2003) Rev Mod Phys 75:473
49. Xiang T, Panagopoulos C, Cooper JR (1998) Int J Mod Phys B 12:1007
50. Tamasaku K, Ito T, Takagi H, Uchida S (1994) Phys Rev Lett 72:3088
51. Hofer J, Schneider T, Singer JM, Willemin M, Keller H, Sasagawa T, Kishio K, Conder K, Karpinski J (2000) Phys Rev B 62:631–639
52. Hussey NE, Abdel-Jawad M, Carrington A, Mackenzie AP, Ballcas L (2003) Nature (London) 425:814
53. Bansil A, Lindroos M, Sahrakorpi S, Markiewicz RS (2005) Phys Rev B 71:012503
54. Sahrakorpi S, Lindroos M, Markiewicz RS, Bansil A (2005) Phys Rev Lett 95:157601
55. Markiewicz RS, Sahrakorpi S, Lindroos M, Lin H, Bansil A (2005) Phys Rev B 72:054519

Kochelaev BI, Waldner F (2007) In: Bussmann-Holder A, Keller H (eds)
High T_c Superconductors and Related Transition Metal Oxides.
Springer-Verlag Berlin Heidelberg, 191–199
DOI 10.1007/978-3-540-71023-3

Skyrmions in Lightly Doped Cuprates?

B. I. Kochelaev[1] (✉) · F. Waldner[2]

[1]Kazan State University, 420008 Kazan, Russia
Boris.kochelaev@ksu.ru

[2]Physics Institute, University of Zürich, CH-8057 Zürich, Switzerland

Abstract We study the stability of skyrmions in anisotropic layered magnets and consider a possible creation of the skyrmions in lightly doped layered cuprates. The picture of thermally excited and induced by the doped quasi-localized holes skyrmions is used to describe magnetic properties and spin kinetics of underdoped cuprates. A consistency of the skyrmion approach and the phase separation phenomenon is discussed.

Keywords High T_c superconductors · Nuclear magnetic resonance · Skyrmion · Two-dimensional magnetism

Introduction

The discovery of high-temperature superconductivity (HTS) by G. Bednorz and K.A. Müller [1] created world-wide enormous efforts to understand the intriguing transformation of non-conducting antiferromagnetic layered cuprates into superconductors by doping their CuO_2 planes with electronic holes. At the same time this discovery gave a great impulse to study general properties of strongly correlated systems, in particular two-dimensional magnets. Many efforts have been undertaken to develop a theory of magnetic properties and spin kinetics of undoped cuprates, which can be represented by a quantum Heisenberg two-dimensional antiferromagnet (QHAF) $S = 1/2$ on a square lattice with a large isotropic exchange coupling constant between the nearest neighbours ($J = 1500$ K in La_2CuO_4). Most of the results were obtained for low temperatures $T \ll J$, using the renormalization group analysis of a non-linear σ-model, the chiral perturbation theory, the $1/N$-expansion method, see for example [2–7].

On the other hand earlier there were given some arguments that topological excitations of the skyrmion type could be important for magnetic properties of cuprates [8]. An existence of skyrmions in the two-dimensional classical ferromagnet was established by Belavin and Polyakov in the frame of the non-linear σ-model [9]. The skyrmion approach for a description of magnetic properties and spin kinetics of QHAF was developed some years ago on the basis of quantum consideration of spin excitations above the skyrmion background [10, 11]. The energy spectrum of spin excitations, local order parameter, average skyrmion radius r_0 and the nuclear spin relaxation rate $1/T_1$ were found for the temperature region $0 < T < J$. These results were consistent with results obtained by other methods in the limit $T \ll J$. Moreover, there was shown previously that in weekly doped cuprates a hole, being quasi-localized on a plaquette of oxygen ions in the CuO_2 plane, induces a three-dimensional spin texture whose topology is that of a single skyrmion [12]. A similar conclusion was made recently about the spin distortion structure of the antiferromagnetic order around the Zhang–Rice singlet [13]. It was quite natural to investigate an evolution of magnetic properties of cuprates with doping taking into account both thermally excited skyrmions and skyrmions created by the doped quasi-localized holes [14, 15]. There the average skyrmion radius was calculated (identified with a spin correlation length as $\xi = 2r_0$) and the nuclear spin relaxation rate as functions of doping and temperature in a qualitative agreement with experimental results.

However, a question could erase, whether the skyrmion, being a topological excitation of the two dimensional isotropic magnet, can survive in the layered cuprates with the anisotropic Heisenberg interaction between the Cu ions. Another question relates to the consistency of the skyrmion approach for the underdoped cuprates with many experimental evidences of a phase separation in these materials (see, in particular [16]). In this contribution we analyze and discuss these problems.

Stability of Skyrmions in Layered Magnets

We model the layered anisotropic magnet by the cubic lattice of spins coupled by the anisotropic Heisenberg exchange interaction in the x, y planes with a small isotropic exchange coupling between the plains:

$$H_{\text{ex}} = \sum_{\langle j,\delta \rangle} \left\{ J_z S_j^z S_{j+\delta}^z + J_{xy} \left(S_j^x S_{j+\delta}^x + S_j^y S_{j+\delta}^y \right) \right\} + J_1 \sum_{\langle j,\rho \rangle} \mathbf{S}_j \mathbf{S}_{j+\rho} \,. \quad (1)$$

Here J_z, J_{xy} are exchange integrals between the nearest neighbors in the x, y planes, whereas J_1 is the coupling constant between the nearest spins of neighboring x, y planes; the sign $\langle j, \ldots \rangle$ means the sum over spins, where the given pair met only once. It was shown previously that in the case of the isotropic exchange interaction in the x, y planes and $J_1 = 0$ the stable spin configurations could be found by the energy minimization in the mean field approximation after the unitary transformation of Eq. 1 to different axes of quantization for all spins [10]. The minima conditions could be reduced in a continuum limit to the set of equations:

$$\begin{aligned} &\Delta\theta - \sin\theta\cos\theta(\nabla\phi)^2 = 0 \\ &\nabla\left(\sin^2\theta \cdot \nabla\phi\right) = 0 \end{aligned} \quad (2)$$

with θ, ϕ as the polar and azimuth angles of the quantization axis direction. These equations are exactly the same as that obtained from the classical non-linear sigma model [9]. It is well known that besides of the trivial homogeneous solution the two simplest one are:

$$\begin{aligned} &\theta(x,y) = 2\mathrm{arctg}(r/r_0)\,, \quad \phi(x,y) = \mathrm{arctg}(y/x)\,; \\ &\theta(x,y) = 2\mathrm{arctg}(r_0/r)\,, \quad \phi(x,y) = \mathrm{arctg}(-y/x)\,. \end{aligned} \tag{3}$$

These two solutions represent the single skyrmion and antiskyrmion, they are shown in Fig. 1 for the square lattice of spins 15×15 (the length of arrows gives the projection of the spins on the x, y plane). The energy of the skyrmion and antiskyrmion $E_0 = 4\pi J$ does not depend on their radius r_0. An analytical solution of Eq. 2 for any number of skyrmions and antiskyrmions was given in [9].

Our search of stable spin configurations in the presence of instability factors $J_1 \neq 0$ and $J_z - J_{xy} \neq 0$ is based on the phenomenological Landau–Lifshitz kinetic equation for the magnetic moment density $\boldsymbol{M}(\boldsymbol{r}, t)$:

$$\frac{\partial \boldsymbol{M}(\boldsymbol{r},t)}{\partial t} = g\,[\boldsymbol{M}(\boldsymbol{r},t) \times \boldsymbol{h}(\boldsymbol{r},t)] + \boldsymbol{R}(\boldsymbol{r},t)\,. \tag{4}$$

Here $\boldsymbol{h}(\boldsymbol{r}, t)$ is the effective magnetic field, $\boldsymbol{R}(\boldsymbol{r}, t)$ is the relaxation term. A structure of the latter was established on the basis of the energy conservation law [17]:

$$\boldsymbol{R}(\boldsymbol{r},t) = \frac{1}{\tau_1}\boldsymbol{h}(\boldsymbol{r},t) - \frac{1}{\tau_2}\big[\boldsymbol{n}(\boldsymbol{r},t) \times [\boldsymbol{n}(\boldsymbol{r},t) \times \boldsymbol{h}(\boldsymbol{r},t)]\big]\,, \tag{5}$$

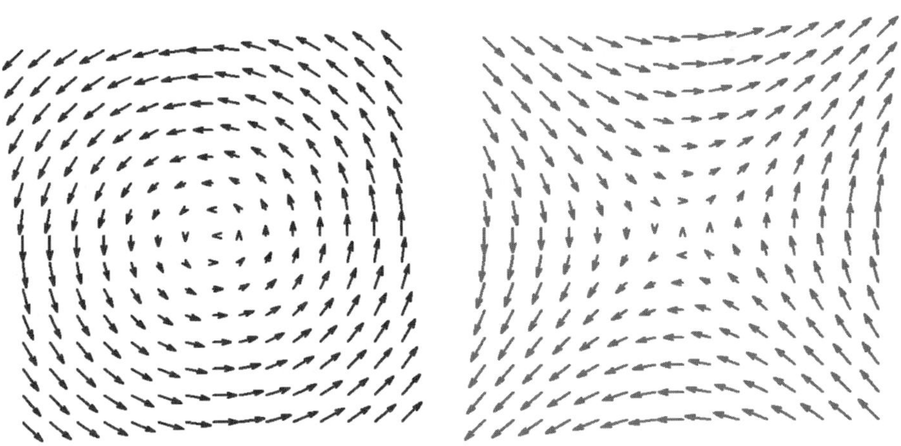

a b

Fig. 1 A projection of the skyrmion (**a**) and antiskyrmion (**b**) on the x–y plane according to Eq. 3

where $n(r,t) = M(r,t)/|M(r,t)|$; τ_1 and τ_2 are the relaxation times of the length and direction of the magnetization vector correspondingly, usually $\tau_1 \gg \tau_2$. As a matter of fact the last term in Eq. 5 is responsible for the final spin configuration in our study. The relaxation term (Eq. 5) can be used in a quantum kinetic equation for a lattice of spins with the Hamiltonian (Eq. 1) in the mean field approximation. This "relaxation method" was suggested and tested in [18].

At first we study the creation of skyrmions and antiskyrmions by the electronic holes localized on the plaquettes of an ideal square lattice of spins with the isotropic exchange interaction. An initial spin configuration is shown in Fig. 2a as four ferromagnetic domains with the total zero magnetization for 16×16 spins. All the spins are coplanar except the corners of the domains, where the spins deviate from the plane modeling the spin disturbance created by the quasilocal motion of the holes on the plaquettes according to [12]. We have found that the relaxation process can lead to different metastable spin configurations depending on the relaxation rate. For every metastable configuration we calculated its energy in order to find the stable state with the lowest energy. The result is presented in Fig. 2b (every spin is rotated by the angle $\phi = \pi/2$ to show a similarity to Fig. 1). The distribution of z-components of the spins is given in Fig. 3. As a matter of fact these numerical results reproduce the analytical solution according to [9] for the elementary unit of a square lattice of neighboring skyrmions and antiskyrmions in the two-dimensional ferromagnet. In the case of the two-dimensional antiferromagnet the results related to staggered sublattices of the spins coincide with described above. So, to study the skyrmion stability we consider further the ferromagnetic case only.

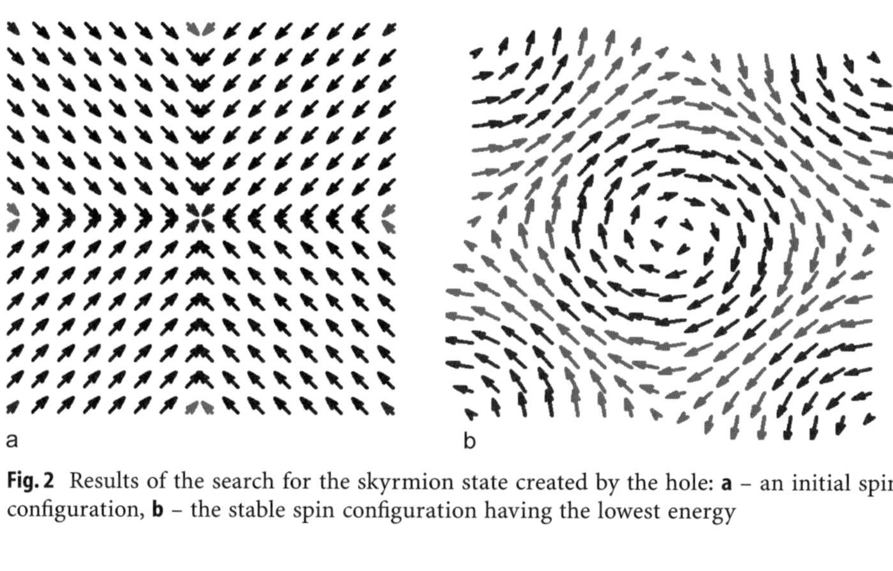

Fig. 2 Results of the search for the skyrmion state created by the hole: **a** – an initial spin configuration, **b** – the stable spin configuration having the lowest energy

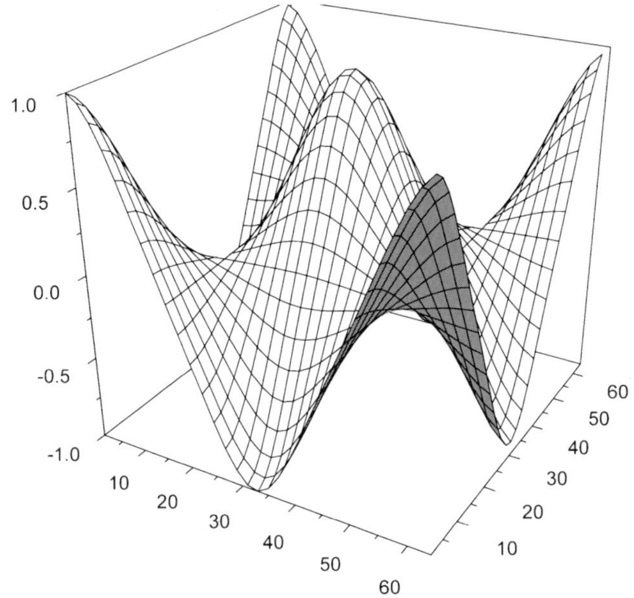

Fig. 3 The distribution of the z-component for the spin configuration shown in Fig. 2b

Now we consider the role of the three-dimensionality of real magnets. In other words we put $J_1 \neq 0$ and $J_z = J_{xy} = J$. We study two cases: two coupled planes of 16×16 spins (having in mind the YBa_2CuO_6 structure) and a quasi-two dimensional ferromagnetic cube of $16 \times 16 \times 16$ spins (the La_2CuO_4 type structure). We start with the same initial spin configuration shown in Fig. 2a for every plane. An evolution of the skyrmion structure with the increasing J_1 value is given in Fig. 4 as the cross section of the skyrmion by the x–z plane for the cube. On can see that the skyrmion structure survives at $J_1 < 0.015\,\mathrm{J}$ and then suddenly collapses. Similar behavior was found for the two coupled planes, but the critical value of the coupling constant was twice as large. It is evident that the actual exchange coupling between the CuO_2 planes in the layered cuprates is many orders of magnitude less than the calculated one above critical values and can not destroy the skyrmion structure.

Next we show consequences of the instability factor due to the anisotropy of the exchange interaction: $\Delta J = J_z - J_{xy} \neq 0$, $J_1 = 0$ (see [18]). Resulting stable spin configurations are shown in Fig. 5. One can see that their structure sufficiently depends on the sign of ΔJ. In the case $\Delta J > 0$ the skyrmion structure deforms in such a way that at $\Delta J / J_{xy} > 0.4$ it was transformed to ferromagnetic domains with spins up and down divided by the relatively narrow walls. In the opposite case $\Delta J < 0$ all the spins intend to lie in the plane and at $|\Delta J| / J_{xy} > 0.3$ the skyrmion structure was transformed into the lattice of vortices. It is important to point out that the critical values of ΔJ for the skyrmion destruction in both cases is

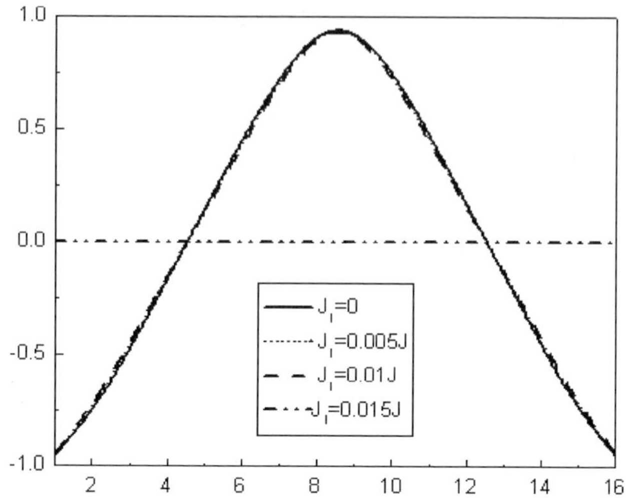

Fig. 4 The cross section of the stable spin configuration by the x–z plane for the layered cube with different exchange coupling J_1 between the x–y planes

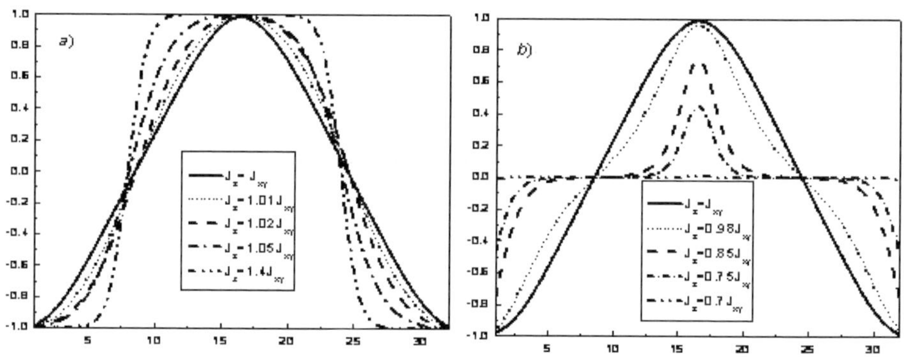

Fig. 5 The cross section of the stable spin configurations by the x–z plane for the single plane with the different anisotropy factors: **a** – the case $J_z > J_{xy}$, **b** – the case $J_z < J_{xy}$

much less in comparison with the actual exchange anisotropy in layered cuprates.

We have evaluated, also, the role of the exchange coupling of the next neighbors in the plane lattice of the spins. We have found that the skyrmion structure can be destroyed only, if the coupling constant has an "incorrect" sign and its absolute value reaches a half of the nearest neighbors coupling constant.

Skyrmions and Magnetic Properties of Cuprates

Our study of the skyrmion stability in the quasi-two-dimensional anisotropic magnets given in the previous section supports the skyrmion approach to describe magnetic properties and spin kinetics of weekly doped cuprates. It was mentioned already that the nuclear spin relaxation rate $1/T_1$ and the spin correlation length were calculated previously in underdoped cuprates taking into account both thermally excited skyrmions and skyrmions created by the doped quasi-localized holes [14, 15]. The key ingredient of these calculations was a quantum consideration of spin waves above the skyrmion background. The calculated temperature dependence of the nuclear spin relaxation rate $1/T_1$ and the spin correlation length are able to describe qualitatively their drastic transformation with doping observed experimentally. However, the predicted critical hole concentrations at which this transformation starts are more then two orders of magnitude less in comparison with the experimental nominal values. This apparent contradiction can be related to a phenomenon of the electronic phase separation. This phenomenon means that at low temperatures the hole distribution in the Cu_2O planes becomes very inhomogeneous, creating hole rich metallic and non-conducting antiferromagnetic regions. Properties of these systems become very peculiar showing different features of these regions. How these properties could be consistent with skyrmions and with experimental study by different methods?

In particular, well known NMR measurements of the nuclear spin relaxation rate $1/T_1$ in $La_{2-x}Sr_xCuO_4$ revealed a dramatic transformation of its temperature dependence on doping [19]. While this rate diverges ($1/T_1 \to \infty$) in pure parent cuprates as the temperature is approaching zero, it goes to zero ($1/T_1 \to 0$) in doped cuprates with $x \geq 0.04$. In the frame of the skyrmion approach described above the beginning of this transformation starts at the concentration of quasilocalized holes $n > 10^{-4}$. This discrepancy can be naturally explained by the electronic phase separation: the most of the holes are collected in the metal regions, which can not contribute to the NMR signal having a too small volume at this level of the nominal doping. We suppose that the observed NMR signal comes actually from the non-conducting antiferromagnetic regions containing a small fraction of holes, which are quasilocalized. A similar situation arises with measurements by neutron scattering of the spin correlation length in the CuO_2 plane. It is well established that in the parent cuprates the spin correlation length diverges when the temperature approaches zero, while it becomes finite in doped cuprate. In particular, in the sample $La_{2-x}Sr_xCuO_4$ with $x = 0.02$ the relative spin correlation length was found to be $\xi/a \approx 40$ [20]. According to the skyrmion approach this value should correspond to the concentration of the quasilocalized holes roughly $n \approx 6 \times 10^{-4}$, which is again much less than the nominal value.

At the same time the metallic regions were revealed in other experiments although they have a very small volume in the underdoped cuprates. In particular, the EPR method allows to measure, due to its high sensitivity in comparison with the NMR, the temperature dependence of the volume of metallic regions [16] in good agreement with the conductivity measurements [21].

Conclusion

We have investigated the stability of the spin topological excitations of the skyrmion structure in the layered crystals with the anisotropic Heisenberg interactions. Our numerical calculations show that skyrmions should survive in the layered underdoped cuprates and their parent compounds. A short description is given how the skyrmion approach gives an opportunity to describe the main features of the transformation of magnetic properties and spin kinetics of cuprates with doping. Starting with the assumption that the quasilocalized holes doped into 2D QHAF induce a formation of skyrmions and using the picture of both thermally excited and hole-created skyrmions the spin correlation length and the nuclear spin relaxation rate were calculated in a qualitative agreement with experiments. At the same time we would like to point out that the skyrmions are still metastable excitations and investigations of the detailed spin structure of underdoped cuprates should depend on the time scale of the observation.

Acknowledgements The authors are very grateful to Alex Müller and Hugo Keller for their interest in this study. B.I.K. acknowledges the support of the Russian Federal Program for the leading scientific schools via contract 02.445.11.7402 and the Swiss National Science Foundation via grant IB7420-110784.

References

1. Bednorz JG, Müller KA (1986) Z Phys B64:189
2. Chakravarty S, Halperin BI, Nelson DR (1989) Phys Rev B 39:2344
3. Tyc S, Halperin BI, Chakravarty S (1989) Phys Rev Lett 62:835
4. Chakravarty S, Orbach R (1990) Phys Rev Lett 64:224
5. Hasenfratz P, Niedermayer F (1991) Phys Lett B268:231
6. Hasenfratz P, Niedermayer F (1993) Z Phys B92:91
7. Chubukov AV, Sachdev S, Ye J (1994) Phys Rev B49:11919
8. Waldner F (1986) JMMM 873:54–57
9. Belavin AA, Polyakov AM (1975) Pis'ma ZhETF 22:503
10. Belov SI, Kochelaev BI (1997) Solid State Commun 103:249
11. Belov SI, Kochelaev BI (1998) Solid State Commun 106:207
12. Gooding RJ (1991) Phys Rev Lett 66:2266
13. Morinari T (22 Feb 2005) arXiv: cond-mat/0502437 v3
14. Belov SI, Ineev AD, Kochelaev BI (2005) Pis'ma ZhETF 81:478
15. Belov SI, Ineev AD, Kochelaev BI (2006) J Superconductivity and Novel Magnetism 19 N 19:1–2

16. Shengelaya A, Bruun M, Kochelaev BI, Safina AM, Conder K, Müller KA (2004) Phys Rev Lett 93:017001
17. Akhiezer A, Baryakhtar V, Peletminskii S (1958) ZhETF 33:474
18. Waldner F (2004) JMMM 981:272–276
19. Imai T, Slichter CP, Yoshimura K, Kosuge K (1993) Phys Rev Lett 70:1002
20. Keimer B, Belk N, Birgeneau RJ et al. (1992) Phys Rev B 49:14034
21. Ando Y et al. (2002) Phys Rev Lett 88:137005

Deng S, Simon A et al. (2007) In: Bussmann-Holder A, Keller H (eds)
High T_c Superconductors and Related Transition Metal Oxides.
Springer-Verlag Berlin Heidelberg, 201–211
DOI 10.1007/978-3-540-71023-3

Lone Pairs, Bipolarons and Superconductivity in Tellurium

S. Deng · A. Simon · J. Köhler (✉)

Max-Planck-Institut für Festkörperforschung, Heisenbergstr. 1, 70569 Stuttgart, Germany
j.koehler@fkf.mpg.de

Abstract We have characterized a new type of bipolaron in the ambient pressure modification of tellurium, Te-I, which originates from Coulomb interactions instead of electron-phonon coupling as for the conventional Anderson bipolaron. The studies at the Hartree-Fock level and the constrained LDA calculations give an estimate ($\sim$ 1 eV) for the stability of the bipolaron. A 3D-tight binding model has been proposed to explain the electronic structure obtained from first-principle calculations. The possible relevance of such bipolarons with superconductivity in the high pressure phase Te-II is discussed.

Keywords Bipolaron · First-principle · Flat steep band · Tellurium · Tight-binding

Introduction

The relevance of preformed pairs of electrons (holes) in superconductors is widely discussed [1–4]. Localized non-bonding electron pairs determine the elemental structures of the chalcogens S, Se, and Te, which are non-metals at ambient pressure and become metals and superconductors under high pressure, e.g. sulphur at 17 K [5]. The association of the non-bonding two-electron object with a bipolaron lies at hand.

The concept of polarons first proposed by Landau [6] proved to be important and useful in the development of condensed matter physics and material sciences, and even the discovery of the high T_c cuprates was led by the concept of Jahn–Teller polarons according to K.A. Müller [7, 8]. The possibility of a bipolaronic mechanism in superconductivity has been suggested early [9]. In the last few years, our search for fingerprints of superconductors left on their normal state properties [10–13] resulted in a flat/steep band scenario. The existence and origin of bipolarons in superconductors is significant for testing the flat/steep band model vs. bipolaron superconductivity. Here we report theoretical studies on tellurium and show the characteristics of bipolarons in ambient pressure Te-I and their possible for superconductivity relevance in high pressure Te-II.

Bipolarons in Te-I

Origin of Bipolarons

The basic electronic structure of chalcogens (S, Se, Te) is ns^2np^4, where the hybridization between the ns and np state is rather small due to the large s–p gap [14]. Because of this the ns state was treated as core state in the earlier vector charge density wave (VCDW) model [15, 16], though it is a too rough approximation as indicated by our recent work [14]. Interestingly the np^4 valence electron configuration is triply degenerate in its atomic state $(p_x^2p_yp_z)$, $(p_xp_y^2p_z)$, $(p_xp_yp_z^2)$. Here the question arises, to what extent this lone-pair configuration or on-site bipolaron is stable in the atomic and solid state, respectively, and, furthermore, what the origin and character of this bipolaron is?

As first suggested by Fukutome [15, 16], here the bipolaron results from the Coulomb interaction (U) among the four p electrons, so it is of a different origin than the negative U center of the Anderson bipolaron in amorphous semiconductors [17]. The origin of this kind of bipolaron can be derived from

$$H\,|\Psi\rangle = E\,|\Psi\rangle\ . \tag{1}$$

At the Hartree–Fock level, one can choose $|\Psi\rangle$ as a Slater determinant for a bipolaron configuration as follows,

$$|\Psi_x\rangle = \left|p_x\alpha p_x\beta p_y\alpha p_z\alpha\right\rangle\ , \tag{2}$$

where α and β indicate spin up and down, respectively. The total energy of this bipolaron configuration is calculated as

$$E = 4h + 6U - 3J\ , \tag{3}$$

where h is the single electron energy, and U and J are the Coulomb and exchange integrals, respectively. In deriving Eq. 3, we have assumed that all of the six Coulomb integrals for the p-states take the same quantity U as well as J for the three exchange integrals. For a bipolaron direction along y or z direction, $|\Psi\rangle = \left|\Psi_y\right\rangle$ or $|\Psi_z\rangle$, one obtains the same result as in Eq. 3.

Now, we assume that there is no such a bipolaron configuration, namely the 4 electrons homogeneously distribute in the 6 spin polarized orbitals as follows,

$$|\Psi_n\rangle = \left|p_x\alpha p_x\beta p_y\alpha p_y\beta p_z\alpha p_z\beta\right\rangle \tag{4}$$

with the occupancy of 4/6 for each spin orbital like $p_x\alpha$ etc. The energy for this configuration is calculated as

$$E = 4h + 20/3U - 8/3J\ . \tag{5}$$

It is obvious that the energy of state $|\Psi_n\rangle$ is higher than that of $|\Psi_x\rangle$ by a value of $2/3U + 1/3J$, because $U \geq J > 0$ [18]. Thus the bipolaron configuration is energetically favorable because of the electron-electron interactions. Beyond the Hartree–Fock approximation using one more determinant, e.g. the following wave function can be considered

$$|\Psi\rangle = 1/\sqrt{3}\left(|\Psi_x\rangle + \left|\Psi_y\right\rangle + |\Psi_z\rangle\right) , \tag{6}$$

and the total energy is then

$$E = 4h + 6U - 3J - C , \tag{7}$$

where C is the correlation integral calculated as

$$C = \left\langle \phi_{p_i}\phi_{p_j} \left| \phi_{p_k}\phi_{p_j} \right.\right\rangle . \tag{8}$$

The subscripts i, j, k in Eq. 8 can be any one of x, y, z, and in Eq. 7 all involved correlation integrals are denoted as C. As the Hartree–Fock energy is an upper bound for the exact energy, C must be positive. So, the inclusion of the configuration interaction (CI) further decreases the energy of the bipolaron state. As demonstrated above, the origin of the bipolaron in electronic configurations of the chalcogens is attributed to the electron-electron repulsion instead of electron–phonon interactions as considered by Anderson [17].

Coulomb Interaction Parameters

It is important to calculate the above parameters, particularly the largest one, U, for testing the stability of the bipolarons in chalcogens. There have been many attempts to calculate the Coulomb parameter U from first-principle methods in the field of high T_c cuprates [19–21] for d electrons. The main postulate is actually the same as we have discussed above, namely to start from a mean-field like total energy

$$E\left(n_\alpha, n_\beta\right) = \frac{n(n-1)}{2}U - \frac{n_\alpha\left(n_\alpha - 1\right)}{2}J - \frac{n_\beta\left(n_\beta - 1\right)}{2}J , \tag{9}$$

where we omitted the single electron energy term, and n_α, n_β indicate the numbers of spin-up and spin-down electrons in the p orbital, respectively. From the definition of the eigenvalue of an orbital of interest [22, 23],

$$\varepsilon_i \equiv \frac{\partial E}{\partial n_i} . \tag{10}$$

One can easily get the following expressions to calculate U and J,

$$\varepsilon_{l\alpha}\left(\frac{n}{2} + \frac{1}{2}, \frac{n}{2}\right) - \varepsilon_{l\alpha}\left(\frac{n}{2} + \frac{1}{2}, \frac{n}{2} - 1\right) = U \tag{11}$$

$$\varepsilon_{l\alpha}\left(\frac{n}{2} - \frac{1}{2}, \frac{n}{2} + \frac{1}{2}\right) - \varepsilon_{l\beta}\left(\frac{n}{2} - \frac{1}{2}, \frac{n}{2} + \frac{1}{2}\right) = J . \tag{12}$$

In Eqs. 11 and 12, the orbital energy $\varepsilon_{l\alpha}$ is calculated with the given number of electrons in α and β spin polarized orbitals, respectively. For the calculation of a solid the orbital energy should be corrected with respect to the corresponding Fermi energy, if one uses a finite size supercell as in our case [20]. Furthermore it should be noted that there are different definitions [24] for the Coulomb interaction U, and for its calculation [19–21].

Calculation Details for the Parameter *U*

For atomic calculations, we have used a full relativistic approach [25] based on the local-density approximation (LDA) formulated by Janak et al. [26]. For Te-I, the experimental crystallographic data reported in reference [27] were used throughout this work. One needs to imbed an impurity Te atom in the Te-I solid where the occupancy of the orbital of interest in the impurity atom must be fixed during the self-consistent calculations. In this work, a supercell which doubles the original unit cell along the c axis was chosen, because the nearest-neighbor distances along the a and b axes are already quite large. This supercell includes the second- and third-nearest neighbors with respect to the impurity Te atom. All Te atoms have the same initial electronic configuration ($5s^2 5p^4$) except for the impurity Te atom, for which the occupancy of the $5p$ orbital is assigned according to Eqs. 11–12. The total number of electrons in the solid remains unchanged. In the self-consistent calculations, the $5p$ orbital of the impurity atom is isolated with respect to the system to avoid the change of its occupancy. The calculations were performed by using the FP-LMTO method [28] on the basis of LDA [26].

The calculated bare parameter U_0 for the Te atom from Eq. 11 is 8.40 eV, which is close to the value of 8.31 eV obtained from the following expression as given in [21] for O

$$U_0 = E\left(p^4 s^0\right) + \left(p^2 s^0\right) - 2E\left(p^3 s^0\right) . \tag{13}$$

Both values agree quite well with the value of 8.59 eV determined semiempirically from experimental atomic spectra [29]. The screened U as calculated from Eq. 11 is 2.75 eV and 1.27 eV for the Te atom and solid Te-I, respectively. This result indicates that the Coulomb interaction between the electrons is significantly screened by other electrons both in the atom and the solid, though in the latter case the screening is more effective. It should be noted that the calculated value of U varies to some extent with the choice of configurations used in Eq. 11. In a non-spin-polarized calculation the occupancy of an orbital of interest can be different to that in spin-polarized calculations. For example, the non-spin-polarized calculations result in a value of 1.10 eV for solid Te-I. In contrast to U, the value of J is much less sensitive to the used configurations. We have taken into account various similar configurations as for the bare and screened U to obtain $J = 1.0 \pm 0.04$ eV for the Te atom. The calculated value of J for solid Te-I is 0.7 eV. As discussed above, the energy gain for the bipolaron configuration is about $2/3U + 1/3J \sim 1.0$ eV, yielding an estimate for the stability of the bipolaron in Te-I. This will be further confirmed by the band structure calculations.

3D-TB Models and Band Structure of Te-I

3D-TB Models for Te-I

From the atomic energies of the 5*s* and 5*p* orbitals, it can be expected that the inner 5*s* orbital will form a bipolaron band when the Te atoms are condensed to form a solid. The two unpaired electrons in the 5*p* shell can form two bonding and antibonding bands. The *p*-type bipolaron band is located in the valence and conduction band gap because of its non-bonding character. This argument is true if there is no *s*–*p* hybridization. However, for the requirement of maximizing the bonding of the two unpaired 5*p* electrons, *s*–*p* hybridization occurs as in much more complicated "hypervalent" compounds [30]. As a result, the bipolaron orbital does not keep perpendicular to the other two p orbitals in contrast to the VCDW model [15, 16] where complete *s*–*p* separation is assumed. Furthermore, to avoid the Coulomb repulsion between the bipolarons and the electron pairs in the bonds, a helical chain with a period of three as shown in Fig. 1 is energetically more favorable than a straight chain.

As a consequence, each of the bands discussed above becomes a triplet due to the helical modulation with indentity after 3 Te atoms. To come to a better understanding, we have derived the tight binding (TB) band structures by

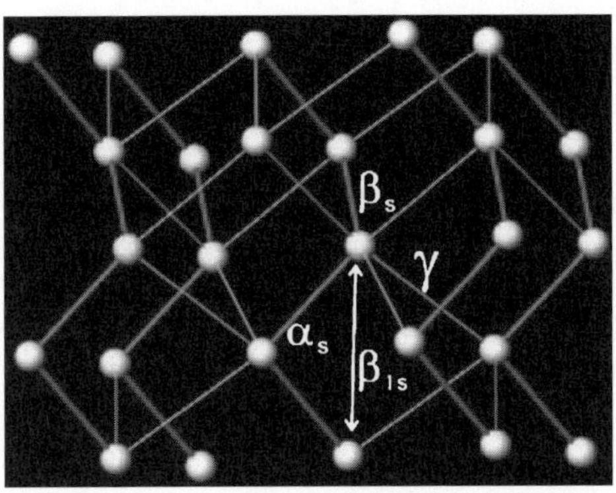

Fig. 1 Crystal structure of Te-I. The red sticks correspond to the short Te – Te distance (283.5 pm), and the thin yellow sticks to a Te – Te distance of 349.1 pm. The Greek letters α_s, β_s, β_{1s} and γ indicate the onsite energies of the *s* orbital, the intra-chain nearest neighbor, the next nearest neighbor hopping integral and the inter-chain hopping integrals, respectively

using the parameters indicated in Fig. 1. The analytic TB band dispersions for one single chain of Se and Te atoms have already been worked out by Joannopoulos et al. [31]. Here we derive a TB model which explicitly includes the inter-chain interactions, which is important in explaining many details of the band structure of Te-I. For the s orbital, we denote the onsite energy by α_s, the nearest neighbor hopping integral by β_s and the inter-chain hopping integral by γ_s. As shown in Fig. 1, each Te atom has four equidistant (349.1pm) inter-chain contacts. By solving the secular equation, one gets the s bands, and e.g. the bands along the Γ-A direction are obtained as follows,

$$\begin{aligned}
E_1 &= \alpha_s + 2\beta_s \cos(2/3\lambda\pi) + 4\gamma_s \cos(2/3\lambda\pi) \\
E_2 &= \alpha_s + 2\beta_s \cos 1/3(2\lambda\pi + 2\pi) + 4\gamma_s \cos 1/3(2\lambda\pi + 2\pi) \\
E_3 &= \alpha_s + 2\beta_s \cos 1/3(2\lambda\pi - 2\pi) + 4\gamma_s \cos 1/3(2\lambda\pi - 2\pi)\,,
\end{aligned} \tag{14}$$

where λ is a parameter ranging from 0 (Γ) to 1/2 (A). The bands along other symmetry directions can be obtained in a similar way. The inclusion of γ_s is important because otherwise two bands along L-H will be degenerate with a constant energy of $\alpha_s + \beta_s$, while the other band has a constant energy of $\alpha_s - 2\beta_s$. When γ_s is included the bands become dispersive and the eigenvalues at H and L points are

$$\begin{aligned}
E(H_1) &= E(H_2) = \alpha_s + \beta_s - \gamma_s \\
E(H_3) &= \alpha_s - 2\beta_s + 2\gamma_s \\
E(L_1) &= \alpha_s - 1/2\beta_s + \gamma_s - 1/2\sqrt{9\beta_s^2 - 4\gamma_s\beta_s + 4\gamma_s^2} \\
E(L_2) &= \alpha_s + \beta_s - 2\gamma_s \\
E(L_3) &= \alpha_s - 1/2\beta_s + \gamma_s + 1/2\sqrt{9\beta_s^2 - 4\gamma_s\beta_s + 4\gamma_s^2}\,.
\end{aligned} \tag{15}$$

For the p-type bonding and antibonding band we have used a different approach in which a bonding and antibonding combination of the basis orbital is included to construct the Hamiltonian matrix [31].

As shown in Fig. 2, the onsite inter-orbital (β) and the other 2nd nearest neighbor hopping parameters ($\beta_i, i = 1-3$) for one chain have been included to derive the Hamiltonian matrix. The parameter β_3 is found to be very small in comparison with other parameters, and therefore we could omit it in the calculation. For the inter-chain interactions we assume an isotropic parameter γ regardless of the different orientations of the respective s–p hybrids. The obtained bonding p-type bands along the Γ-A direction are then as follows,

$$\begin{aligned}
E_1 &= \alpha_p + \beta_p + 2\gamma_p + 2(t + 2\gamma_p)\cos 2/3\lambda\pi + 2\gamma_p \cos 4/3\lambda\pi \\
E_2 &= \alpha_p + \beta_p + 2\gamma_p + 2(t + 2\gamma_p)\cos 1/3(2\lambda\pi + 2\pi) + 2\gamma_p \cos 1/3(4\lambda\pi - 2\pi) \\
E_3 &= \alpha_p + \beta_p + 2\gamma_p + 2(t + 2\gamma_p)\cos 1/3(2\lambda\pi - 2\pi) + 2\gamma_p \cos 1/3(4\lambda\pi + 2\pi)
\end{aligned} \tag{16}$$

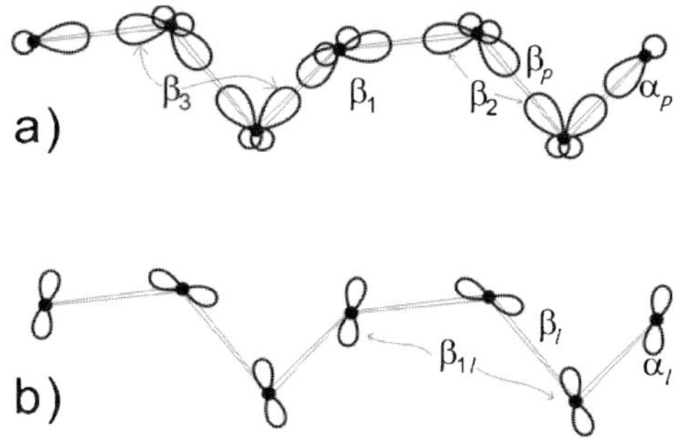

Fig. 2 **a** Schematic representations for the various energy integrals used to derive the 3D-TB model for *p*-type bonding and antibonding bands; **b** for *p*-type bipolaron bands

In Eq. 16 t is a combined parameter, $t = 1/2(\beta_1 + 2\beta_2)$, λ is a parameter ranging from 0 to 1/2. As can be seen from the following expressions, the dispersion relation along the L-H direction is simpler than the above ones,

$$\begin{aligned} E_1 &= \alpha_p + 2\beta_p - 2t \\ E_2 &= \alpha_p + \beta_p + t - 3\gamma_p \cos 2\lambda\pi \\ E_3 &= \alpha_p + \beta_p + t + 2\gamma_p \cos 4\lambda\pi - \gamma_p \cos 2\lambda\pi \,. \end{aligned} \tag{17}$$

It is obvious that without the inter-band interactions ($\gamma_p = 0$) the three bands will become completely flat with two degenerate bands along the L-H direction. The dispersion of the antibonding band can be derived from the antibonding combinations of the sp-hybrid as shown in Fig. 2. The result along the Γ-A direction is given by

$$\begin{aligned} E_1 &= \alpha_p - \beta_p - 2\gamma_p + 2(t_1 + 2\gamma_p) \cos 1/3(2\lambda\pi + 2\pi) - 2\gamma_p \cos 1/3(4\lambda\pi - 2\pi) \\ E_2 &= \alpha_p - \beta_p - 2\gamma_p + 2(t_1 + 2\gamma_p) \cos 1/3(2\lambda\pi - 2\pi) - 2\gamma_p \cos 1/3(4\lambda\pi + 2\pi) \\ E_3 &= \alpha_p - \beta_p - 2\gamma_p + 2(t_1 + 2\gamma_p) \cos 2/3\lambda\pi - 2\gamma_p \cos 4/3\lambda\pi \,, \end{aligned} \tag{18}$$

where $t_1 = 1/2(-\beta_1 + 2\beta_2)$. Due to the non-bonding character of the bipolarons only the nearest and next-nearest neighbor hopping integrals along the chain direction together with the inter-chain hopping integral are considered for deriving the bipolaron band. The results along the Γ-A direction are as follows,

$$\begin{aligned} E_1 &= \alpha_l + 2\beta_l \cos 1/3(2\pi\lambda + 2\pi) + 2\beta_{1l} \cos 1/3(4\pi\lambda - 2\pi) \\ &\quad + 4\gamma_l \cos 1/3(2\lambda\pi + 2\pi) \\ E_2 &= \alpha_l + 2\beta_l \cos 1/3(2\pi\lambda - 2\pi) + 2\beta_{1l} \cos 1/3(4\pi\lambda + 2\pi) \\ &\quad + 4\gamma_l \cos 1/3(2\lambda\pi - 2\pi) \\ E_3 &= \alpha_l + 2\beta_l \cos 2\pi\lambda/3 + 2\beta_{1l} \cos 4\pi\lambda/3 + 4\gamma_l \cos 2\lambda\pi/3 \,. \end{aligned} \tag{19}$$

The analytic expressions for the energy bands of Te-I in the entire first Brillouin zone can be derived within the methods discussed above, though some of them may be rather complicated. To test the 3D-TB model and to find the most important interactions, we have also calculated the band structure of Te-I by using both TB-LMTO [32] and FP-LMTO methods [28]. The computational details are the same as given in [14].

Band Structure of Te-I

The band structure calculated from first-principle methods and that derived from the 3D-TB model along the Γ-A direction are shown in Fig. 3. The good agreement between them indicates that the 3D-TB model contains the essentials of the electronic structure for Te-I. The 3D-TB model clearly reveals the *s*-, *p*-bonding, *p*-bipolaron and *p*-antibonding characters of the various triplets centered at – 12, – 3, – 2, and 1 eV, respectively. From the above Eqs. it is obvious, that the inter-chain interactions add extra dispersions to the one-dimensional model and break the degeneracy at some symmetry points in the 1D model. For example, an extra dispersion of $-2\gamma_s$ is added from Γ to A for the second *s*-band. For the *p*-bonding band, the dispersion along both, the Γ-M and L-H directions, are also a consequence of the inter-chain interactions. Through fitting only the bands along symmetry directions the obtained parameter values are – 13.07, – 0.76 and – 0.24 eV for α_s, β_s and γ_s, respectively, for the *s*-triplet. The corresponding values for the *p*-bands are – 2.47, – 1.83, 0.0, 0.62 and – 0.24 eV for α_p, β_p, β_1, β_2, and γ_p, respectively. For the *p*-type bipolaron band the values are – 3.1, 0.21, 0.0 and – 0.05 eV for α_l, β_l, β_{1l}, and γ_l, respectively. Though relatively small, the inter-band interaction parameter γ is significant even for the *s*-band, for which the first-principle bands are a little more dispersive than those derived from the 3D-TB model shown in Fig. 3. The rather small values of the β and γ parameters for the bipolaron bands show their non-bonding character. The anisotropy for the *p*-bands

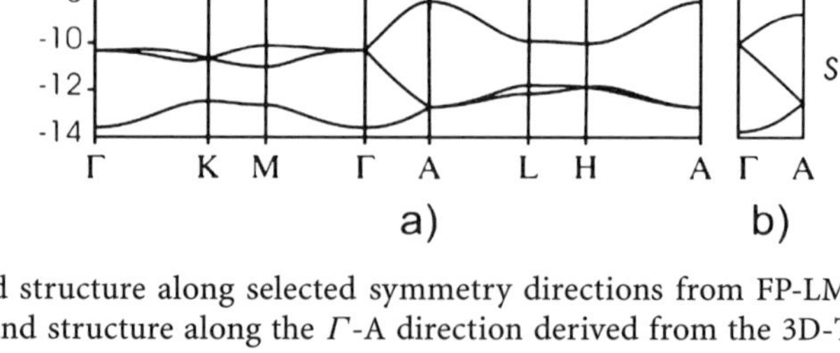

Fig. 3 **a** Band structure along selected symmetry directions from FP-LMTO calculations for Te-I; **b** band structure along the Γ-A direction derived from the 3D-TB model

requires s–p hybridization enhancing the bond strength along the chain direction. On the other hand, the s character favors the inter-chain interactions due to its symmetry. Our calculations indicate that without s–p hybridization, the band structure is dramatically changed, and Te-I develops metallic character. As it is obvious from Fig. 3, the 3D-TB model is not able to reproduce every detail of the bands obtained from first-principle calculations, particularly for the antibonding p-type bands. To improve this model, more hopping integrals need to be included. Particularly the anisotropic inter-chain hopping integrals need to be considered for the p-type bands.

Bipolarons and Superconductivity

In addition to the above studies, Te-I has been demonstrated to be a bipolaron crystal [14] by calculating the electron localization function [33–36]. With the characterization of these bipolarons, which are just below the Fermi level, the question arises, whether superconductivity occurs in such a system, and what will be the role of the bipolarons. The first possibility is that the bipolarons remain stable and get the hopping mobility due to hole doping in the bipolaron bands. Then superconductivity is the consequence of Bose–Einstein condensation of these bipolarons. The main difficulty of this approach is the stability of the bipolarons after holes being doped in the system. For Te-I, as we discussed above the bipolarons are stabilized through Coulomb interactions with an energy gain of ~ 1 eV. This is just the estimated upper bound as it is well known that the Hartree-Fock approximation often overestimates the energy gap. Our first-principle calculations reveal a gap of ~ 0.34 eV between the bipolaron band and the antibonding band. This value agrees very well with the experimental result of ~ 0.34 eV [37].

The second possibility is described by the two-component model [38] where a bipolaron can change directly into a Cooper pair state as shown in Fig. 4. In this model the bipolarons remain relatively stable as preformed pairs even above the superconducting transition temperature (T_c), which depends on the relative concentrations of the bipolarons and the wide band electrons.

The third possibility follows the flat/steep band model, where the bipolarons are part of the flat band electrons in the low energy electronic structure. In contrast to the former two possibilities the flat band electrons obey Fermi–Dirac statistics instead of Bose–Einstein statistics, and the bipolaron is not stable against decomposition into two steep band electrons. In Te-I the inter-chain interactions γ between bipolarons have bonding character. Hence, one can expect a breakage of the bipolarons upon a compression of the inter-chain distances. This occurs with the transition into Te-II under pressure, when the Te chains condense into puckered layers [39].

Our earlier studies [14] indicate that the bipolaron bands indeed overlap with the antibonding p-type band in Te-II, which thus creates a flat/steep

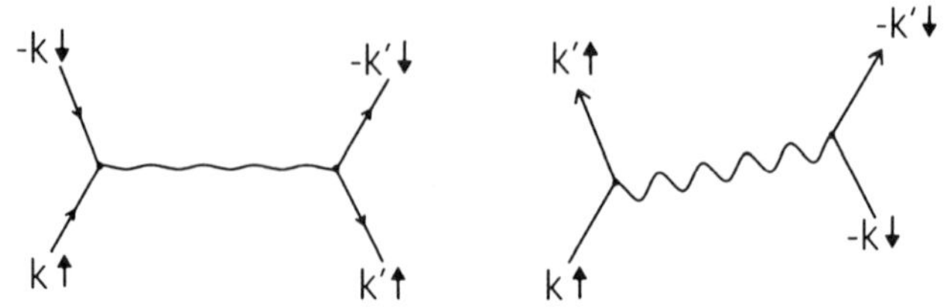

Fig. 4 **a** Schematic representation for a process of a bipolaron decaying into a Cooper pair; **b** Two electrons coupled into a Cooper pair by retarded electron-phonon interactions

band feature for the normal state electronic structure. Te-II is a superconductor with a $T_c \sim 5$ K [40]. Further theoretical studies on the electron-phonon coupling will be needed to clarify the role of bipolarons identified in Te-I as also leading to superconductivity in Te-II. We suggest an experiment to determine the dependence of the correlation length on pressure. If the bipolarons in Te-I play the key role for superconductivity of Te-II according to the bipolaron model or two-component model, the correlation length will increase with pressure in contrast to the prediction within the flat/steep band model.

Conclusions

In summary, we have quantitatively characterized a p-type bipolaron in Te-I through calculating the total energy of the various configurations at the HF and HF + CI level. The constrained LDA calculations result in an estimated upper bound ($\sim$ 1 eV) for the stability of such bipolarons. Our analysis reveals that the bipolaron originates from Coulomb interactions in contrast to the conventional Anderson bipolaron, which is a result of electron-phonon coupling. The 3D-TB model gives a simple and clear explanation of the bands obtained from first-principle calculations, which agrees with the experimental results. The possible role of bipolarons in leading to superconductivity in Te-II together with a suggestion for the verification via calculations and experiment is given.

References

1. Micnas R, Ranninger J, Robaszkiewicz S (1990) Rev Mod Phys 62:113
2. Friedberg R, Lee TD (1989) Phys Rev B 40:6745
3. Gor'kov LP, Sokol AV (1987) JETP Lett 46:421
4. Ioffe LB, Millis AJ (1996) Phys Rev B 54:3645
5. Struzhkin VV, Hemley RJ, Mao H-K, Timofeev YA (1997) Nature 390:382
6. Landau L (1933) Phys Z Sowjetunion 3:664
7. Bednorz JG, Müller KA (1986) Z Phys B 64:189
8. Müller KA (2000) J Supercond 13:863

9. Alexandrov AS (1981) J Ranninger Phys Rev B 24:1164
10. Simon A (1997) Angew Chem Int Ed Engl 36:1788
11. Deng S, Simon A, Köhler J (1998) Angew Chem Int Ed Engl 37:640
12. Deng S, Simon A, Köhler J (2002) J Amer Chem Soc 124:10712
13. Deng S, Simon A, Köhler J (2005) Struct Bonding 114:103
14. Deng S, Simon A, Köhler J (2006) Angew Chem Int Ed Engl 45:599
15. Fukutome H (1984) Prog Theor Phys 71:1
16. Oka T, Fukutome H (1988) Prog Theor Phys 79:608
17. Anderson PW (1975) Phys Rev Lett 34:953
18. Roothaan CCJ (1951) Rev Mod Phys 23:69
19. Gunnarsson O, Andersen OK, Jepsen O, Zaanen J (1989) Phys Rev B 39:1708;
20. Anisimov VI, Gunnarsson O (1991) Phys Rev B 43:7570;
21. McMahan AK, Martin RM, Satpathy S (1988) Phys Rev B 38:6650
22. Slater JC (1974) Quantum Theory of Molecules and Solids. McGraw-Hill, New York, Vol. 4
23. Janak JF (1978) Phys Rev B 18:7165
24. Herring C (1966) in Magnetism, edited by Rado GT and Suhl H. Academic, New York, Vol. IV
25. Liberman DA, Cromer DT, Waber JT (1971) Comp Phys Comm 2:107
26. Janak JF, Moruzzi VL, Williams AR (1975) Phys Rev B 12:1257
27. Adenis C, Langer V, Lindqvist O (1989) Acta Cryst C 45:941
28. Savrasov SY (1996) Phys Rev B 54:16470
29. Fraga S, Karwowski J, Saxena KMS (1979) Atomic energy levels, Elsevier, New York, Vol. 4
30. Papoian GA, Hoffmann R (2000) Angew Chem Int Ed 39:2408
31. Joannopulos JD, Schlüter M, Cohen ML (1975) Phys Rev B 11:2186
32. Tank R, Jepsen O, Burkhardt A, Andersen OK, TB-LMTO-ASA (version 4.7) 1998, MPI für Festkörperforschung, Stuttgart, Germany
33. Becke AD, Edgecombe KE (1990) J Chem Phys 92:5397
34. Savin A, Jepsen O, Andersen OK, Preuss H, von Schnering HG (1992) Angew Chem 31:187
35. Silvi B, Savin A (1994) Nature 371:683
36. Burdett JK, McCormick TA (1998) J Phys Chem A 102:6366
37. Guillaume CB, Debever JM (1965) Solid State Comm 3:19
38. Micnas R, Robaszkiewwicz S, Bussmann-Holder A (2005) Struct Bonding 114:13
39. Hejny C, McMahon MI (2004) Phys Rev B 70:184109
40. Gregoryanz E, Struzhkin V, Hemley RJ, Eremets MI, Mao H-K, Timofeev YA (2002) Phys Rev B 65:064504

Kremer RK, Kim JS et al. (2007) In: Bussmann-Holder A, Keller H (eds)
High T_c Superconductors and Related Transition Metal Oxides.
Springer-Verlag Berlin Heidelberg, 213–226
DOI 10.1007/978-3-540-71023-3

Carbon Based Superconductors

R. K. Kremer (✉) · J. S. Kim · A. Simon

Max-Planck-Institut für Festkörperforschung, Heisenbergstr. 1, 70569 Stuttgart, Germany
rekre@fkf.mpg.de

Abstract We review the characteristics of some carbon based novel superconductors which emerged in the past two decades since the discovery of superconductivity in the high-T_c oxocuprates. In particular, we summarize the properties of ternary layered carbide halides of the rare rarth metals with composition $RE_2C_2X_2$ (RE = Y, La; X = Cl, Br, I) and of the rare earth di- and sesquicarbides, YC_2, LaC_2 and La_2C_3. Finally, we briefly discuss the properties of the recently discovered Ca and Yb intercalated graphite superconductors, CaC_6 and YbC_6.

Introduction

The discovery of high-T_c superconductivity by Bednorz and Müller [1] in 1986 marks the beginning of a period of a vivid search for – chemically and physically partly extremely complex – new oxocuprates and for theoretical approaches to understand their puzzling properties, quite a few of which remained controversial even until today. The advent of this completely unexpected class of new superconductors also revived the interest in more conventional – "low-T_c" – superconductors. In due course, a number of new systems were found, and already known superconductors were reinvestigated with improved and refined experimental and theoretical tools. These activities led to surprising new discoveries as that of the 40 K superconductor MgB_2 by Nagamatsu et al. [2] Apart from its T_c, MgB_2 is special primarily for two reasons: Compared e.g. to the high-T_c oxocuprates its crystal structure is of remarkable simplicity allowing electronic and phononic structure calculations of high precision, and MgB_2 is the first system for which *multigap* superconductivity has independently been evidenced by several experimental techniques [3–5].

Until the discovery of MgB_2, doped fullerenes had shown the highest T_c values after the high-T_c oxocuprates. With large enough quantities of purified C_{60} available, [6] Hebard et al. prepared superconductors with a T_c of 18 K by doping polycrystalline C_{60} and C_{60} films with alkali metals [7]. Subsequently, by adjusting the separation of the C_{60} molecules using a proper composition of different alkali metals, T_c's up to $\sim$ 33 K were reached [8].

Superconductivity in doped fullerenes also redraw attention to carbon based superconductors in general. Especially, binary and quasibinary transition metal carbides have a long history in showing T_c's which were among the highest found before the discovery of the high-T_c oxocuprates [9]. Later borocarbides of composition REM_2B_2C, with RE = Y or Lu and M = Ni or Pd, with T_c's up to 22 K attracted considerable interest [10–12].

Superconductivity in graphite intercalation compounds (GICs) is another early field of research which recently was revived. The discovery of superconducting GICs dates back to the pioneering work of Bernd Matthias' group in the 1960s, however, the T_c's of these early GICs remained well below 1 K [13, 14]. Subsequently, the T_c's of alkali metal intercalated GICs could be raised by intercalation under pressure with e.g. Li and Na, but T_c did not significantly exceed the boiling point of liquid helium [15, 16]. It was not until recently that T_c of the GICs could be significantly enhanced by intercalating divalent alkaline earth metals like Ca and Yb [17].

Finally, after graphite and C_{60}, diamond was also converted into a superconductor by hole doping induced by a substitution of about 3% B into C sites. Ekimov et al. showed that such a boron-doped diamond is a bulk, type-II superconductor below $T_c \sim 4$ K with superconductivity surviving in a magnetic field up to $H_{c2}(0) \geq 3.5$ T [18].

In our search for complex metal-rich rare earth halides we found a series of new superconducting layered carbide halides of the rare earth metals with T_c's up to ~ 10 K [19–21]. For a deeper understanding of the chemistry and physics of these we in turn reinvestigated also the properties of binary dicarbides and sesquicarbides of composition REC_2 and RE_2C_3, with R = Y, La. Superconductivity in binary carbides of rare earth metals had been an intensively investigated topic in the sixties and seventies of the last century. In this family of compounds T_c values peaked with $(Y_{0.7}Th_{0.3})_2C_3$ at 17 K [22].

Superconductivity in rare earth metal sesquicarbides recently regained considerable attention after the reports by Amano et al. and Nakane et al. about the successful synthesis of binary Y_2C_3 under high pressure conditions (~ 5 GPa) [23, 24]. The reported T_c's reached 18 K and the upper critical field exceeded 30 T. In the following we will summarize some of the characteristic properties of the ternary layered rare earth metal carbide halides and the binary di- and sesquicarbides. We conclude with some remarks on our results on the recently discovered alkali earth GICs.

Superconductivity in Rare Earth Carbide Halides and Rare Earth Carbides

Ternary Layered Cabide Halides of the Rare Earth Metals

The carbide halides of the rare earth metals, $RE_2C_2X_2$ (X = Cl, Br, I and RE being a rare-earth metal) crystallize with layered structures which contain

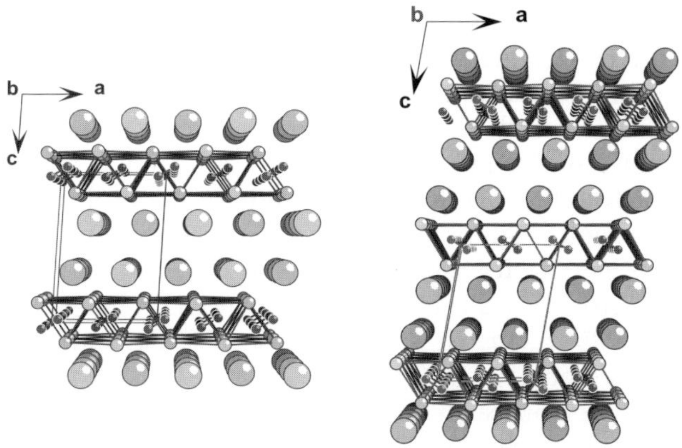

Fig. 1 (*Left*) Crystal structure of $Y_2C_2I_2$ (1*s* stacking variant) and (*right*) crystal structure of $Y_2C_2Br_2$ (3*s* stacking variant) projected along [010] with the unit cells outlined. C, Y, and (I,Br) atoms are displayed with increasing size

double layers of close-packed metal atoms sandwiched by layers of halogen atoms to form X – RE – C_2 – RE – X slabs as elementary building blocks. These connect via van der Waals forces in stacks along the crystallographic *c*-axis. Different stacking sequences (1*s* and 3*s* stacking variants) have been found. The carbon atoms form C – C dumbbells which occupy the octahedral voids in the close-packed metal atom doublelayers (cf. Fig. 1) [25, 26].

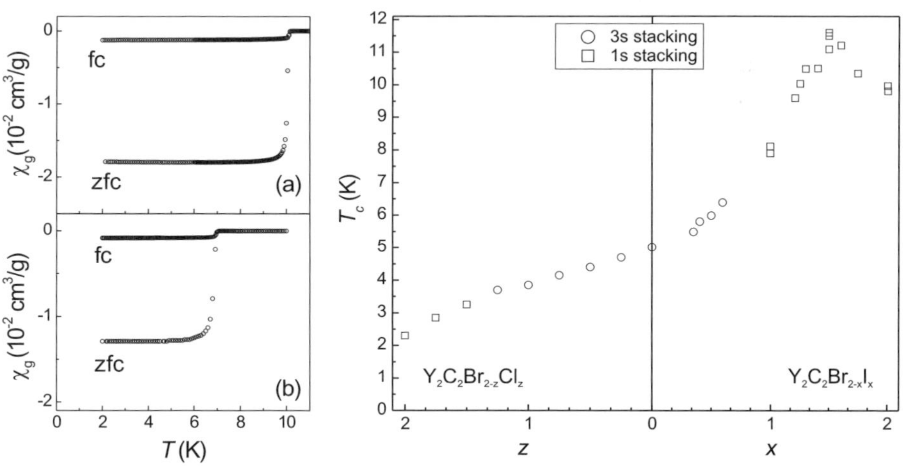

Fig. 2 (*Left*) field-cooled (fc) and zero field-cooled (zfc) magnetic susceptibilities of **a** $Y_2C_2I_2$ (after [28]) and **b** $La_2C_2Br_2$ (after [30]). (*Right*) T_c's of a series of quasiternary mixtures of $Y_2C_2Br_{2-x}I_x$ and $Y_2C_2Br_{2-z}Cl_z$. Different stacking variants of the compounds are indicated by different symbols (after [20])

Compounds containing the nonmagnetic rare-earth metals Y and La are superconductors (Fig. 2). The maximum T_c of 11.6 K which was achieved by adjusting the composition in the quasi-ternary phases $Y_2C_2(X,X')_2$ [20]. The variation of $T_c(x)$ across the transition of the 3s and the 1s stacking variant indicating that superconductivity is essentially a property of the configuration of an individual $X-RE-C_2-RE-X$ slab rather than of the stacking details in the crystal structure. The transition temperatures of all known superconducting phases $RE_2C_2X_2$ are compiled in Table 1.

The heat capacity of $Y_2C_2I_2$ shows a sharp anomaly, however with a jump height $\Delta C_P(T_c)/\gamma T_c \approx 2$ which is considerably larger than the value 1.43 expected from weak coupling BCS theory [21, 28]. A fit of the heat capacity anomaly with the empirical α-model [31] indicates strong coupling with $2\Delta(0)/k_B T_c \approx 4.2$, the superconducting gap being enhanced by about 20% over the BCS value, similar to the σ-gap in MgB_2 [32]. There is, however, no indication from the temperature dependence of the heat capacity anomaly for a multiple gap scenario. Using approximate equations for strong coupling superconductors which relate $2\Delta(0)/k_B T_c$ and $\Delta C_P(T_c)/\gamma T_c$ to the logarithmic average over the phonon frequencies ω_{ln} [33, 34] one estimates the typical phonon frequency range for $Y_2C_2I_2$ to be $\sim 80-100\,cm^{-1}$. In this range A_g modes have been discerned by Raman spectroscopy in which Y and halogen atoms vibrate in-phase parallel and perpendicular to the layers [35, 37].

C stretching and tilting vibrations have considerably higher energies, and their role for electron-phonon coupling, particularly in the case of the tilting modes, could be important [38]. The electronic structure in close neighborhood to the Fermi energy, E_F, is characterized by bands of low dispersion which are reminiscent of the quasimolecular character of the HOMO and LUMO orbitals of an isolated $C-C$ dumbbell [37, 39]. These together with highly dispersive bands establish a flat/steep band scenario which in our view is a prerequisite of superconductivity in a more general sense [36].

The low-dispersive bands give rise to two peaks in the electronic density of states, DOS, each about 100 meV above and below the Fermi energy which

Table 1 Transition temperatures and upper critical fields, $\mu_0 H_{c2}$, of the known superconducting phases $RE_2C_2X_2$ (RE = Y, La; X = Cl, Br, I)

Compound	T_c (K)	$\mu_0 H_{c2}$ (T)	Refs.
$Y_2C_2Cl_2$	2.3	–	[20]
$Y_2C_2Br_2$	5.04	3	[20, 21, 27]
$Y_2C_2I_2$	10.04	12	[20, 27–29]
$Y_2C_2Br_{0.5}I_{1.5}$	11.6	–	[20]
$La_2C_2Br_2$	7.03	–	[30]
$La_2C_2I_2$	1.72	–	[30]

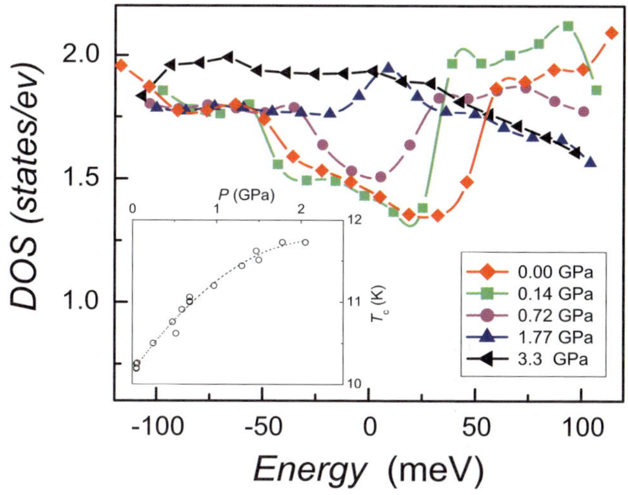

Fig. 3 Electronic density of states, DOS, of $Y_2C_2I_2$ in the close vicinity to E_F. The *inset* shows the pressure dependence of T_c of $Y_2C_2I_2$. The *dotted line* is a guide to the eye (after [41, 42])

enclose a "pseudogap" at E_F [37, 39]. Deviations from the linear temperature dependence of the Korringa relaxation of ^{13}C nuclei probed by ^{13}C NMR are a clear manifestation for the proposed structure in the DOS close to E_F [40].

The electronic structure and the dispersion of the bands in the vicinity of E_F is very sensitive to slight structural variations and can be very effectively tuned e.g. by hydrostatic pressure to increase the DOS and maximize T_c [41]. When hydrostatic pressure is applied to $Y_2C_2I_2$ T_c increases, and a maximum of about 11.7 K is reached at 2 GPa, similar to the maximum T_c found in the quasi-ternary mixtures [29, 41, 42]. The increase of T_c with pressure in $Y_2C_2I_2$ and also $La_2C_2Br_2$ is remarkable and parallels the findings observed for the Hg based oxocuprates but also for *fcc*-La for which similar values for the relative increase $1/T_c \cdot dT_c/dP$, have been detected [43–45].

Binary Dicarbides and Sesquicarbides of the Rare Earth Metals

YC_2 crystallizes with the body centered tetragonal CaC_2 structure type (Fig. 4) with C – C dumbbells centering Y metal atom octahedra which are slightly elongated along [001] [46]. YC_2 had been found to be a superconductor with a $T_c \sim 3.88$ K [47]. Proper heat treatment of stoichiometric YC_2 samples results in superconductors with a sharp transition and onset T_c's up to 4.02(5) K, somewhat increased over those previously reported [48, 49]. LaC_2 shows a T_c of about 1.6 K [47].

Heat capacity measurements (Fig. 5) nicely reveal the anomaly at the transition to superconductivity which follows closely the BCS weak-coupling

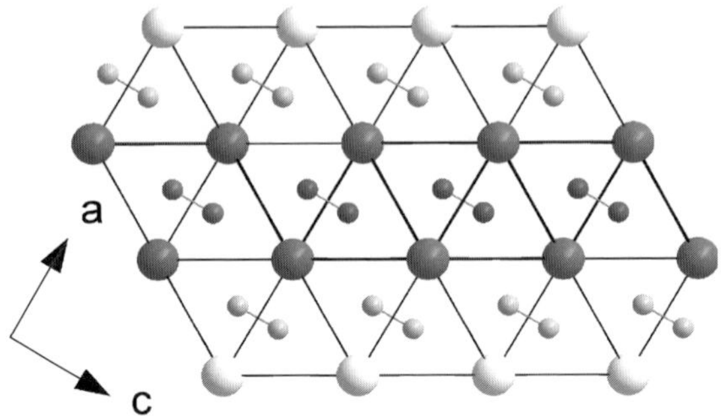

Fig. 4 Crystal structure of YC_2 along [010]. Y and C atoms are drawn with decreasing size. An $Y-C_2-Y$ doublelayer as found in the ternary carbide halides of the rare earths metals, $RE_2C_2X_2$ (RE = Y, La; X = Cl, Br, I), is highlighted in *dark grey*

predictions but already indicate significantly decreased critical fields as compared to those of the layered carbide halides [27].

Electronic structure calculations for YC_2 reveal strongly dispersive bands in planes perpendicular to the *c*-direction originating from Y $d_{x^2-y^2}$ orbitals and also strongly dispersive bands in the *c*-direction emerging from combinations of Y d_{xz}, d_{yz}, and C p_x, p_y orbitals [48]. As a consequence the electronic density of states close to the Fermi level is to a large extent featureless with a slight positive slope. Doping with Th or Ca (10% to 20%) decreases T_c [48].

As compared to the layered yttrium carbide halides, the critical fields of YC_2 (< 0.1 T, cf. Fig. 6 and Table 1) are reduced by up to two orders of mag-

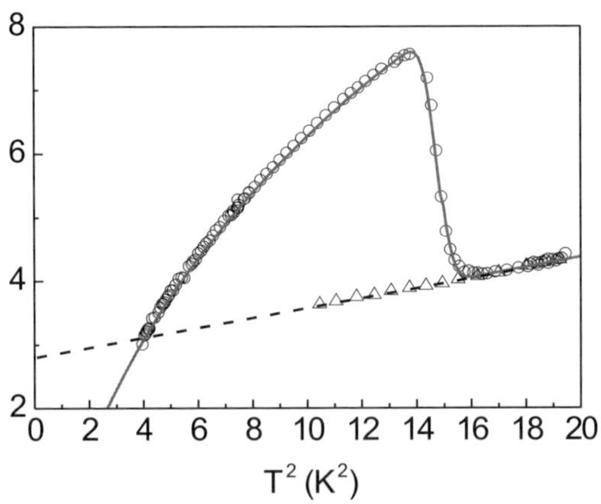

Fig. 5 Superconducting anomaly in the heat capacity of YC_2 ($B_{ext} = 0$, ○). The normal state △ has been reached by applying an exernal field of 0.4 T (after [48]). The (*red*) *solid line* represents a fit to the predictions of the BCS theory with a slight smearing of T_c being included

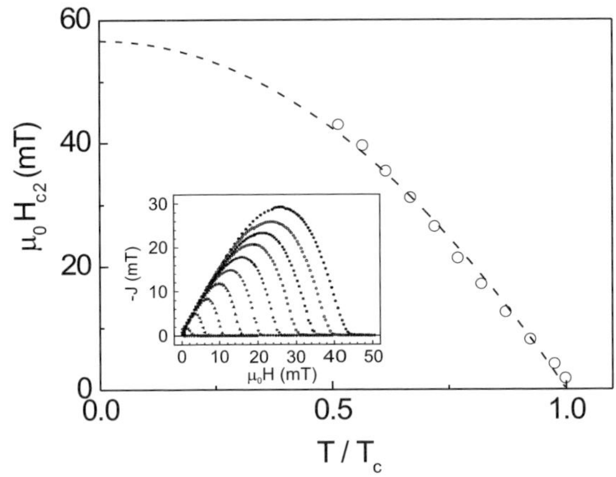

Fig. 6 Upper critical field, $\mu_0 H_{c2}$, determined from the isothermal magnetization measured of a spherical sample of YC_2. The *inset* displays the isothermal magnetizations measured at constant temperatures of 2 K, 2.2 K, ..., 3.8 K, 4 K, in increasing order (after [27])

nitude. The significant difference in the upper critical fields between the layered carbide halides and the dicarbides as well as the marked increase in the anisotropy of the coherence lengths ($\xi_{\parallel}/\xi_{\perp} \approx 5$ [27]) supports Ginzburg's suggestion that from the point of view of possibilities to enhance T_c promising materials are layered materials and dielectric-metal-dielectric sandwich structures [50–52]. In fact, by comparing the crystal structures of the dicarbides and the carbide halides of the rare earths (Figs. 1 and 4) one realizes that the $R-C_2-R$ doublelayers carrying the superconductivity in the ternary carbide halides can be considered as sections of the three dimensional structure of the dicarbides which are sandwiched by dielectric halogen layers. In this respect, the dicarbides and the carbide halides of the rare earth metals are interesting examples to test Ginzburg's conjecture.

La_2C_3 (like Y_2C_3) crystallizes with the cubic Pu_2C_3 structure in the space group $I\bar{4}3d$ which belongs to the tetrahedral crystallographic class T_d with no center of symmetry [53]. The structure contains $C-C$ dumbbells in a distorted dodecahedral coordination ("bisphenoid") formed by 8 La atoms (cf. Fig. 7). For a more detailed discussion of the problems of C deficiency and the problem of the aniostropy of the thermal ellipsoids of the C atoms see [61]. A recent study of the crystal structure up to high pressures could not detect any structural phase transitions up to 30 GPa [54].

In non-centrosymmetric systems with significant spin-orbit coupling superconducting order parameters of different parity can be mixed. A recent system which attracted particular interest in this respect, is the heavy fermion superconductor $CePt_3Si$ which shows unconventional properties, as e.g. antiferromagnetism and superconductivity at $T_N \sim 2.2$ K and $T_c \sim 0.75$ K,

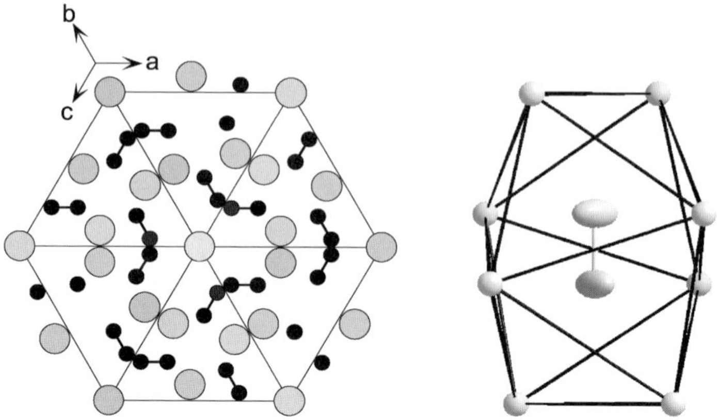

Fig. 7 Crystal structure of La_2C_3 projected along [111] (after [49]). (*Left*) Unit cell with La atoms indicated by the large spheres. (*Right*) La atom environment of a C – C dumbbell. The thermal ellipsoids of the C atoms are shown

respectively, and an upper critical field which considerably exceeds the paramagnetic limit [55]. With no 4*f* electrons present and the high atomic mass of La (as compared to Y) La_2C_3 is therefore an interesting system to study the effects of non-centrosymmetry on superconductivity. Possible multi-gap superconductivity is another interesting issue which has been proposed for Th doped Y_2C_3 and La_2C_3 [65]. Recently, Harada et al. from ^{13}C NMR measurements reported multi-gap superconductivity for Y_2C_3 [66].

In contrast to Y_2C_3 which requires high-pressure synthesis methods [23, 24], samples of La_2C_3 are readily accessible by arc-melting of the constituents. Early on, La_2C_3 was reported to have a T_c of ~ 11 K [56–58]. Subsequently, it has been shown that these samples were not stoichiometric, as anticipated, but exhibit a range of homogeneity from 45.2% to 60.2% atom-% carbon content [49, 59, 60]. Investigations of a series of samples $La_2C_{3-\delta}$ with $0.3 \geq \delta \geq 0$ indicate a separation into two superconducting phases with rather sharp T_c's of ~ 6 K and 13.3–13.4 K (Fig. 8). The high T_c values are attributed to stoichiometric La_2C_3, viz. negligible C deficiency, which was assured individually for the samples by neutron powder diffraction ([28, 61], Kim et al., unpublished data).

Our electronic structure calculations show a splitting of the bands near E_F indicating that the spin degeneracy is lifted due to a sizable spin-orbit coupling in addition to the non-centrosymmetry in the structure (Kim et al., unpublished data). However, the band splitting in La_2C_3 is much smaller than those found for other non-centrosymmetric superconductors like $CePt_3Si$, Li_2Pt_3B or $Cd_2Re_2O_7$ [62–64]. For Li_2Pd_3B, another non-centrosymetric superconductor, where the band splitting is comparable with that of La_2C_3, conventional BCS type behavior with an isotropic superconducting gap has been established via μSR experiments [67]. Based on our heat capacity meas-

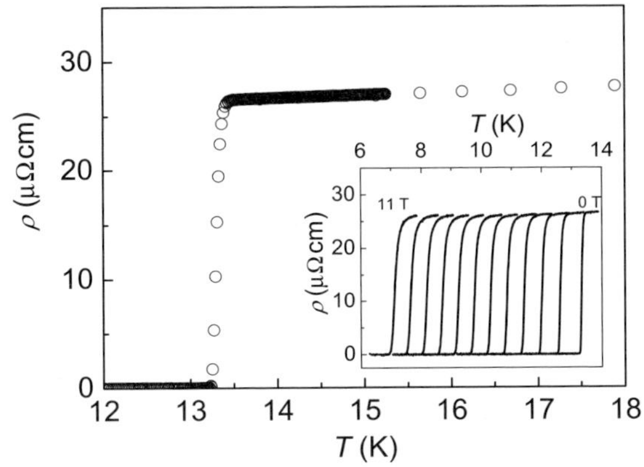

Fig. 8 Low temperature electrical resistivity of La_2C_3 showing the superconducting transition at 13.4 K. The *inset* demonstrates the decrease of the superconducting transition with external magnetic fields ranging from 0 T, 1 T, ..., 11 T

urements we similarly conclude that La_2C_3 is a system with strong electron-phonon coupling with a single gap of isotropic *s*-wave symmetry.

The upper critical field was determined by various methods and reaches a value of ~ 20 T at $T \to 0$ K (Fig. 9). $\mu_0 H_{c2}$ is clearly enhanced over the Werthamer–Helfand–Hohenberg predictions [68], but it does not exceed the paramagnetic limit. Therefore, even though band splitting effects due the non-centrosymmetric structure are present, they appear to be not significant in case of La_2C_3.

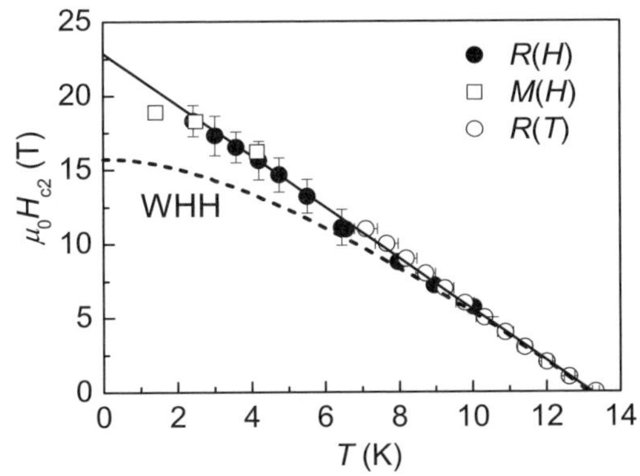

Fig. 9 Upper critical field of La_2C_3 determined by various experimental methods, magnetoresistance ($R(H)$ and $R(T)$) and magnetization ($M(H)$) measurements (after Kim et al., unpublished data)

Superconductivity in Alkaline Earth Intercalated Graphite

The recent discovery of superconductivity in Ca- and Yb-intercalated graphite has refocused considerable interest onto graphite intercalated compounds (GICs) [17, 69]. The superconducting transition temperatures for Ca- and Yb-intercalated graphite are 11.5 and 6.5 K (cf. Fig. 10), respectively, significantly higher than those of the alkali-metal intercalated graphite phases studied before.

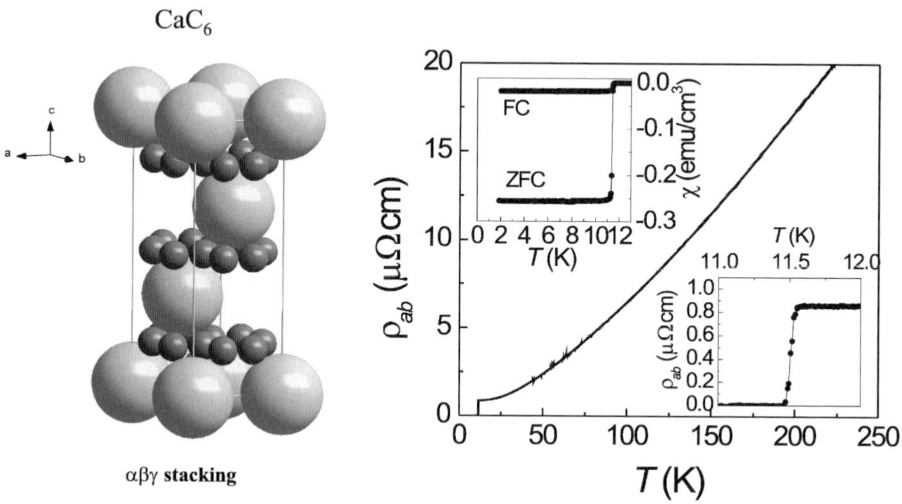

Fig. 10 (*Left*) Crystal structure of CaC_6 (after [69]) and (*right*) in-plane electrical resistivity and magnetic susceptibility (*inset*) of CaC_6 (after [70])

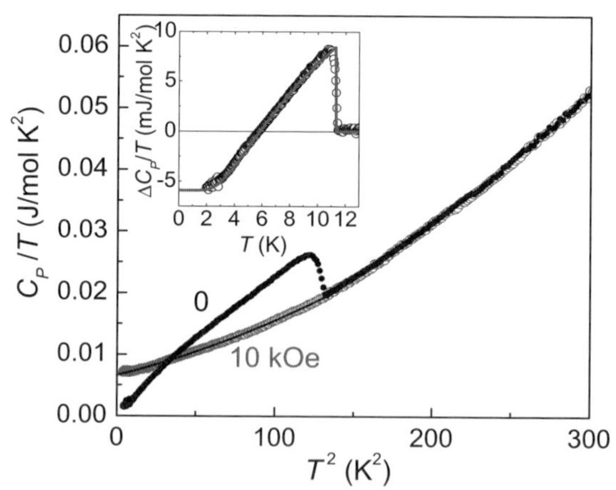

Fig. 11 Temperature dependence of the specific heat of CaC_6 at $B = 0$ and 1 T. The *inset* shows the temperature dependence of $\Delta C_P/T = C_P/T(B = 0) - C_P/T(B = 1\ \mathrm{T})$. The (*red*) *solid line* is the best fit assuming an isotropic *s*-wave BCS gap (after [70])

Apart from the significant enhancement of T_c, two other aspects immediately attracted attention: In case of of YbC_6 it was initially speculated that $4f$ electrons may play a role and that superconductivity might be mediated by valence fluctuation. This possibility, however, could be ruled out and it was found that Yb, like Ca, is divalent and the f electrons provide no essential contributions to the electronic structure at E_F [71]. The second interesting aspect concerned the role of the so-called "interlayer band", i.e. a three-dimensional nearly-free electron band emerging from electrons localized in the intercalant plane, and its relation to superconductivity together with the conjecture of an unconventional electronic pairing mechanism involving excitons [72].

This view was questioned based on the results of the first heat capacity study on CaC_6. It showed that the anomaly at T_c can be clearly resolved indicating the bulk nature of the superconductivity. In particular, both the temperature and magnetic field dependence of C_P strongly evidence a fully gapped, intermediate-coupled, *phonon-mediated* superconductor without essential contributions from alternative pairing mechanisms [70].

Linear response calculations provide the following picture of the electron–phonon coupling (cf. Fig. 12): There are three distinct groups of modes, one at $\omega \sim 10$ meV, another around $\omega \sim 60$ meV, and the third located at $\omega \sim 170$ meV which contribute ~ 0.4, ~ 0.3, and ~ 0.1 to the total coupling constant λ. These three groups are mainly composed of the Ca, out-of plane and in-plane C vibrations, respectively. The observed positive pressure de-

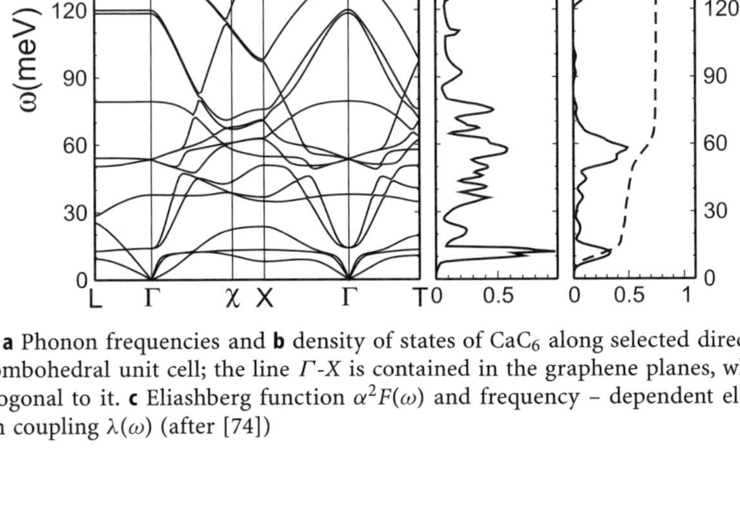

Fig. 12 **a** Phonon frequencies and **b** density of states of CaC_6 along selected directions in the rhombohedral unit cell; the line Γ-X is contained in the graphene planes, while L-Γ is orthogonal to it. **c** Eliashberg function $\alpha^2F(\omega)$ and frequency – dependent electron – phonon coupling $\lambda(\omega)$ (after [74])

pendence of T_c can be understood within this electron–phonon coupling scheme due to a softening of the Ca in-plane phonon modes [73, 74].

Conclusions

The broad chemical bonding abilities of carbon allowing to realize highly anisotropic chemical structures which together with the low atomic mass of carbon make the modifications of carbon as well as carbon-derived compounds to a wide and rewarding playground to search for new and unusual superconductors.

References

1. Bednorz JG, Müller KA (1986) Z Phys B 64:189
2. Nagamatsu J, Nakagawa N, Muranaka T, Zenitani Y, Akimitsu J (2001) Nature 463:401
3. Bouquet F, Fisher RA, Phillips NE, Hinks DG, Jorgensen JD (2001) Phys Rev Lett 87:047001
4. Crabtree G, Kwok W, Canfield PC, Budko SL (2003) Physica C 385:1
5. Dolgov O, Kremer RK, Kortus J, Golubov AA, Shulga SV (2005) Phys Rev B 72:024504
6. Krätschmer W, Lamb DL, Fostiropulos K, Huffman DR (1990) Nature 347:354
7. Hebard AF, Rosseinsky MJ, Haddon RC, Murphy DW, Glarum SH, Palstra TTM, Ramirez AP, Kortan AR (1991) Nature 350:600
8. Tanigaki K, Hirosawa I, Ebbesen TW, Mizuki J, Shimakawa Y, Kubo Y, Tsai JS, Kuroshima S (1992) Nature 356:419
9. Roberts BW (1976) J Phys Chem Ref data 5:581
10. Nagarajan R, Mazumdar C, Hossain Z, Dhar SK, Gopalakrishnan KV, Gupta LC, Godart C, Padalia BD, Vijayaraghavan R (1994) Phys Rev Lett 72:274
11. Cava RJ, Takagi H, Zandbergen HW, Krajewski JJ, Peck WF Jr, Siegrist T, Batlogg B, van Dover RB, Felder RJ, Mizuhashi K, Lee JO, Eisaki H, Uchida S (1994) Nature 367:252
12. Cava RJ, Takagi H, Batlogg B, Zandbergen HW, Krajewski JJ, Peck WF Jr, van Dover RB, Felder RJ, Siegrist T, Mizushashi K, Lee JO, Eisaki H, Carter SA, Uchida S (1994) Nature 367:146
13. Hannay NB, Geballe TH, Matthias BT, Andres K, Schmidt P, MacNair C (1965) Phys Rev Lett 14:225
14. Dresselhaus MS, Dresselhaus G (2002) Adv Phys 51:1
15. Belash IT, Bronnikov AD, Zjarikov OV, Palnichenko AV (1990) Synthetic Metals 36:283
16. Belash IT, Bronnikov AD, Zjarikov OV, Palnichenko AV (1989) Solid Sate Commun 69:921
17. Weller TE, Ellerby M, Saxena SS, Smith RP, Skipper NT (2005) Nature Phys 1:39
18. Ekimov EA, Sidorov VA, Bauer ED, Mel'nik NN, Curro NJ, Thompson JD, Stishov SM (2004) Nature 428:542
19. Simon A, Mattausch HJ, Eger R, Kremer RK (1991) Angew Chem Int Ed 30:1188
20. Simon A, Yoshiasa A, Bäcker M, Henn RW, Kremer RK, Mattausch Hj, Felser C (1996) Z anorg allg Chem 622:123
21. Henn RW, Schnelle W, Kremer RK, Simon A 1996 Phys Rev Lett 77:374
22. Krupka MC, Giorgi AL, Krikorian NH, Szklarz EG (1969) J Less-Common Met 17:91

23. Amano G, Akutagawa S, Muranaka T, Zenitani Y, Akimitsu J (2004) J Phys Soc Jpn 73:530
24. Nakane T, Mochiku T, Kito H, Nagao M, Itoh J, Kumakura H, Takano Y, cond-mat/0311561
25. Schwanitz-Schüller U, Simon A (1985) Z Naturforsch 40b:710
26. Mattausch HJ, Kremer RK, Eger R, Simon A (1992) Z Anorg Allg Chem 609:7
27. Henn RW (1996) PhD Thesis, University Karlsruhe (TH)
28. Kremer RK, Simon A (2004) Curr Appl Phys 4:563
29. Ahn K, Grzechnik A, Kremer RK, Simon A (2003) J Solid State Chem 171:367
30. Ahn K, Gibson BJ, Kremer RK, Mattausch HJ, Stolovits A, Simon A (1999) J Phys Chem B 103:5446
31. Padamsee H, Neighbor JE, Shifman CA (1973) J Low Temp Phys 12:387
32. Golubov AA, Kortus J, Dolgov OV, Jepsen O, Kong Y, Andersen OK, Gibson BJ, Ahn K, Kremer RK (2002) J Phys Condens Matter 14:1353
33. Carbotte JP (1990) Rev Mod Phys 62:1027
34. Allen PB, Dynes RC (1975) Phys Rev B 12:905
35. Henn RW, Strach T, Kremer RK, Simon A (1998) Phys Rev B 58:14364
36. Deng S, Simon A, Köhler J (2005) In: Müller KA, Bussmann-Holder A (eds) Superconductivity in Complex Systems. Structure and Bonding, Springer, Berlin, Heidelberg, 114:103
37. Puschnig P, Ambrosch-Draxl C, Henn RW, Simon A (2001) Phys Rev B 64:024519
38. Simon A (1997) Angew Chem Int Ed Engl 36:1788
39. Miller GJ, Burdett JK, Schwarz C (1986) Inorg Chem 25:4437
40. Herrling E, Dormann E, Henn RW, Kremer RK, Simon A (1998) Synth Metals 92:13
41. Ahn K, Kremer RK, Simon A, Marshall WG, Puschnig P, Ambrosch-Draxl C (2005) J Phys Condens Matter 17:S3121
42. Henn RW, Ahn K, Krug von Nidda H-A, Kremer RK, Simon A (2000) Physica C 719:341–348
43. Gao L, Xue YY, Chen F, Xiong Q, Meng RL, Ramirez D, Chu CW (1994) Phys Rev B 50:4260
44. Chu CW, Gao L, Chen F, Huang ZJ, Meng RL, Xue YY (1993) Nature 365:323
45. Schilling JS, Klotz S (1992) In: Ginsberg DM (ed) Physical Properties of High Temperature Superconductors III. World Scientific
46. Atoji M (1961) J Chem Phys 35/6:1950
47. Giorgi AL, Szklarz AL, Krupka MC, Wallace TC, Krikorian NH (1968) J Less-Common Met 14:247
48. Gulden TH, Henn RW, Jepsen O, Kremer RK, Schnelle W, Simon A, Felser C (1997) Phys Rev B 56:9021
49. Gulden TH (1997) PhD Thesis, Universität Stuttgart
50. Ginzburg VL (1964) Phys Lett 13:101;
51. Ginzburg VL, Kirznits DA (1964) Sov Phys JETP 19:269
52. Ginzburg VL, Kirznits DA (eds) (1982) High-Temperature Superconductivity. Consultant Bureau, Plenum, New York
53. Zachariasen WH (1965) Acta Crystallogr 5:17
54. Wang X, Loa I, Syassen K, Kremer RK, Simon A, Hanfland M, Ahn K (2005) Phys Rev B 72:064520
55. Bauer E, Hilscher G, Michor H, Paul CH, Scheidt EW, Gribanov A, Seropegin YU, Noël H, Sigrist M, Rogl P (2004) Phys Rev Lett 92:027003–1
56. Giorgi AL, Szklarz EG, Krupka MC, Krikorian NH (1969) J Less-Common Met 17:121
57. Giorgi AL, Szklarz EG, Krikorian NH, Krupka MC (1970) J Less-Common Met 22:131

58. Francavilla TL, Carter FL (1976) Phys Rev 14:128
59. Spedding FH, Gschneidner KA Jr, Daane AH (1959) Trans AIME 215:192
60. Simon A, Gulden TH (2004) Z Anorg Allg Chem 630:2191
61. Kim JS, Kremer RK, Jepsen O, Simon A (2006) Curr Appl Phys 6:897
62. Samokhin KV, Zijlstra ES, Bose SK (2004) Phys Rev B 69:094514
63. Lee K-W, Pickett WE (2005) Phys Rev B 72:174505
64. Eguchi R, Yokoya T, Baba T, Hanawa M, Hiroi Z, Kamakura N, Takata Y, Harima H, Shin S (2002) Phys Rev B 66:012516
65. Sergienko IA (2005) Physica B 581:359–361
66. Harada A, Akutagawa S, Miyamichi Y, Mukuda H, Kitaoka Y, Akimitsu J (2006) cond-mat/0612578
67. Khasanov R, Landau IL, Baines C, La Mattina F, Maisuradze A, Togano K, Keller H (2006) Phys Rev B 73:214528
68. Werthamer NR, Helfand E, Hohenberg PC (1966) Phys Rev 147:295
69. Emery N, Hérold C, d'Astuto M, Garcia V, Bellin CH, Marêché JF, Lagrange P, Loupias G (2005) Phys Rev Lett 95:087003
70. Kim JS, Kremer RK, Boeri L, Razavi FS (2006) Phys Rev Lett 96:217002
71. Mazin II, Molodtsov SL (2005) Phys Rev B 72:172504
72. Csanyi G, Littlewood PB, Nevidomskyy AH, Pickard CJ, Simons BD (2005) Nature Physics 1:42
73. Calandra M, Mauri F (2005) Phys Rev Lett 95:237002
74. Kim JS, Boeri L, Kremer RK, Razavi FS (2006) Phys Rev B 74:214513

Lee WS, Cuk T et al. (2007) In: Bussmann-Holder A, Keller H (eds)
High T_c Superconductors and Related Transition Metal Oxides.
Springer-Verlag Berlin Heidelberg, 227–236
DOI 10.1007/978-3-540-71023-3

Band Renormalization Effect in $Bi_2Sr_2Ca_2Cu_3O_{10+\delta}$

W. S. Lee[1] (✉) · T. Cuk[1] · W. Meevasane[1] · D. H. Lu[1] · K. M. Shen[1] · Z.-X. Shen[1] · W. L. Yang[2] · X. J. Zhou[2] · Z. Hussain[2] · C. T. Lin[3] · J.-I. Shimoyama[4] · T. P. Devereaux[5]

[1]Department of Physics, Applied Physics,
and Stanford Synchrotron Radiation Laboratory, Stanford University,
Stanford, CA 94305, USA
leews@stanford.edu

[2]Advanced Light Source, Lawrence Berkeley National Lab, Berkeley, CA 94720, USA

[3]Max-Planck Institute für Festkörperforschung, 70569 Stuttgart, Germany

[4]Department of Applied Chemistry, University of Tokyo, 113-8656 Tokyo, Japan

[5]Department of Physics, University of Waterloo, Waterloo, Ontario, N2L 3G1, Canada

Abstract The band renormalizations of optimally-doped Bi2223 system were studied by angle-resolved photoemission spectroscopy. In the superconducting state, we found that the band renormalizations are dominated by one bosonic mode. In the normal state, the band renormalizations are weak, but still observable. To better visualize the renormalization features in the normal state, we extracted the real part of the self-energy, $\mathrm{Re}\Sigma$, in which we found a shoulder at about 35 meV in addition to the maximum at about 70 meV. This result suggests that multiple bosonic modes contribute to the band renormalizations, instead of just one single mode. These modes are likely oxygen related optical phonon mode. Compared to the extracted $\mathrm{Re}\Sigma$ in the superconducting state, we show that the renormalization effect exhibits an interesting and subtle temperature dependence at the nodal region.

The observation of a kink in the band dispersion by ARPES [1–10] has drawn a lot of attention in the field of high-T_c superconductor. This sudden band renormalization is generally interpreted as electrons coupled to collective modes; however, the nature of the modes is still a subject of debate. So far, most studies focused on double layer Bi-based cuprates, Bi2212, and relatively less experimental work exist on its brother compound, the trilayer crystal, Bi2223. Previous studies [11, 12] on the Bi2223 system concluded that the antinodal renormalization below T_c is due to the magnetic resonance mode and has *different* origin from the nodal kink. Because the band renormalization effect in the antinodal region disappears in the normal state, while the kink in the nodal dispersion still remains. However, it is a puzzle why the kink in the nodal direction did not exhibit a corresponding shift due to the opening of a superconducting gap when goes through a phase transition from the normal to superconducting states.

In this paper, with improved data quality, we show that the extracted $\mathrm{Re}\Sigma$ exhibit fine structures suggesting a multiple modes coupling interpretation.

In contrast to previous studies, our data suggest that below T_c, the prominent and anisotropic renormalization effects of the node and the antinode are of the *same* origin. However, the apparent kink in the nodal dispersion for the normal state and the superconducting state is dominated by different modes. Furthermore, our data show that the coupling of the lower energy mode, likely the 35 meV B_{1g} oxygen phonon [13], is quite different between Bi2201 and that of Bi2212 and Bi2223, which are superconductors with substantially higher T_c.

High quality $Bi_2Sr_2Ca_2Cu_3O_{10+\delta}$ (Bi2223) single crystals were grown by the floating zone method as described in [14]. After annealing under oxygen flow, the samples reach near optimal doping with an onset transition temperature of 110 K. ARPES measurements were performed at Beamline 10.0.1 of the Advanced Light Source (ALS) and Beamline 5-4 of the Stanford Synchrotron Radiation Laboratory (SSRL). In our experiments, 27 eV photons were used to excite the photoelectrons. The energy resolution was set to 15 meV and the momentum resolution is about 0.012 Å^{-1} at this photon energy. Samples were cleaved in situ with a base pressure better than 4×10^{-11} Torr. Consistent with the previous observation [15, 16], the near-E_F spectra are dominated by a feature that forms a single hole like Fermi surface (FS), without evidence of FS splitting associated with multiple CuO_2 layers in each unit cell.

Figure 1a shows intensity plots of the raw data at different positions along the Fermi surface when the system is in the normal (120 K) and superconducting states (25 K); dramatic differences of the band dispersion between the normal state and superconducting state is observed, especially near the antinodal region. Figure 1b displays the Energy Distribution Curves (EDCs) in the superconducting state. A peak-dip-hump (PDH) structure can be clearly observed in the antinodal region ("c3" and "c4"). Furthermore, the band dispersion separates into two branches as shown by the dotted lines in the corresponding image plots. In the normal state (120 K), the PDH structure disappears and only one continuous branch of band dispersion can be observed (lower paneal of Fig. 1a). In the nodal region ("c1" and "c2"), a kink around 70 meV in the dispersion derived from Momentum Distribution Curves (MDCs) [3] was observed in both the normal state and superconducting state. We note that the kink is enhanced in the superconducting state; in addition, the EDCs around the kink position in the superconducting state clearly show a two-peak feature suggesting that the band separates into two branches at the kink energy (thick EDCs in "c1" and "c2" of Fig. 1b).

The observed momentum and temperature dependence is very similar to that of the Bi2212 system, which has been interpreted as an effect of electron-boson coupling [8, 9, 17, 18]. In this interpretation, the band dispersion is separated into two branches at an asymptotic energy of $\Omega + \Delta_0$, where Ω and Δ_0 are the energy of the bosonic mode and the the maximum of the $d_{x^2-y^2}$ superconducting gap, respectively. In the antinodal region, EDCs evolve into the peak-dip-hump (PDH) structures with the dip energy at roughly the same

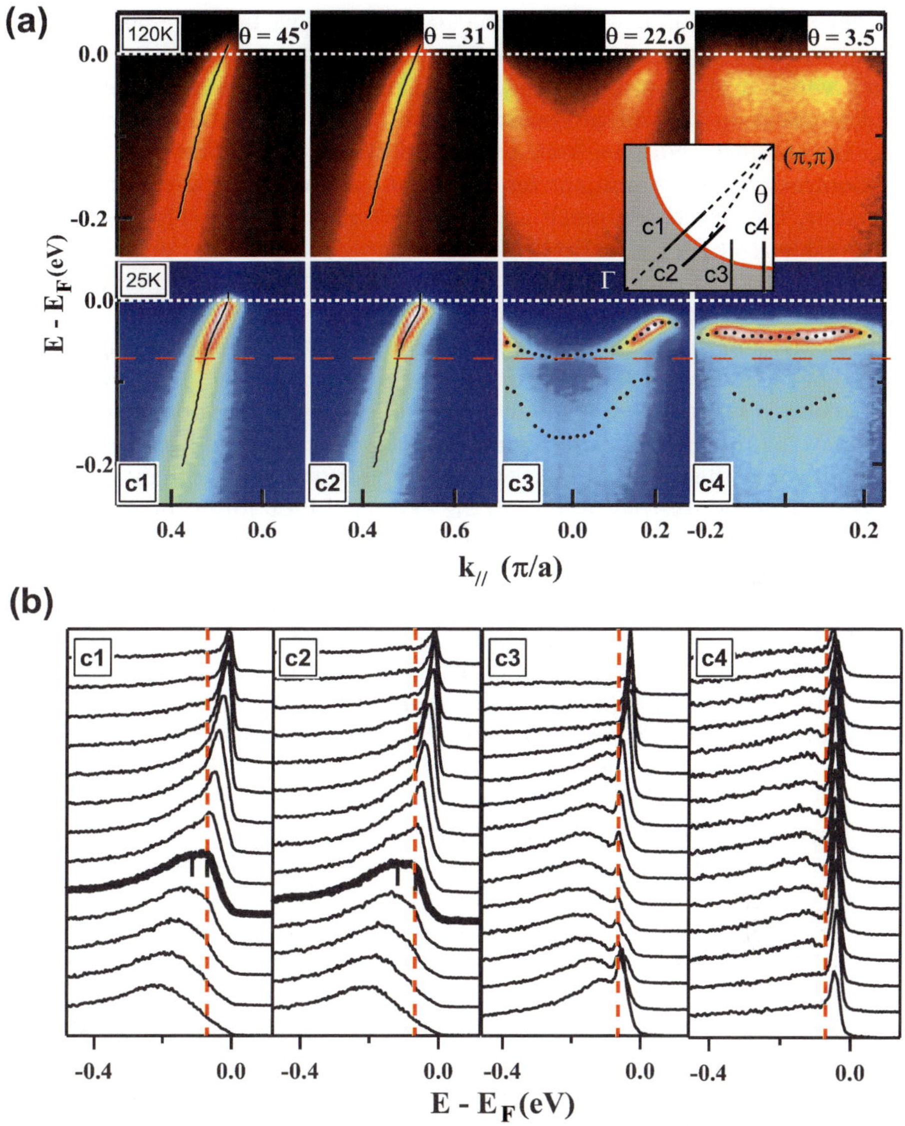

Fig. 1 **a** Intensity plots of the normal (120 K) and superconducting state (25 K) data from nodal to antinodal region as indicated in inset with the *red curve* presenting the FS. The *solid black curves* are the band dispersions extracted from the MDCs analysis, and the *dotted curves* are the positions of peak and hump in EDCs. **b** Stack plots of the EDCs in the superconducting state. The *red dashed lines* are all located at 70 meV below the Fermi energy. The EDCs plotted by *thicker curves* in "c1" and "c2" exhibit two features

energy scale. In Fig. 1b, we mark the asymptotical value of the peak dispersion for "c1", "c2" and "c3" and the "dip" position for "c4" by the red dashed lines, which are all around 70 meV suggesting that the observed prominent and momentum dependent renormalization effect for the entire Fermi surface in the *superconducting state* is induced by the same mode.

Contrary to the prominent renormalization effects in the superconducting state, the effects are much weaker in the normal state (Fig. 1), but still observable. To better visualize the subtle changes in the band dispersion, we extracted the real-part of the self-energy ($\mathrm{Re}\Sigma$) by subtracting an assumed liner bare band determined by connecting the Fermi crossing point and the end of MDC-derived band dispersion from the MDC-derived band dispersion. For the $\mathrm{Re}\Sigma$ data shown in this paper, the end of the MDC-derived band dispersion is set at 200 meV below the Fermi energy for the nodal region and 150 meV for the intermediate region between the node and antinode, which are well above the apparent kink energy (at least twice larger than the apparent kink energy). We plot in Fig. 2a the extracted $\mathrm{Re}\Sigma$ for both normal state and superconducting state in the nodal region. We found that the maximum position of the $\mathrm{Re}\Sigma$, which corresponds to the apparent kink position in the MDC-derived band dispersion, does not shift with the opening of a superconducting gap, but remains at about the same energy ($\sim$ 70 meV). Here, we use the term, "real part of self-energy", loosely, since we can not determine the "true" bare band experimentally. Nevertheless, the position of the kink in the extracted $\mathrm{Re}\Sigma$, is robust against the choice of bare band. This no-shift

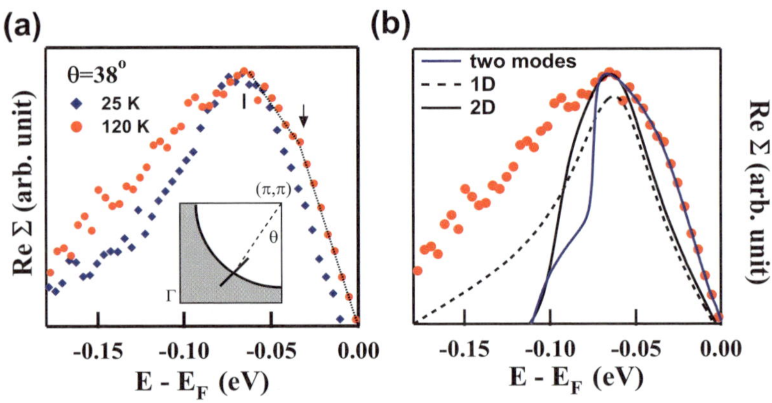

Fig. 2 **a** The real part of the self-energy, $\mathrm{Re}\Sigma$, obtained by subtracting the MDC-derived band dispersion from a linear bare band. The data were measured near the nodal region along the *black line* in the *inset*. The magnitude of the $\mathrm{Re}\Sigma$ has been normalized for a better illustration of the shape differences between the normal and superconducting state. The *vertical bar* and the *arrow* indicate the positions of the 70 meV "kink" and the 35 meV "subkink", respectively. The *dotted lines* serve as a guide to the eye only. **b** A comparison of the simulated $\mathrm{Re}\Sigma$ with the normal state data shown in **a**

behavior is paradoxical: given the superconducting gap of ~ 40 meV at 25 K in Bi2223 [12, 15], the kink position is expected to shift by about the same amount when the system make a phase transition from the normal to superconducting states [8, 9, 18].

To investigate the origin of this paradox, we first address the shape difference of the ReΣ in the normal and superconducting states (Fig. 2a). While the data at 25 K show a linear behavior from the peak to the gap edge, the normal state data reveal a shoulder near 35 meV as indicated by the arrow. Such a shoulder would not exist if the electrons coupled only to a single mode near 70 meV. In Fig. 2b, we plot simulated ReΣs of electronic states coupled to a 65 meV dispersionless bosonic mode for both one dimensional and two dimensional case at 120 K as a comparison to our normal state results[1]. As can be seen in Fig. 2b, the curvature of the normal state ReΣ is *convex* for the lower binding energy (< 65 meV), while that of the simulated ReΣ is *concave* in the same range. This difference suggests that the shape of the normal state self-energy cannot be explained by a single bosonic mode coupling scenario.

The experimental reproducibility of this convex curvature in the ReΣ is demonstrated in Fig. 3a in which we plot the extracted ReΣ in the nodal region from four independent measurements. All of the extracted ReΣs show a maximum around 70 meV corresponding to the apparent "kink" position in the MDC-derived dispersion, which is consistent with the previous study on the same material [11]. In addition to this apparent 70 meV maximum, a convex curvature of ReΣ at lower binding energy, or more specifically, a "slope break" at about 35 meV, are observed in all extracted ReΣ. This reproducibility of our data indicates that this 35 meV "subkink" is intrinsic, rather than caused by noise. Since these four sets of data were collected from two ARPES systems located in different synchrotron radiation centers, we have eliminated possible systematic errors of the experimental set-up. Furthermore, as shown in Fig. 3b, these 35 and 70 meV structures in ReΣ persist more than half-way between the nodal and antinodal region, where the MDCs can still fit reasonably to a Lorentzian function. The coexistence of the "kink" 70 meV and "subkink" (35 meV) suggests that the renormalization effects in Bi2223 system are caused by multiple bosonic modes instead of a single bosonic mode coupling.

Recently, renormalization features caused by multiple mode coupling has also been found in Bi2201 [20], Bi2212 [10], and LSCO [19]. These results cast doubts to the single bosonic mode interpretation, such as spin resonance mode interpretation [6, 8, 9, 11], but can be naturally explained by coupling to multiple phonon modes. As the phonon density of states is not available for this compound, we compare our results with the real part of the

[1] The simulations were calculated by using the finite temperature Nambu-Eliashberg electron-phonon self-energy formalism [18] with a constant electron–phonon coupling vertex. For the one dimensional case, a parabolic band is taken as the bare band; while in the two dimensional case, a six-parameter tight binding model [8, 9] fitted to our Bi2223 data were adopted in our calculation

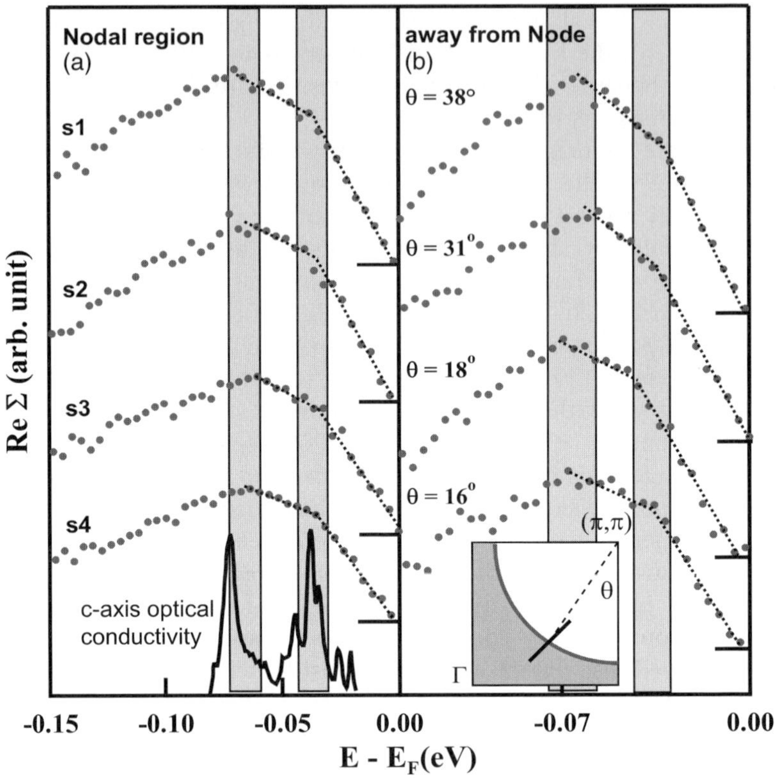

Fig. 3 ReΣ extracted from the data taken in **a** the nodal region, and (**b**) different positions along the Fermi surface in the normal state (120 K) from four independent measurements performed at SSRL and ALS. The definition of the angle is illustrated in the inset with the *black line* representing the measurement direction and *red curve* representing the Fermi surface. The *shade areas* serve as a guide to the eye for the position of the kinks around 35 meV and 70 meV. The *curve*, shown at the bottom of (**a**), is the real part of the *c*-axis optical conductivity, which was reproduced from Fig. 3 of [26]

c-axis optical conductivity [26] in which the phonon features are still prominently detectable even in the highly doped region. We note that the energy positions of the two most prominent phonon features observed in *c*-axis optical conductivity coincide with those of the kink and subkink as shown in Fig. 3. Motivated by this agreement of the energy scales between our data and the optical data, we calculated the ReΣ in a model consisting two Einstein phonons with energies of 35 meV and 70 meV. As illustrated by the blue curve in Fig. 2b, the simulated ReΣ can reproduce the observed convex curvature in the low energy part of our data[2]. This lends support to the proposals by

[2] We focus only on the phonon induced structures in the dispersion in the lower binding energy and ignore other renormalizations at a broader energy range, which may account for the discrepency between the simulation and data in the higher binding energy part

W. Meevesana et al. [21] and T.P. Devereaux et al. [22], which state that the electrons couple strongly to *c*-axis phonons because of the poor screening effect along the *c*-axis. We remark that at a relatively high temperature (120 K) as in our case, only the modes coupled most strongly to the electronic structure could be resolved as a slope change, while the fine structures produced by other phonons with weaker couplings will be smeared out by the temperature broadening. Nevertheless, it is reasonable to conclude that *at least* two phonons with energies of around 35 meV and 70 meV couple strongly to the electronic states.

With this multiple phonon mode coupling interpretation, we address the paradox regarding the absence of temperature dependence of the nodal kink position in Fig. 4, in which we summarize the positions of the kinks and sub-kinks in both normal and superconducting states. We found that the kinks around 70 meV exist in both the superconducting state (25 K) and normal state (120 K). With the 35 meV mode identified, we propose that the normal state 35 meV kink shifts to around 70 meV in the superconducting state because of the opening of the superconducting gap ($\sim$ 40 meV). It happens that the 35 meV mode plus the gap size is very close to the energy of the 70 meV mode, giving us an *illusion* that the kink position does not shift upon the opening of a superconducting gap. In other words, the 70 meV kink in normal state and superconducting state is caused by different bosonic modes. In principle, the normal state 70 meV kink should also shift to around 105 meV, but it is hard to confirm from our data because MDCs at an energy higher than the apparent kink energy (70 meV) become significantly broader and thus noisier causing larger error bars for the fitted peak positions. We note that we have found a similar effect in Bi2212 with a more detailed temperature dependence study, which will be elaborated in a separate paper [23].

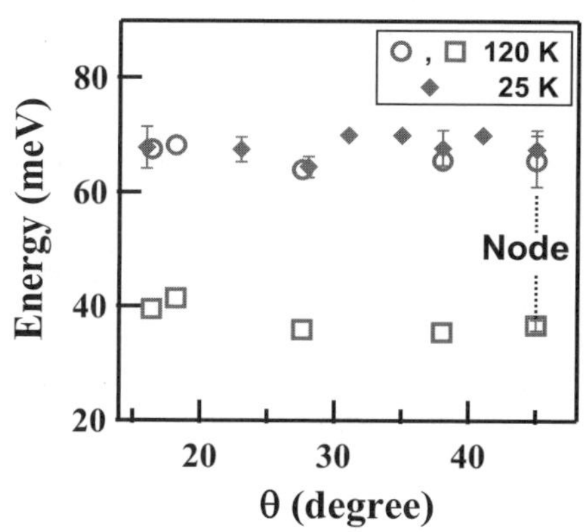

Fig. 4 A summary of the energy positions of the fine structures in ReΣ along the Fermi surface in the normal state and superconducting state. The angle, θ, is defined in the *inset* of Fig. 3

What are these phonons, especially the one dominates the renomrmalization effect in the superconducting state? With the energy scales of the phonon modes determined by ARPES, we could narrow down to some candidates along with other phonon sensitive techniques, such as the Raman spectroscopy, IR spectroscopy, and neutron spectroscopy. These techniques have shown that several phonon modes around these two energies, such as the buckling and breathing modes of in-plane oxygen, and the apical oxygen modes, exhibit strong coupling to the electronic state in the high T_c cuprates [13, 24, 26]. While the breathing and apical oxygen modes are close to 60–80 meV, the B_{1g} buckling phonon near 35 meV exhibits a very anisotropic coupling strength, and has been shown able to explain the temperature and momentum dependence of the band renormalizations in the Bi2212 system [10, 17]. Since the electronic structures of Bi2223 and Bi2212 system are qualitatively similar, this 35 meV mode in the Bi2223 system, which causes the dramatic band renormalization effect in the superconducting state, is likely the B_{1g} buckling phonon.

This B_{1g} mode could also explain the the difference of the antinodal spectra between the single layer and multilayer systems. Figure 5 shows antinodal

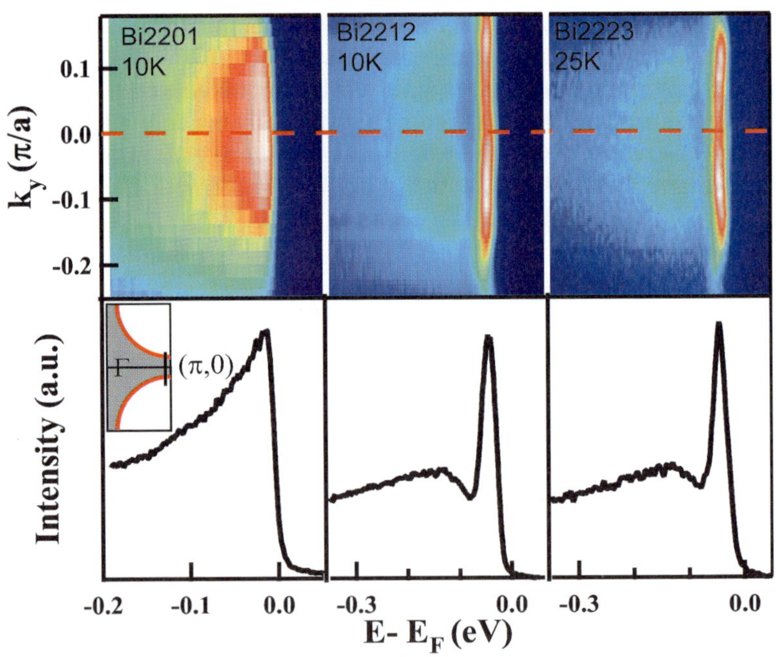

Fig. 5 *Upper panels*: the intensity plot the data from optimally-doped BSCCO family taken in the antinodal region as indicated by the *black line* in the *inset*. The T_c's of the samples are 33 K, 94 K, and 110 K for Bi2201, Bi2212, and Bi2223, respectively. *Lower panels*: corresponding EDCs near $(\pi, 0)$, the *red dashed line*

spectral in superconducting state from optimally doped BSCCO single layer (Bi2201), bilayer (Bi2212), and trilayer (Bi2223) system. Clearly, the band dispersion separates into two branches for both bilayer and trilayer systems, but remains to be a single branch for the single layer system. In other words, the strong PDH structure in the antinodal region only exists in the multilayer systems (the lower panel of the Fig. 5), suggesting a weak B_{1g} mode coupling in the single layer system. This can be explained by considering the mechanism for electronic coupling to the B_{1g} phonon mode; a first order coupling can be produced by breaking the mirror symmetry of the copper-oxygen plane [13]. For multilayer systems, the outmost layers have only one apical oxygen atom on one side in each unit cell; thus, this asymmetry generates a net electric field perpendicular to the the CuO_2 plane which enhances the coupling between the electrons and buckling B_{1g} phonon. For the single layer system, on the contrary, this local asymmetry induced electric field does not exist because of the presence of apical oxygen atoms on both sides of the CuO_2 plane. Therefore, the coupling to the B_{1g} phonon is weak and the renormalization effect due to B_{1g} mode is not observable. We note that the B_{1g} phonon may enhance T_c in the Bi-based multilayer systems because the B_{1g} mode coupling has a *d*-wave form factor [13] and thus favors a *d*-wave paring [25]. However, in the single layer systems, other many-body processes and phonons may be more relevant to the high T_c superconductivity, since one does not expect the existence of a strong electronic coupling to the B_{1g} phonon. This example of B_{1g} mode coupling demonstrates a material dependent renormalization effects; therefore, it is desirable to conduct a detailed analysis of other families of cuprates to gain further insight into the material dependence of the band renormalizations and its relation to the high T_c superconductivity.

Acknowledgements We would like to thank N. Nagaosa for inspiring discussion. SSRL is operated by the DOE Office of Basics Energy Science, Divisions of Chemical Science and Material Science. The research at Stanford is supported by the SLAC/XLAM program through DOE contract: DE-AC02-76SF00515. The ARPES measurements at Stanford were also supported by NSF DMR-064701.

References

1. Damascelli A, Hussain Z, Shen Z-X (2003) Rev Mod Phys 75:473
2. Bogdanov PV et al. (2000) Phys Rev Lett 85:2581
3. Kaminski A et al. (2001) Phys Rev Lett 86:1070
4. Johnson PD et al. (2001) Phys Rev Lett 87:177007
5. Lanzara A et al. (2001) Nature 412:510
6. Kim TK et al. (2003) Phys Rev Lett 91:167002
7. Gromko AD et al. (2003) Phys Rev B 68:174520
8. Eschrig M, Norman MR (2003) Phys Rev B 67:144503
9. Eschrig M, Norman MR (2000) Phys Rev Lett 85:3261
10. Cuk T et al. (2004) Phys Rev Lett 93:117003

11. Sato T et al. (2003) Phys Rev Lett 91:157003
12. Matsui H et al. (2003) Phys Rev B 67:060501(R)
13. Devereaux TP et al. (1995) Phys Rev B 51:505
14. Liang B et al. (2002) Physica C 383:75
15. Feng DL et al. (2002) Phys Rev Lett 88:107001
16. Mori M et al. (2002) Phys Rev B 66:064502
17. Devereaux TP et al. (2004) Phys Rev Lett 93:117004
18. Sandvik AW et al. (2004) Phys Rev B 69:094523
19. Zhou XJ et al. (2004) Cond-Mat/0405130
20. Meevasana W et al. (2006) Phys Rev Lett 96:157003
21. Meevasana W, Devereaux TP, Nagaosa N, Shen Z-X, Zaanen J (2006) Phys Rev B 74:174524
22. Devereaux TP, Shen Z-X, Nagaosa N, Zaanen J, preprint
23. Lee WS et al., in preparation
24. Limonov M et al. (2002) Phys Rev B 66:054509
25. Nazarenko A, Dagotto E (1996) Phys Rev B 53:R2987
26. Kovaleva NN et al. (2004) Phys Rev B 69:054511

Mannhart J, Kopp T et al. (2007) In: Bussmann-Holder A, Keller H (eds)
High T_c Superconductors and Related Transition Metal Oxides.
Springer-Verlag Berlin Heidelberg, 237–242
DOI 10.1007/978-3-540-71023-3

How Large is the Intrinsic Flux Noise of a Magnetic Flux Quantum, of Half a Flux Quantum and of a Vortex-Free Superconductor?

J. Mannhart[1] (✉) · T. Kopp[1] · Y. S. Barash[2]

[1]Center for Electronic Correlations and Magnetism, EP6, Universität Augsburg, 86135 Augsburg, Germany
jochen.mannhart@physik.uni-augsburg.de

[2]Institute of Solid State Physics, Russian Academy of Sciences, Chernogolovka, 142432 Moscow District, Russia

Abstract This article addresses the question whether the magnetic flux of stationary vortices or of half flux quanta generated by frustrated superconducting rings is noisy. It is found that the flux noise generated intrinsically by a superconductor is, in good approximation, not enhanced by stationary vortices. Half flux quanta generated by π-rings are characterized by considerably larger noise.

Standard electronic devices are based on the manipulation of electrodynamic quantities such as electric charge, current, and magnetic fields. It is obvious that under practical conditions, which require contacts to the devices and operation at a finite temperature T, all these devices are subject to noise which arises usually from several sources. Thermal noise and shot noise are prime examples. Digital superconducting devices operate by processing quanta or half quanta of magnetic flux. For some devices, in particular for qubits, their noise needs to be vanishingly small. Exploring whether for superconducting devices a fundamental, ultimate noise limit exists, we ask the following questions: Consider a magnetic flux quantum that is hold stationary in a hole of a superconductor cooled to $T \ll T_c$ (Fig. 1). The flux of the vortex penetrates a pick-up coil with an area A_{coil} of a diameter that is much larger than the London penetration depth λ_L. Is the flux Φ in A_{coil} noisy? Or is the flux of a superconducting vortex under practical operation conditions completely noise-free? If it is noise-free, is it possible to realize devices that operate with flux quanta in a noise-free mode? Of course, being quantized, the fluxoid [1] exactly equals $\Phi_0 = h/2e$ at all instants. But does the flux fluctuate?

To tackle this question, we call to mind that in the superconductor the flux line generates a screening current with density $\vec{J}(\vec{r})$ circling the hole on a path with inductance L. The length scale of its penetration into the superconductor is given by λ_L [1, 2]. Although these Meissner currents are supercurrents, they fluctuate with $\Delta\vec{J}(\vec{r})$. This noise of the screening currents is caused, in the presence of the magnetic field of the vortex, by thermal and quantum fluctu-

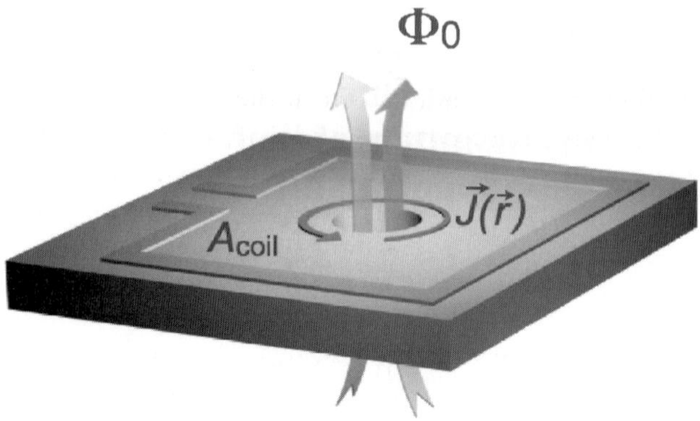

Fig. 1 Sketch of the device configuration. A vortex Φ_0 is pinned stationary by a hole in a superconductor and penetrates a coil (*blue*) with an area A_{coil}. The Meissner screening currents $\vec{J}(\vec{r})$ are sketched in *red*

ations of the gauge invariant phase, as well as by fluctuations of the density of the condensate coexisting with the noisy quasiparticle system.

Consequently, the magnetic flux penetrating the hole is noisy, expressed by the vector potential noise $\Delta\vec{A}(\vec{r})$. The noisy screening current and the noisy magnetic field give rise to a gradient of the phase of the superconducting order parameter that is given by:

$$\hbar\vec{\nabla}\varphi = \frac{m}{n2e}\vec{J} + 2e\vec{A} ,$$

where m is the Cooper pair mass and $\vec{A}$ is the vector potential [2]. The phase φ and the number of Cooper pairs n are conjugate variables, and therefore

$$\Delta\varphi\Delta n \gtrsim 1 ,$$

an uncertainty relation which holds for large n – the error being of the order of $1/n$ [2, 3].

Because the typical order of magnitude of n is 10^{23} and $\Delta n/n \sim n^{-1/2} \sim 3 \times 10^{-12}$, for any loop closed around a vortex, at any instant, the integrated phase gradient has to amount exactly to an integer number of 2π, the precision being of the order of 10^{-12}.

Is the total flux penetrating A_{coil} noisy? We start to answer this question by disregarding current fluctuations at the boundary of A_{coil}, having in mind to return to this boundary effect later. Also we postpone discussions of effects induced by intrinsic equilibrium noise of the current taking place in the absence of the vortex. Then, the topological constraint described yields the requirement that despite of the noise of the local current density $\Delta\vec{J}(\vec{r})$, the noise on the total current $\Delta\vec{I}$, and the noise of the magnetic field $\Delta\vec{A}(\vec{r})$, the total flux through the loop has to equal exactly $h/2e$ in a large spectral range. This refers to all loops that comprise the vortex completely. The noise of the current density and the magnetic field does therefore not generate noise in the total flux through the loop. The noise is quenched by the topological constraint resulting from the uniqueness of the phase. This noise quenching occurs by the following microscopic mechanism: if, for example, at position $\vec{r}$ the screening current fluctuates to exceed its average value, the resulting induction changes the path of $\vec{J}(\vec{r})$ such that its inductance is lowered to the value that keeps the induced flux constant – and noise-free. This process specifically reduces the noise of flux quanta, noise arising from other sources is not necessarily affected.

The noise is quenched over a large spectral range. At high frequencies, the range is limited by the gap frequency $\Delta/\hbar$: the condensate does not respond sufficiently to faster fluctuations. At low frequencies, rare events triggered by highly energetic fluctuations occur when the flux line approaches the boundary of A_{coil} or even crosses it. Such events are heralded by their large magnetic flux changes. Because the time interval between such jumps scales exponentially with the energy barrier for entry of flux lines into the superconductor, these events are rare and leave correspondingly long, noise-free periods. For many applications, such jumps do not jeopardize device performance. Should a quantum jump, the device has to be reset and started afresh.

Thus we conclude that under the assumptions made, the screening current and the phase fluctuations do not cause flux noise for a superconducting vortex. Due to the macroscopic character of the superconducting flux quantum and the topological requirement of the order parameter phase, the noise generated by a vortex has a very low value, even under practical experimental conditions. Therefore, the frequently practiced encoding of data as separate quantum states presents a route to low noise-data processing; the system can quench low-energy noise at every processing step.

It has to be mentioned that a significant difference exists between the flux noise in a weak-link free superconducting loop and of a loop interrupted with weak-links such as Josephson junctions. In the latter case only the sum of the magnetic flux contribution and the phase difference across the junctions is equal to a multiple of 2π. Under these conditions, noise in the phase difference across the junctions usually controls the flux noise of the loop. In case the junctions are in the zero-voltage state, the phase difference noise is caused via the first Josephson relation by critical current fluctuations. If the junctions are in the voltage state, the phase difference noise is associated via the second Josephson relation with the voltage noise. Therefore, as described by the effective electric circuit of the device (including, in particular, resistive elements) several noise sources are coupled to the flux noise. Such important effects have been thoroughly studied [4–6]. They also take place in loops with odd numbers of π-junctions, for example, in tricrystal rings [7]. Keeping the phase difference of π across the junction leaves almost a half value of magnetic flux quanta for the flux through the ring. Due to noise in the junctions, the flux noise is significantly enhanced as compared to a loop devoid of weak-links. Nevertheless, also in this case the fluxoid is of course exquisitely noise-free. Half flux quanta generated by frustrated π-loops can provide a highly precise and stable flux bias for quantum interference devices that also can be rapidly switched (Fig. 2) [8], the flux generated by these loops is, however, not free of noise.

In contrast, the results for weak-link free loops seem to suggest the principal possibility to build noise-free superconducting devices that manipulate flux quanta. Is this possible, indeed? Because the flux of a vortex is, to a good approximation, noise-free, this problem boils down to the question whether under practical conditions also a vortex-free superconductor is free of flux noise.

Does a vortex-free superconductor induce noise currents in a coil with a diameter $\gg \lambda_L$ as shown in Fig. 3? Yes, the superconductor generates flux noise and, associated with the flux noise, also voltage noise. Supercurrents and quasiparticle currents that are activated by thermal or by quantum fluctuations even in the absence of the vortex, generate fluctuating, local magnetic fields. These fields form closed loops. Loops that are closed within A_{coil} do not alter the flux in A_{coil}. Flux noise is, however, generated by fluctuating loops that straddle the boundary of A_{coil} (Fig. 3). By this intrinsic process, superconductors induce a small but finite magnetic flux noise at all parts of their surfaces. Therefore, above the surface of a superconductor, a cloud of minute flux and voltage noise is generated.

This flux noise is closely related to the electromagnetic fluctuations present in condensed matter [9]. It is controlled by the temperature-dependent quasiparticle density and phase fluctuations and thus represents a basic property

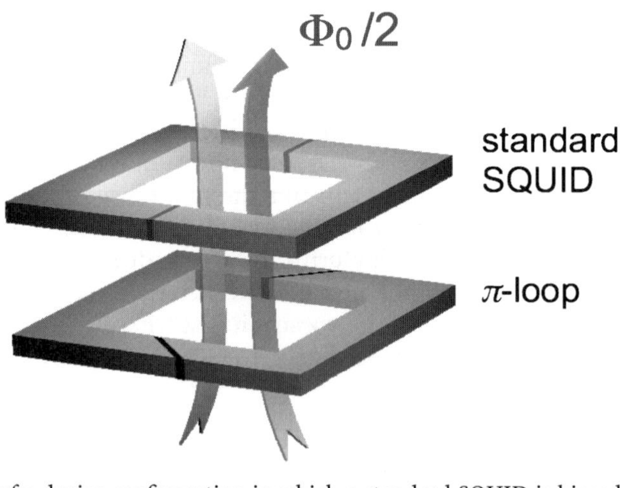

Fig. 2 Sketch of a device configuration in which a standard SQUID is biased by half a flux quantum $\Phi_0/2$ generated by a frustrated superconducting loop (π-loop), formed, e.g., by a tricrystal ring

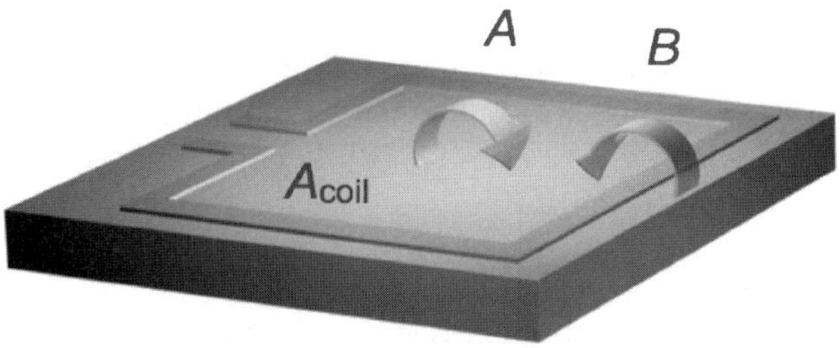

Fig. 3 Sketch of a sample configuration to illustrate the flux noise intrinsically generated by a superconductor. Phase fluctuations generate loops of magnetic flux (A, B) that penetrate a detector coil (*blue*) with an area A_{coil}. Loops that straddle the boundary of A_{coil} (B) generate magnetic flux noise in the detector coil, loops that are closed well within A_{coil} (A) do not

of the material. The flux noise varies as a function of the distance to the surface of the superconductor. We are not aware that this intrinsic flux noise has been determined for, say Al at 100 mK, to give an example, or that its spectral density has ever been calculated. Nevertheless it is obvious that the noise is smallest for fully gapped, *s*-wave superconductors. The flux noise causes a minute, temperature dependent, attractive contribution to the force that acts between two closely spaced superconductors. In superconducting qubits, at finite temperature this flux noise provides an ultimately small, yet intrinsic source of decoherence that cannot be overcome.

The effects of a vortex on the superconducting order parameter and also the magnetic field of the vortex are screened in the depth of the superconducting ring. Because the intrinsic flux noise is controlled deep inside the superconductor by current fluctuations on the contour of integration, the intrinsic flux noise and the vortex can interact only slightly. There are several possible mechanisms to cause a small interaction. Fluctuation-induced quasiparticles can, for example, interact with the screening current of the flux line, or be subject to Aharonov-Bohm type phase shifts when circling the vortex. Due to these interactions, the contribution of a stationary vortex to the flux noise is finite – minute, but not exactly zero.

In summary, we conclude that intrinsic flux noise generated by a superconductor is stronger for superconductors with gap nodes. A stationary vortex does, in good approximation, not produce additional noise. Its noise is suppressed by the topology and the macroscopic nature of the vortex. The flux of half vortices generated by frustrated π-rings is characterized by a significantly larger amount of noise.

Acknowledgements The authors are grateful to A.J. Leggett for helpful discussions and to A. Herrnberger for drawing the figures. This work was supported by the DFG (SFB 484), by the ESF (THIOX), and by the Nanoxide program of the EC.

References

1. London F (1950) Superfluids. Vol. 1, Wiley, Inc., New York
2. Tinkham M (1996) In: Introduction to Superconductivity. McGraw-Hill, Inc, New York
3. Nagaosa N (1999) In: Quantum Field Theory in Condensed Matter Physics. Springer, Berlin, Heidelberg, New York
4. Jackel LD, Buhrman RA (1975) J Low Temp Phys 19:201
5. Clarke J (1993) SQUIDs: Theory and Practice. In: Weinstock H, Ralston RW (eds) The New Superconducting Electronics. Kluwer
6. Larkin A, Varlamov A (2005) Theory of Fluctuations in Superconductors. Oxford University Press
7. Tsuei CC, Kirtley JR (2000) Rev Mod Phys 72:969
8. Mannhart J, Mayer B, Allenspach R, Tsuei CC (1995) IBM Tech Discl Bull 38:185
9. Lifshitz EM, Pitaevsky LP (1986) In: Statistical Physics, Part 2. Pergamon Press, Oxford

Mihailovic D, Kusar P (2007) In: Bussmann-Holder A, Keller H (eds)
High T_c Superconductors and Related Transition Metal Oxides.
Springer-Verlag Berlin Heidelberg, 243–251
DOI 10.1007/978-3-540-71023-3

Lattice and Magnetic Excitations in Relation to Pairing and the Formation of Jahn–Teller Polaron Textures in Cuprates

D. Mihailovic (✉) · P. Kusar

Dept. of Complex Matter, Jozef Stefan Institute, SI-1000 Ljubljana, Slovenia
dragan.mihailovic@ijs.si

Abstract Dynamic nanoscale textures such as pairs and stripes in charge-doped cuprates can form as a result of interactions which may be of polaronic or of magnetic origin. We discuss new experiments on quasiparticle relaxation dynamics in the high photoexcitation intensity regime with the aim of determining which of the two interactions is responsible for the functional behaviour in cuprates. On the basis of concise energy conservation considerations in quasiparticle relaxation, we deduce that lattice excitations can easily mediate quasiparticle recombination and pseudogap formation in $La_{2-x}Sr_xCuO_4$. We also find that magentic excitations are incapable of absorbing all the energy released upon recombination, and on their own thus cannot be responsible for pairing in the cuprates.

Introduction

The Jahn–Teller (JT) polaron was originally proposed as a possible origin of superconductivity in $La_{2-x}Ba_xCuO_4$ by Bednorz and Müller [1]. Since then, the JT effect has been discussed by a number of authors in the context of polaron formation and superconductivity [2–8]. Recently, Kabanov and Mihailovic [9, 10] considered the case of Jahn–Teller electron-phonon coupling at finite wavevector – where hybridized electronic states at finite wavevector are coupled to $k \neq 0$ phonons. The Jahn–Teller interaction together with long-range Coulomb interaction V (the Jahn–Teller-Coulomb or JTC model) was shown to lead to the formation of diverse textures such as the one shown in Fig. 1 [11, 12]. Relevance to superconductivity is evident from the clustering dynamics: the density of clusters of different size shown in the right panel of Fig. 1 as a function of doping x shows that pairs (clusters with $j = 2$) are dominant over a large part of the phase diagram, coexisting with larger clusters and single particles. It is implicit in the JTC model that the pairing boson, which is the cause for the formation of such textures, is a high-frequency JT-active lattice vibration. In order to gain wider acceptance of the JTC model and the importance of the JT-effect in general, we need to show that its predictions – particularly the involvement of the lattice in pairing and pseudogap formation – can be observed in experiments.

The driving force for spatial inhomogeneity in the JTC model arises from the competition of long and short range forces, but this can also arise from

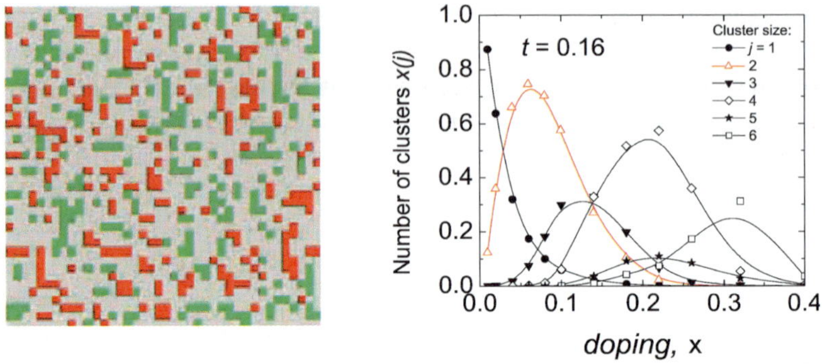

Fig. 1 **a** The JT polaron texture state predicted by the JTC model at high doping. ($x = 0.3$). The two colours correspond to the two types of states with orthorhombic distortion axes along x and y, respectively. **b** Cluster size statistics calculated on the basis of Monte-carlo simulations of the JTC model show that pairs (i.e. clusters of size 2) are present over a large part of the phase diagram, shown here at low temperature $t = 0.16$, where $t \sim T/T^*$ and T^* is the pseudogap temperature

magnetic interactions. Notably, Low, Emery, Fabricius and Kivelson [13] and many others since then have considered a mechanism for stripe formation based on short-range ferromagnetic interactions and long-range Coulomb interactions, which also naturally leads to phase separation, stripe formation and similar textures as shown in Fig. 1. Thus, since both magnetic and Jahn–Teller interactions in combination with long-range Coulomb repulsion can lead to phase separation, in order to determine which interaction is responsible for the phase separation in the cuprates, we must turn to further experiments. In this paper we begin by briefly examining some of the existing evidences that the model is appropriate for describing the functional behaviour of the cuprate superconductors and then proceed to the more fundamental issue of the pairing boson. We argue that the energy released upon recombination of pairs or the formation of the pseudogap state cannot be absorbed by spin excitations, but must be absorbed by the lattice, directly implying that the pairing boson is a lattice excitation.

Evidence for Jahn–Teller Interactions

Since the first review on the experimental evidence for phase separation in 1996 [14], the two-component paradigm associated with the existence of electronic and lattice inhomogeneity has proven quite robust, and in the last ten years it has gained wider acceptance. Magnetic [15], optical [16–18], EPR [19–21], NMR [22, 23], and more recently scanning tunneling microscopy (STM) [24–26] experiments have shown the presence of electronic inhomogeneity, while local lattice probes such as XAFS [27, 28] and PDF [29,

30] have shown that the local structure – presumably due to lattice polarons and stripes – differs from the average structure. Thus one aspect of the model, namely phase separation, appears to have been firmly established.

A more microscopic indication that the JTC model is relevant for describing the cuprates comes from the symmetry breaking phenomena which it predicts. The bipolarons and stripe clusters which form as a result of the JTC model (Fig. 1) do not have the tetragonal (or quasi-tetragonal) crystal symmetry, but transform according to an irreducible representation of the little group at finite k [9]. The symmetry breaking caused by local pair or stripe formation is directly observable – and has indeed been reported in numerous experiments [9, 10].

The observation of an isotope effect on both T_c [31] and T^* [32] firmly establishes a role of the lattice in the pairing mechanism. However, in spite of a large body of experimental evidence, still a large part of the community appears to be in favour of a magnetic-mediated superconductivity mechanism. A common argument for spin-mediated superconductivity is that only HTS cuprates have low spin $S = 1/2$ copper states, which makes them unique for achieving high T_c's. A counter-argument is that there are quite a few other oxide superconductors which do not have magnetic ions, notable examples being doped $SrTiO_3$, $LiTi_2O_4$, $BaBiO_3$. The very high $T_c = 34$ K of $Ba_{1-x}K_xBiO_3$ shows that the presence of magnetic ions is not essential to achieve moderately high T_c's, and the presence of the pseudogap and incipient structural instability in this compound is very similar to the cuprates. However, neither argument is proof that the glue is mediated by spins or the lattice. Thus the problem which is still outstanding is that there are so far none – or very few – experiments which relatively unambiguously exclude the possibility of either lattice or spin as the origin of the pairing interaction and superconductivity.

Quasiparticle Recombination and Emission of the Glue Boson

To study the "glue boson" involved in the recombination of photoexcited charge carriers in $La_{2-x}Sr_xCuO_4$, we have conducted a study of the ultrafast QP dynamics by laser excitation in the high-excitation regime. The availability of nearly 4 orders of magnitude of photoexcited QP densities ranging from $n_p \ll n_0$ to $n_p \geq n_0$ enables us to continuously cover the crossover from a low- to a high-density regime. In the high intensity limit, a single laser pulse can easily break all pairs within the excited volume, effectively destroying the condensate and enabling us to determine an upper limit to the condensation energy U_c. Furthermore, at such high photoexcited carrier densities, the Cu spins are expected to be entirely quenched, which enables us to verify whether the pseudogap is of magnetic origin. We concentrate on the energy conservation issues, showing that the electronic and magnetic degrees of freedom cannot take up the energy released upon QP recombination, leaving emission of lattice excitations as the only possibility.

Time resolved photoinduced reflectivity experiments were performed using the standard pump-probe technique [33] on $La_{2-x}Sr_xCuO_4$ ($T_c = 30$ K). (The details of the experiments are reported elsewhere [34].) In these experiments, the photoinduced reflectivity $\Delta R(t)$ is proportional to the quasiparticle (QP) density n_p at any given time t [33, 35] enabling us to monitor the QP density in real time as a function of temperature, doping and photoexcitation carrier density. The measured photoexcited QP density at the superconducting gap Δ_s is estimated [33] to range from $\sim 10^{18}$ to 10^{22} cm^{-3}, easily exceeding the normal-state carrier density, $n_0 \simeq 10^{21}$ cm^{-3} (for $x = 0.1$) in La_{2-x} Sr_xCuO_4.

The time-dependence of the normalized photoinduced reflectivity change $\Delta R/R$ for different excitation intensities are shown in the left panel of Fig. 2 below T_c (4.4 K) and above T_c (100 K). Below T_c we can clearly identify two relaxation processes on short timescales, which we label as A and B. Immediately above T_c only signal B remains and is visible up to T^*. On the basis of previous extensive systematic experiments on many different materials [35, 36], components A and B are assigned to recombination across the superconducting gap $\Delta_s(T)$ and the "pseudogap" Δ_p respectively [33, 34].

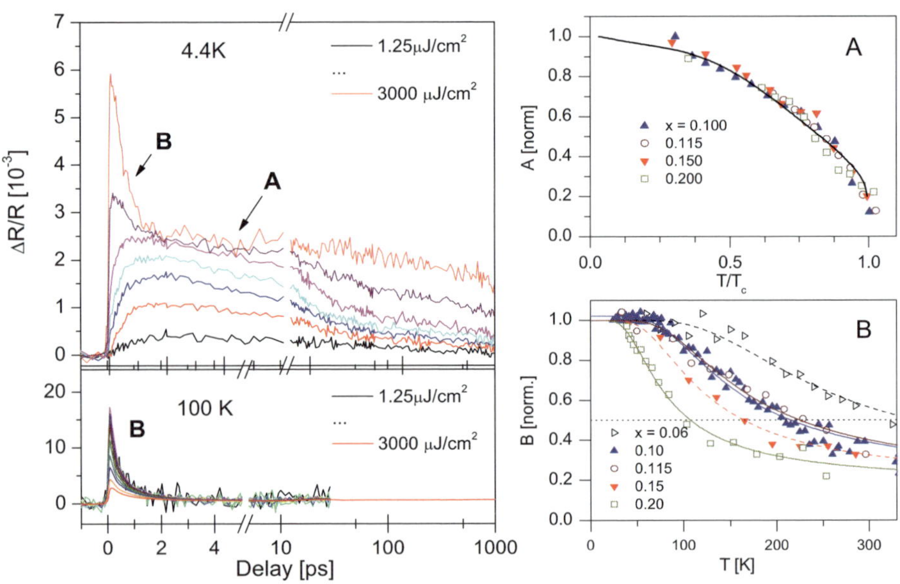

Fig. 2 *Left*: The photoinduced reflectance $\Delta R/R$ as a function of time below and above T_c at different excitation fluences $\mathcal{F}$. Note, that signal A is present only in the superconducting state, while signal B is present both, in the superconducting state and the pseudogap state up to T^*. *Right*: The amplitude of signals A (*top*) and B (*bottom*) as a function of T for different doping levels x. The lines are a fit to the data using the RTK model [34]

The two signals *A* and *B* corresponding to the QP density $n_s(T)$ and $n_p(T)$ have very distinct temperature dependences, associated with T_c and T^* respectively as shown in Fig. 2 (right panels). The data for *A* (top-right panel in Fig. 2) can be fitted very well with the established Rothwarf–Taylor–Kabanov (RTK) model [33, 37, 38] for the superconducting response, while the component *B* (bottom-right panel in Fig. 2) shows a characteristic pseudogap response and is fit using a two-level system in the RTK bottleneck regime [34]. As can be seen in Fig. 3, the pseudogap temperatures T^* and the pseudogap values from the RTK fits of Δ_p agree well with other experiments such as magnetic susceptibility [15, 39]. The same is true for the superconducting gap Δ_s, unambiguously establishing the assignment of the two signals *A* and *B* to the superconducting gap recombination and the pseudogap recombination respectively [34].

The RTK model, while derived for a general case of a superconductor with emission of phonons is perfectly applicable – at least in principle – to magnons as the pairing glue. Thus, in the pairing process, one might suggest that a zone boundary magnon is emitted instead of a high-frequency phonon. Since the amount of energy that the magnetic subsystem can absorb in the recombination process is much smaller than the lattice subsystem (given by the ratio of the lattice and magnetic specific heats C_L/C_M [40]), the decisive question which the experiments can try to answer is what happens at high photoinduced densities where a large amount of energy is released very quickly.

To answer this question, we examine the behaviour of the superconducting QP density n_s and the QP n_p density in the pseudogap state under conditions

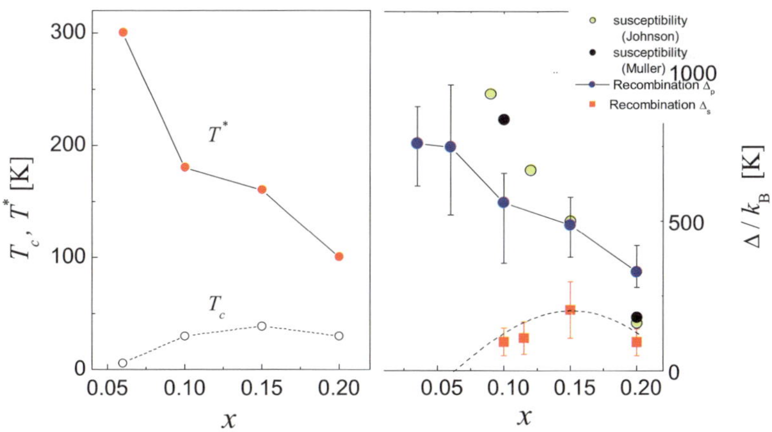

Fig. 3 The dependence of T^* and Δ_s and Δ_p as a function of doping. While T^* is read out directly from the T-dependence as shown in Fig. 2, the values of Δ are obtained from the fits to the same data using the RTK model. Data on the pseudogap are compared to susceptibility data from Muller et al. [15] and from Johnson et al. [39]

of high photoexcitation intensity. In Fig. 4 we have plotted the QP densities of n_s and n_p corresponding to signals *A* and *B* respectively as a function of excitation fluence $\mathcal{F}$. The left panel shows the behaviour of both n_s and n_p in the superconducting state at low $\mathcal{F}$, while the right panel shows the behaviour at very high $\mathcal{F}$. The saturation of n_s (signal *A*) at $\mathcal{F} \simeq 12\,\mu J/cm^2$ occurs when all the QPs are excited across the superconducting gap. Remarkably, in the PG state, signal *B* remains present up to $\mathcal{F} > 1500\,\mu J/cm^2$, indicating that the PG state is very robust with respect to photoexcitation. Even more remarkably, the *T*-dependence of n_p (Fig. 5) is apparently virtually unchanged at high $\mathcal{F}$ compared to low-density excitation, meaning that Δ_p is virtually unchanged by the high density of photoexcited carriers. This means that a large amount of energy is rapidly dissipated out of the electronic system on the timescale of a few 100 femtoseconds.

The optical energy deposited in the excitation volume corresponding to the condensate vaporization threshold $\mathcal{F} = 12\,\mu J/cm^2$ at 4.4 K (Fig. 4) is $U_p \simeq 3.4\,K/Cu$. This value is 5.6 times larger than the BCS value $U_{BCS} = N(E)\Delta_s^2/2 \sim 0.6\,K/Cu$ and much larger than the measured condensation energy $U_c = 0.12\,K/Cu$ extracted from specific heat data [41], where we have taken Δ_s from Fig. 3 and $N(E) \simeq 1\,eV^{-1}\,Cu^{-1}$. This indicates that a substantial amount of energy (approximately 5 U_{BCS}) is released within the 10 ps time window [38]. However, U_p is significantly smaller than the energy $U_L \simeq 9.3\,K/Cu$ required to heat the lattice degrees of freedom form 4.4 K to 30 K in $La_{2-x}Sr_xCuO_4$ calculated by integrating the experimentally measured specific heat $c_p(T)$ [42]. Thus, the lattice can easily take up all the energy re-

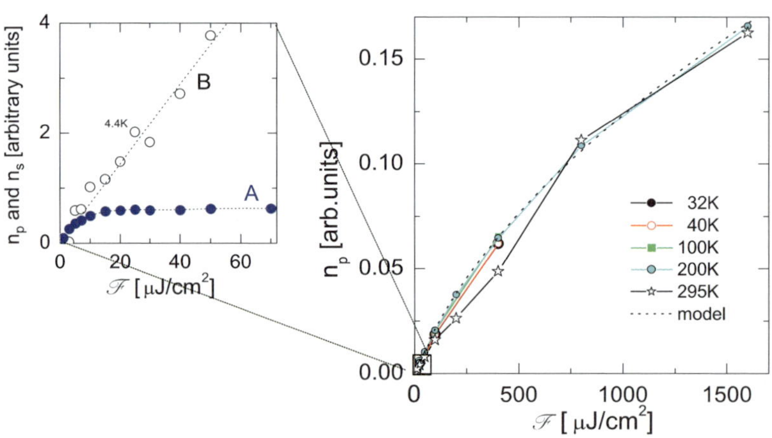

Fig. 4 The intensity dependence of the QP density in the superconducting state (*left*) and in the pseudogap state (*right*). The point when all the QPs in the superconducting condensate are excited is clearly visible at $12\,\mu J/cm^2$, which corresponds to the superconducting condensation energy of 3.4 K/Cu. Remarkably, the PG state appears not to show signs of saturation up to $> 1500\,\mu J/cm^2$

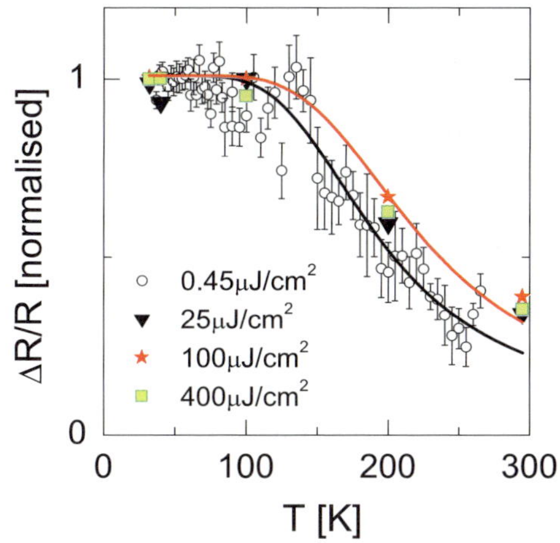

Fig. 5 The temperature dependence of the QP density in the pseudogap state for different fluences. A remarkable stability of the PG state with respect to high-intensity photoexcitation can be inferred from the relatively weak dependence of T^* (or Δ_p) on $\mathcal{F}$ up to $1500\,\mu J/cm^2$

leased in the recombination process. Comparing with the maximum available magnetic reservoir energy $U_M \simeq 0.01$ K/Cu, (using the magnetic specific heat in undoped La_2CuO_4 [40]), we see that the magnetic system is incapable of relaxing the photoexcited QPs in the superconducting state, its heat capacity being too small by a factor of $U_p/U_M \simeq (3.4\ K/Cu)/(0.01\ K/Cu) = 340$.

The situation is even more convincingly in favour of lattice interactions causing the formation of the *pseudogap state*. In Fig. 4 we see that the PG is robustly present up to $\mathcal{F} > 1500\,\mu J/cm^2$. The QP density at this point is approximately $n_p^p = \frac{\mathcal{F}}{\Delta_p \lambda_{op}} \frac{e-1}{e} \simeq 10^{22}\ cm^{-3} \simeq 1.2/Cu$. At such densities, the Cu atoms would be expected to be in a zero spin ($S = 0$) 3 + state, so any magnetic state based on interactions between Cu^{2+} $S = 1/2$ spins would be quenched. It is hard to envisage that the pseudogap state would remain intact and Δ_p unchanged under such conditions.

We end by returning to a discussion of the recombination dynamics in the context of the JTC model [11]. The process of recombination formally describes the pairing event (or cluster-formation event) by considering the dynamics of the transition from an unbound state to a bound state. The clusters shown in Fig. 1 of course statistically fluctuate, but each time a pair or a cluster is formed, the event is accompanied by the release of energy Δ per particle. This energy is relatively accurately measurable in the time-resolved experiments and is given by Δ_p and Δ_s for pseudogap excitations and superconducting state excitations, respectively. The recombination dynamics experiments appear to unambiguously imply that the glue involved is a lattice excitation of energy $\hbar\omega > \Delta$, lending support to electron-phonon interaction driven pairing mechanisms, such as the JTC model. Indeed, in previous work on ultrafast QP dynamics in the cuprates we have been able to describe the entire dynamics quantitatively and self-consistently using the RTK [33, 37] model. While the RTK model itself does not define the pairing boson, we have now been able to identifying the "glue" by carefully considering model-independent energy conservation arguments.

References

1. Bednorz JG, Müller KA (1986) Z Phys B 64:189
2. Gor'kov LP, Sokol A (1987) JETP Lett. 46:420
3. Markiewicz RS (1992) Physica C 200:65
4. Englman R, Halperin B, Weger M (1990) Physica C 169:314
5. Weger M, Englman R (1999) Physica A 168:324
6. Bersuker GI, Goodenough JB (1997) Physica C 274:267
7. Kamimura H, Suwa Y (1993) J Phys Soc Jpn 62:3368
8. Kamimura H et al. (1996) Phys Rev Lett 77:723
9. Mihailovic D, Kabanov VV (2001) Phys Rev B 63:054505
10. Kabanov VV, Mihailovic D (2002) Phys Rev B 65:212508
11. Mertelj T, Kabanov VV, Mihailovic D (2005) Phys Rev Lett 94:147003
12. Mertelj T, Kabanov VV, Mihailovic D, to be published
13. Low U, Emery VJ, Fabricius K, Kivelson SA (1994) Phys Rev Lett 72:1918
14. Mihailovic D, Müller KA (1996) In: High-T_c Superconductivity: Ten Years after the Discovery, edited by Kaldis E, Liarokapis E, Müller KA (Kluwer Academic Publishers, Dordrecht, 1997), p 243
15. Müller KA (1989) Physica C 159 Issue 6:717–726
16. Sugai S (1990) Sol Stat Comm 76:371
17. Kabanov VV, Mihailovic D (2002) Phys Rev B 65:212508
18. Misochko OV, Georgiev N, Dekorsy T, Helm M (2002) Phys Rev Lett 89:067002
19. Shengelaya A, Bruun M, Kochelaev BI, Safina A, Conder K, Müller KA (2004) Phys Rev Lett 93:017001
20. Kochelaev BI et al. (1997) Phys Rev Lett 79:4274
21. Kochelaev BI et al. (1994) Phys Rev B 49:13106
22. Teplov MA et al. (1999) J Superconductivity 12:113
23. Teplov MA et al. (1997) In: Kaldis E, Liarokapis E, Müller KA (eds) High-T_c Superconductivity 1996: Ten Years after the Discovery. Kluwer Academic Publishers, Dordrecht, p 531
24. Derro DJ et al. (2002) Phys Rev Lett 88:097002
25. Lang KM (2002) Nature 415:412
26. Pan SH, Hudson EW, Lang KM, Eisaki H, Uchida S, Davis JC (2000) Nature 403:746
27. Bianconi A, Saini NL, Lanzara A, Missori M, Rossetti T, Oyanagi H, Yamaguchi H, Oka K, Ito T (1996) Phys Rev Lett 76:3412
28. Haskel D, Stern EA, Dogan F, Moodenbaugh AR (2000) Phys Rev B 61:7055
29. McQueeney RJ, Petrov Y, Egami T, Yethiraj M, Shirane G, Endoh Y (1999) Phys Rev Lett 82:628
30. Bozin E, Billinge S, Kwei GH, Takagi H (1999) Phys Rev B 59:4445
31. Faltens TA et al. (1987) Phys Rev Lett 59:915

32. Lanzara A, Guo-meng Zhao, Saini NL, Bianconi A, Conder K, Keller H, Müller KA (1999) J Phys Condens Matter 11 48:L541
33. Kabanov VV, Demsar J, Podobnik B, Mihailovic D (1999) Phys Rev B 59:1497
34. Kusar P, Demsar J, Mihailovic D, Sugai S (2005) Phys Rev B 72:014544
35. Dvorsek D, Kabanov VV, Demsar J, Kazakov SM, Karpinski J, Mihailovic D (2002) Phys Rev B 66:020510
36. Demsar J, Hudej R, Karpinski J, Kabanov VV, Mihailovic D (2001) Phys Rev B 63:054519
37. Rothwarf A, Taylor BN (1967) Phys Rev Lett 19:27
38. Kabanov VV, Demsar J, Mihailovic D (2005) Phys Rev Lett 95 72:147002
39. Johnston DC (1989) Phys Rev Lett 62:957
40. Singh MR, Barrie S (1998) Phys Stat Sol b 205:611
41. Matsuzaki T, Momono N, Oda M, Ido M (2004) J Phys Soc Jpn 73:2232
42. Wen HH et al. (2004) Phys Rev B 70:214505

Oyanagi H (2007) In: Bussmann-Holder A, Keller H (eds)
High T_c Superconductors and Related Transition Metal Oxides.
Springer-Verlag Berlin Heidelberg, 253–258
DOI 10.1007/978-3-540-71023-3

Lattice Effects in High-Temperature Superconducting Cuprates Revealed by X-ray Absorption Spectroscopy

H. Oyanagi

Photonics Research Institute, AIST, 1-1-1 Umezono, Tsukuba, 305-8568 Ibaraki, Japan
h.oyanagi@aist.go.jp

Abstract In contrast to conventional superconductivity where phonons lead to the formation of Cooper pairs, in high-temperature superconductivity (HTSC), the role of electron–phonon coupling has long been neglected. The in-plane Cu – O bonds in HTSC cuprates show unconventional broadening at low temperature as carriers are doped. Here we focus on the high-quality polarized X-ray absorption spectroscopy (XAS) data for a model HTSC system, $(La,Sr)_2CuO_4$ (LSCO). Thin film single crystal samples were prepared by state-of-the-art MBE, precisely controlling compositions. High-quality data was obtained by use of a segmented X-ray detector. The in-plane Cu – O radial distribution function (RDF) in LSCO ($x = 0.15$) shows broadening as temperature is lowered, which shows a sharp drop at the critical temperature which is followed by a gradual increase (disorder). Comparing the data with resistivity, we find a remarkable coincidence between the sharpening and the onset of superconductivity. Since the sharpening of RDF is interpreted as correlated motion of oxygen atoms (phase coherence due to superconductivity), the results demonstrate that the superconducting state directly relates to the unconventional oxygen displacements in a bond stretching mode. The results will be discussed in relation to local models of distortion of the different nature (metallic vs. insulating) that is strongly influenced by strain.

Keywords $(La,Sr)_2CuO_4$ · Electron–phonon coupling · High-temperature superconductivity · X-ray absorption spectroscopy

Strong coupling between electrons and lattice vibrations (phonons) in high temperature superconducting (HTSC) [1] materials was recently demonstrated by angle-resolved photoemission spectroscopy (ARPES) [2]. X-ray absorption spectroscopy (XAS) revealed the local displacement of oxygens as carriers are doped indicating intimate relation between lattice effects and the pairing mechanism [3]. Isotope [4] and strain (pressure) [5–7] effects are considered as direct experiments that establish the role of lattice in HTSC. Under epitaxial strain, T-$(La,Sr)_2CuO_4$ (LSCO) thin film single crystals exhibit significant strain dependence of the superconducting critical temperature and metallic transport properties [5–7]. In order to probe local lattice of thin film single crystals, highly sensitive XAS technique is needed. For this purpose, we have developed a state-of-the-art pixel array detector (PAD) which minimizes contribution of substrates. Here we report that the results of polarized XAS investigating the local structure of LSCO under strain [8]. The results indicate that the in-plane Cu – O bonds undergo temperature dependent disorder and superconductivity-induced phase coherence which correlates with the onset of superconductivity. We suggest that the pairing mechanism involves the bond-stretching mode local lattice distortion.

Thin films of LSCO single crystal were grown in a molecular-beam epitaxy (MBE) chamber from metal sources using multiple electron-gun evaporators with accurate stoichiometry control of the atomic beam fluxes [9]. The samples were grown on $LaSrAlO_4$ (LSAO) and $SrTiO_3$ (STO) substrates (10 mm × 10 mm). The temperature dependence of resistivity measured by a four probe method showed superconductivity at higher temperature than that of bulk single crystal by ca. 24% ($T_c = 43.4$ K with $\Delta T < 1.0$ K) for LSAO and lower temperature by ca. – 46% ($T_c = 19$ K with $\Delta T < 9.0$ K) for STO substrates. The oxygen concentration was carefully controlled by post-growth oxidation by ozone gas annealing. The lattice parameters of thin films were obtained from X-ray diffraction patterns. Sample (10 mm × 10 mm) is mounted on an aluminum holder and attached to a closed-cycle helium refrigerator which rotates on a high precision goniometer (Huber 420) to change the incidence angle. Samples are 100 nm thick LSCO grown on $LaSrAlO_4$ (LSAO) and $SrTiO_3$ (STO) [10]. The fluorescence signal is recorded over a cone-like solid angle perpendicular to the incidence beam away from scattering plane. This arrangement was used to measure EXAFS with the electric field vector $\boldsymbol{E}$ parallel with the *ab* plane ($\boldsymbol{E}//ab$), while a sample is rotated by 90 degrees for probing along the *c*-axis ($\boldsymbol{E}//c$). XAS measurements were performed in a fluorescence detection mode at BL13B, Photon Factory. The beam energy and maximum electron current were 2.5 GeV and 400–450 mA, respectively. A directly water-cooled silicon (111) double-crystal monochromator was used. Careful alignment of sample geometry on a precision goniometer was repeated until artifacts are not present in the region of interest. The sample was attached to an aluminum holder with a strain-free glue and cooled down using a closed cycle He refrigerator.

Normalized in-plane polarized Cu K-EXAFS oscillations were obtained from the fluorescence yield $F(E)$ subtracting the contribution from atomic absorption μ_0. The radial distribution function (RDF) for the Cu – O and Cu – Cu pairs were separately analyzed using a single scattering formula and theoretical phase shift functions calculated by FEFF7 [11]. The experimental EXAFS data for the nearest-neighbor (NN) in-plane oxygens (O_p) were fitted in *k*-space by theoretical curves. The determined $R_{Cu\text{–}O_p}$ values for LSCO/LSAO and LSCO/STO are 1.88 ± 0.01 Å and 1.90 ± 0.01 Å, respectively. The in-plane Cu – O distances ($R_{Cu\text{–}Op}$) in strained LSCO compared with the crystallographic spacings ($a_0/2$) show that the CuO_6 octahedra are tetragonally deformed in accordance with the variation of the lattice constants.

In Fig. 1, the temperature dependence of the in-plane oxygen displacement $\sigma^2_{Cu\text{–}O}$ is shown in (a) as a function of T while the same plot in an expanded

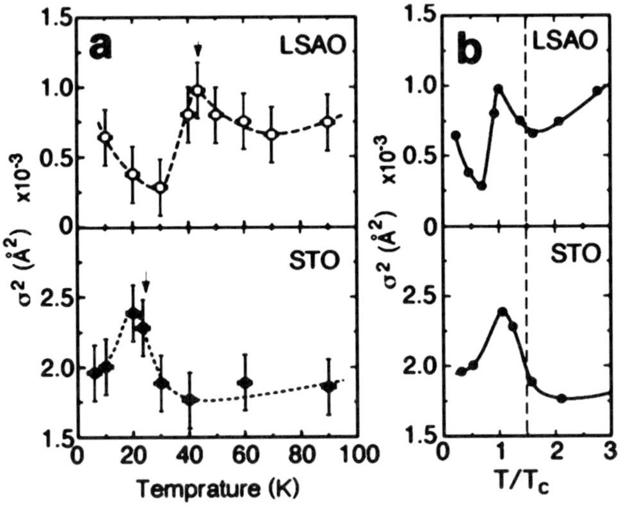

Fig. 1 **a** Mean-square relative displacement $\sigma^2_{Cu-O_p}$ plotted as a function of temperature for the in-plane $Cu-O_p$ distance in $(La, Sr)_2CuO_4$ (LSCO) thin film single crystals grown on $SrTiO_3$ (STO) and $LaSrAlO_4$ (LSAO) substrates. The arrows indicate the positions of the critical temperatures. The *dashed line* is an eye-guide showing the evolution of strain-induced inhomogeneity. **b** $\sigma^2_{Cu-O_p}$ plotted as a function of normalized temperature (T/T_c) for the in-plane $Cu-O_p$ distance in LSCO thin film single crystals grown on STO and LSAO substrates

horizontal scale is shown in (b) against T/T_c. The in-plane oxygen disorder becomes prominent below a much higher temperature than T_c $(T/T_c \sim 1.5)$ as oxygens take a two-site distribution (undistorted and distorted configuration giving the long and short R_{Cu-O_p} values, respectively) [3]. A distinct σ^2_{Cu-O} peak, indicating a maximum fraction of distorted configuration, appears at T_c, which sharply drops, correlating well with the change of resistivity (expressed by arrow). The development of dynamic disorder (incommensurate inhomogeneity) at low temperature is considered as an universal feature for doped HTSC cuprates [12]. We interpret the increase of disorder as increasing volume fraction of distorted oxygen sites which disppears as electron pairs are formed and phase coherence is achieved, in agreement with the observation that the split RDF merges at T_c (Mustre de leon J et al., unpublished results). The nearest neighbor (NN) EXAFS contribution for LSCO/STO ($T_c = 19$ K) is not reproduced by a single-domain model (undistorted homogeneous CuO_2 planes). This is due to inhomogeneity in lattice or the coexistence of different local structures (domains). The NN FT magnitude function can be fitted by a two-domain model having two different $Cu-O_p$ distances without broadening. The temperature-dependent oxygen displacement at low temperature is overwhelmed by the dynamic disorder (distortion) as seen by local probes. The existence of two different Cu – O distances ascribed to undistorted domains (UD) and distorted domains (DD) have been reported [3, 9]. Details of the strain-dependence of the local structure is elsewhere [9].

Figure 2 (top row) illustrates the possible local distortions in LSCO where ΔR indicates magnitude of oxygen displacement under a specific distortion type, i.e., LTT [3] or Q_2 type Jahn–Teller (JT) [13]. The magnitude of distortion is obtained by fitting models to the experimental data. In the right square panels, two-dimensional views of distorted CuO_2 plane are schematically illustrated for the two cases of distortion. As shown theoretically [13],

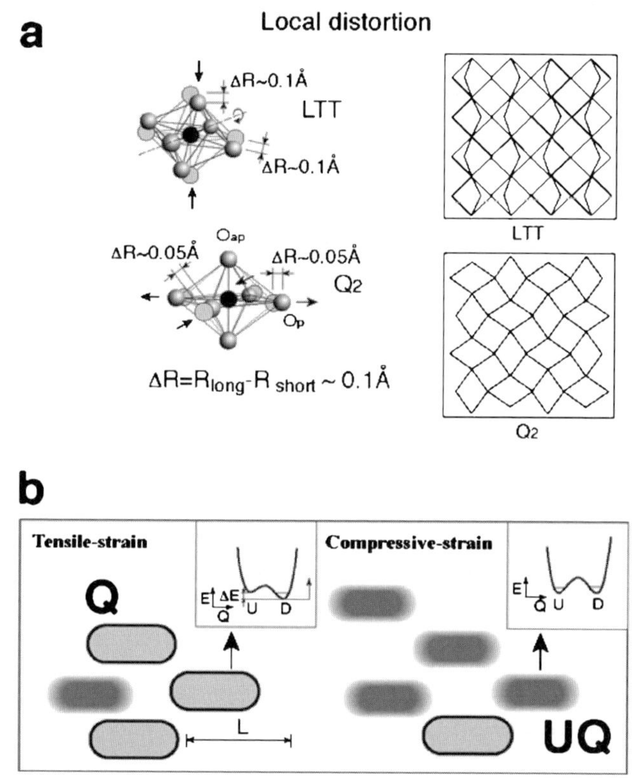

Fig. 2 **a** Schematic illustration of the local distortions in $(La, Sr)_2CuO_4$ (LSCO), i.e., low-temperature tetragonal (LTT), and Q_2-type pseudo Jahn–Teller (JT) distortions (*left row*), and schematic displacement of the CuO_2 plane (*right row*). **b** Evolution of distorted domains (inhomogeneity) under strain. In the top, the insets show the double-well potentials of displaced oxygens under strain. As temperature is lowered across about $1.5T_c$, clusters of dynamic local distortion (unquenched disorder, UQ) start coalescing, increasing their dimensions. Under tensile strain, static distortions (quenched disorder, *Q*) are stabilized. When the length of coalesced domains (*L*) exceed the critical length of coherence length ζ_0 ($L > \zeta_0$), mobile domains (UQ) turn into phase-coherent superconducting state

dynamical distortions are driving insulating antiferromagnetic (AF) domains metallic. In contrast, we find that tensile strain stabilizes domains with bond-stretching type distortion that is unfavourable to superconductivity. This static distortion is quenched disorder (expressed as Q in (b)) and immobile while the dynamic one (expressed as unquenched, UQ). The distorted configuration (Q) which localizes electron states is stabilized under tensile strain. In superconducting doped cuprates, dynamical disorder domain (UQ) is dominant and mobile domains of dynamic distortion coalesce. When the length scale (L) exceeds that of coherence length ζ_0 ($L > \zeta_0$), the UQ domains become uniform as phase coherence develops starting tunnelling between them. As shown Fig. 1, upon cooling, the in-plane oxygen displacement increases below T^* ($\sim 1.5T_c$) gives a maximum at T_c and sharply drops at T_c accompanied with a change in resistivity. As schematically shown in the insets of Fig. 2b, the evolution of oxygen displacement under tensile strain pulls down the energy changing the character of distortion from dynamical to static one. Instantaneous local distortion apparently disappears once phase coherence is formed.

In summary, we used X-ray absorption spectroscopy (XAS) to probe the local structure of a model HTSC system, $(La, Sr)_2CuO_4$ (LSCO) single-crystal thin films to investigate lattice effects under two-dimensional strain. Here we show that thein-plane lattice distortion strongly correlates with the superconducting critical temperature. The results indicate that first, the local lattice distortion is involved in the pairing mechanism and second, it is dependent on strain. Temperature-dependent radial distribution of the in-plane Cu – O bonds in strained LSCO shows that lattice inhomogeneity occurs as a result of nonuniform local distortion involving the Cu – O bond stretching, forming nanometer-scale patterns or domains reported by local probes. Local lattice inhomogeneity could be favorable to superconductivity with dynamical character but it changes its nature to quenched disorder under tensile strain which unfavorable. The fact that uniaxial strain controls the critical temperature (superfluid density) imposes constraints on the realistic models.

Acknowledgements The authors express their greatest thanks to K.A. Müller, A. Bussmann-Holder, H. Kamimura, A. Bianconi, T. Egami, and A. Bishop for their continuous encouragement and fruitful discussions.

References

1. Bednorz JG, Müller KA (1986) Z Phys B 64:189
2. Lanzara A, Bogdanov PV, Zhou XJ, Keller SA, Feng DL, Lu ED, Yoshida T, Eisaki H, Fujimori A, Kishio K, Shimoyama J-I, Noda T, Uchida S, Hussain Z, Shen Z-X (2001) Nature 412:510
3. Bianconi A, Saini NL, Lanzara A, Missori M, Rossetti T, Oyanagi H, Yamaguchi H, Oka K, Ito T (1996) Phys Rev Lett 76:3412

4. Khasanov R, Eshchenko DG, Luetkens H, Morezoni E, Prokscha T, Suter A, Garifanov N, Mali M, Roos J, Conder K, Keller H Phys Rev Lett 92:057602-I
5. Sato H, Naito M (1997) Physica C 280:178
6. Locquet J-P, Perret J, Fompeyrine J, Machler E, Seo JW, Van Tendeloo G (1998) Nature 394:453
7. Bozoic I, Logvenov G, Belca I, Narimbetov B, Sveklo I (2002) Phys Rev Lett 89:107001
8. Oyanagi H, Tsukada A, Naito M, Saini NL (2007) Phys Rev B 75:024511
9. Naito M, Sato H (1995) Appl Phys Lett 67:2557
10. Tsukada A, Greibe T, Naito M (2002) Phys Rev B 66:184515
11. Rehr J, Zabrinsky SI, Albers RC (1992) Phys Rev Lett 69:3397
12. Saini NL, Lanzara A, Oyanagi H, Yamaguchi H, Oka K, Ito T, Bianconi A (1997) Phys Rev B 55:12759
13. Bussmann-Holder A, Keller H (2005) Eur Phys J B 44:487

Oeschler N, Fisher RA et al. (2007) In: Bussmann-Holder A, Keller H (eds)
High T_c Superconductors and Related Transition Metal Oxides.
Springer-Verlag Berlin Heidelberg, 259–268
DOI 10.1007/978-3-540-71023-3

Specific Heat of $Na_{0.3}CoO_2 \cdot 1.3H_2O$, a Novel Superconductor with Structural and Electronic Similarities to the High-T_c Cuprates

N. Oeschler[1] · R. A. Fisher[1] · N. E. Phillips[1] (✉) · J. E. Gordon[2] · M.-L. Foo[3] · R. J. Cava[3]

[1]Lawrence Berkeley National Laboratory and Department of Chemistry, University of California, Berkeley, CA 94720, USA
nephill@cchem.berkeley.edu

[2]Physics Department, Amherst College, Amherst, MA 01002, USA

[3]Department of Chemistry, Princeton University, Princeton, NJ 08544, USA

The specific heats of three samples of $Na_{0.3}CoO_2 \cdot 1.3H_2O$ are dramatically different, even though they were prepared in the same way and show none of the evidence of the poor sample quality that accounted for the sample dependence of the specific heat in early measurements on the cuprates. The differences are the result of a non-magnetic pair breaking action that progresses with increasing sample age and acts preferentially in one of two electron bands, the one with the smaller superconducting-state energy gap. For the two superconducting samples the difference in the specific heat is a combination of the change in the residual DOS associated with the pair breaking, and the shift in the relative contributions of the two bands to the superconducting condensate. For the other, non-superconducting sample, the pair breaking has weakened the electron pairing of the superconducting state to the degree that it has given way to a competing order. Time-dependent changes in structural parameters and critical temperature, which have been reported recently by others, support, and extend, the interpretation of the specific-data.

The 1986 discovery [1] of high-critical-temperature superconductivity in layered perovskite cuprates, for which critical temperatures (T_c) have reached 133 K [2], opened a new era in research on superconductivity. It also raised the question of whether Cu was unique, or whether high-T_c superconductivity might be found in similar systems with transition-metal ions replacing the Cu. Although Co was recognized as an interesting candidate almost immediately, the very fragile superconductivity of $Na_{0.3}CoO_2 \cdot 1.3H_2O$, with T_c only ~ 4.5 K, was not discovered until 2003 [3]. (The extreme sensitivity of the superconductivity to sample composition and history suggests that even then the discovery was a matter of chance.) Most of the structural and electronic features thought to be important for the superconductivity in the cuprates occur in $Na_{0.3}CoO_2 \cdot 1.3H_2O$, but there is one significant difference. In the cuprates, spin-1/2 Cu ions are in *corner-sharing* O octahedra or pyramids, in a two-dimensional approximately *square* array. In the parent undoped Mott insulator the Cu ions are ordered antiferromagnetically, and antiferomagnetic spin fluctuations are thought to play a role in the electron pairing in the doped superconducting phases. In $Na_{0.3}CoO_2 \cdot 1.3H_2O$, spin-1/2 Co ions are at the centers of *edge sharing* O octahedra, in a two-dimensional *triangular* array, which frustrates the magnetic interaction and may affect the superconductivity. Both experimental and theoretical work suggest that the supereconductivity is certainly unusual, and possibly unique.

Theoretical models, some based on particular structural or electronic features of $Na_{0.3}CoO_2 \cdot 1.3H_2O$, e.g., the triangular Co arrangement [4], and Fermi-surface nesting [5], have led to an unusual variety of proposed mechanisms for the electron pairing. In principle, the theoretical predictions could be tested by comparisons with properties such as the specific heat (C) that give information on the symmetry of the order parameter (OP), specifically, the existence of nodes in the energy gap. However, meaningful comparisons have been precluded by a "sample dependence" of thc experimental results that has prevented identification of the relevant "intrinsic" properties. The superconducting-state electron contribution to C (C_{es}) is qualitatively different in different samples, suggesting differences in the superconductivity. It also includes a sample-dependent normal-state-like contribution that corresponds to a "residual" density of states (DOS), which, in analogy with the cuprates, has been attributed to an incomplete transition to the superconducting state and the presence of "normal material". Mazin and Johannes [6] have made a systematic comparison of experimental results with the possible OP. They note that the *experimental* results have "led to an unprecedented number of proposals for the pairing symmetry"; they suggest that the superconductivity may be unique, but they recognize that "experimental reports are often contradictory and solid evidence for any particular pairing state remains lacking"; they have noted ambiguity in the interpretation of the specific heat in particular.

The specific-heat measurements that we describe here show that the sample-to-sample variations in the specific heat of $Na_{0.3}CoO_2 \cdot 1.3H_2O$ are related to sample age. The residual DOS is not associated with normal material, but with non-magnetic pair breaking that progresses with increasing sample age and acts preferentially in one of two electron bands to produce a shift in their relative contributions to the superconducting-state condensate. This shift, together with the increase in the residual DOS, accounts for the sample dependence of the specific heat. This interpretation of the specific-heat data is consistent with, and supported by, other time-dependent changes [7], which also suggest a structural basis for the differences, and the identification of the pair-breaking scattering centers with O vacancies. The structural changes related to the pair breaking are inextricably associated with the *occurrence* of the superconductivity [7], and in this sense the measured properties of different samples *are* the intrinsic properties – the intrinsic properties of slightly different materials with different levels of pair-breaking. The ambiguity in the specific-heat results is explained, but the seemingly difficult problem of accounting for substantial differences in the macroscopic properties on the basis of relatively subtle structural changes takes its place. The evidence for line nodes in the energy gap is ambiguous but pair breaking in the absence of paramagnetic centers does require a sign-changing *component* of the OP. The existence of two gaps implies the existence of a second sheet of the Fermi surface that contributes to the superconductivity, and demonstrates a similarity to another recently discovered unusual superconductor MgB_2 [8]. (In the case of MgB_2, the unusual feature is the 40-K T_c, which is thought to be the extreme upper limit for the phonon mechanism, but is comparable to that of some cuprates. The discovery of the superconductivity also seems to have been a matter of chance, although not for the same reason as for $Na_{0.3}CoO_2 \cdot 1.3H_2O$.) The evidence for two bands is itself of interest in connection with the controversy over the occurrence of ε_g' hole pockets in $Na_{0.3}CoO_2 \cdot 1.3H_2O$, the existence of which is central to several of the proposed pairing theories (see, e.g. [5]).

We have made measurements of C for three samples, including, for two samples, measurements of $C(B)$ in fields (B) to 9 T. The samples were prepared in the same way [9], but differed in sample age (storage time at ambient temperature before cooling for the specific-heat measurement), 3, 5, and 40 days, for Samples 2, 3, and 1, respectively. The concentration of paramagnetic centers, which are associated with the residual DOS in the cuprates, was negligibly small for Sample 2 and below the limit of detection for Samples 1 and 3. Shortly after preparation Sample 1 was superconducting but at the time of the measurements of C there was no superconductivity. There was, however, a weakly field-dependent specific-heat anomaly associated with a CDW transition near 7 K [10]. Samples 2 and 3 were superconducting with $T_c \sim 4.5$ K and showed no trace of the 7-K anomaly. The results are described in more detail elsewhere [11].

The general nature of the results for the superconducting samples is illustrated by $C(0)$ and $C(9T)$ for Sample 2 in Fig. 1a and b, which show the specific-heat anomaly at T_c and the absence of other transitions below 30 K. The low-T "upturn" in $C(9T)$ is in good agreement with the calculated hyperfine contribution (C_{hyp}). After subtraction of C_{hyp}, $C(9T)$ to 12 K was combined with $C(0)$ from 6 to 12 K, and fitted as the sum of the normal-state electron contribution ($C_{en} = \gamma_n T$) and the lattice contribution ($C_{lat} = B_3T^3 + B_5T^5 + B_7T^7$) to obtain C_{lat} and $\gamma_n = 16.1$ mJ K^{-2} mol^{-1}. For $B \leq 1$ T, small T^{-2} terms, e.g., an upturn in $C(0)$, which is barely perceptible in Fig. 1a, are evidence of a contribution associated with paramagnetic centers (C_{mag}) at a concentration of $\sim 10^{-3}$ mol/mol sample, which make no other observable contribution to $C(B)$ [11]. The conduction-electron contribution is $C_e(B) = C(B) - C_{lat} - C_{hyp}(B) - C_{mag}(B)$. $C_e(0)$, which is C_{es} for

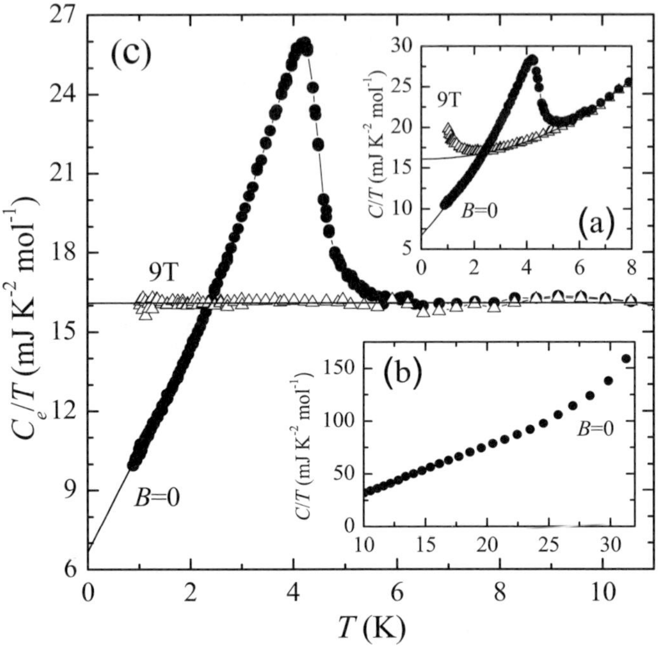

Fig. 1 **a,b** The total specific heat. **c** The electron contribution in the normal state (9 T) and in the superconducting state ($B = 0$)

$T \leq T_c$, and $C_e(9T) = C_{en}$ are shown in Fig. 1c. Below 2 K, C_{es}/T is linear in T: C_{es} is the sum of T-proportional and T^2 terms. The 0-K intercept, the coefficient of the T-proportional term and a measure of the residual DOS, is $\gamma_r = 6.67$ mJ K^{-2} mol^{-1}. The T^2 term, a signature of line nodes in the energy gap (but see below), has also been identified by Yang et al. in a similar sample [12]. For Sample 3, $\gamma_n = 15.7$ mJ K^{-2} mol^{-1}; there is no evidence of a T^2 term in C_{es}; the lowest-T data can be fitted as the sum of the exponential term expected for a "fully-gapped" superconductor and $\gamma_r T$, with $\gamma_r = 11.0$ mJ K^{-2} mol^{-1}.

All specific-heat measurements on $Na_{0.3}CoO_2 \cdot 1.3H_2O$ that permit reasonably unambiguous extrapolations to 0 K give finite values of γ_r. By analogy with *some* of the early specific-heat results on the cuprate superconductors, the finite values of γ_r have been attributed to a volume fraction γ_r/γ_n of normal material. In this "normal-material" model, the electron contribution to the specific heat for one mole of superconducting material is $C'_{es} = [(C_{es} - \gamma_r T)]/(1 - \gamma_r/\gamma_n)$, and specific-heat data for $Na_{0.3}CoO_2 \cdot 1.3H_2O$ have usually been displayed in this form. C'_{es} for Samples 2 and 3 is shown in Fig. 2, which emphasizes the dramatic difference in the superconductivity for the two samples. The entropy-conserving constructions in Fig. 2 give $T_c = 4.52$ K and $\Delta C'_{es}(T_c)/\gamma_n T_c = 1.28$ for Sample 2; $T_c = 4.65$ K and $\Delta C'_e(T_c)/\gamma_n T_c = 2.08$

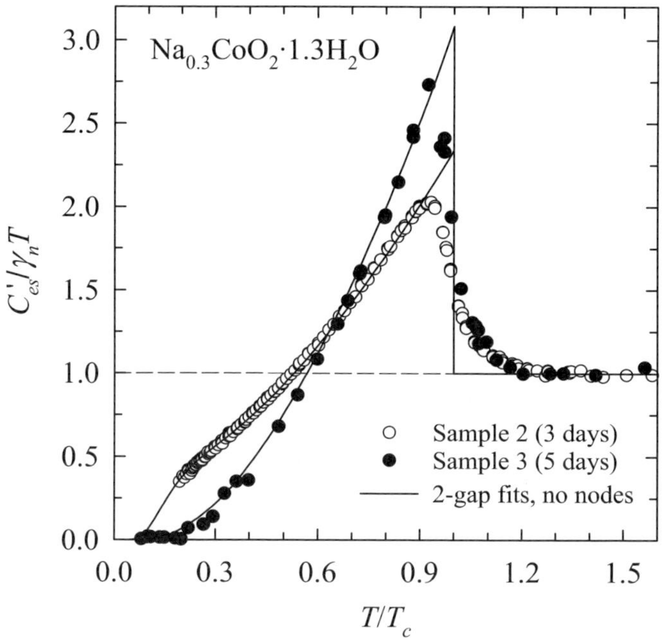

Fig. 2 The superconducting-state electron specific heat "for one mole of superconducting material", C'_{es}, obtained from C_{es} by subtracting $\gamma_r T$ and scaling by $1/(1-\gamma_r/\gamma_n)$ (see text for explanation). The *solid curves* are 2-gap fits, without nodes, for Samples 2 and 3

for Sample 3. Vortex-state data for Sample 2 are shown in the same form, i.e., as $C'_e(B) = [(C_e(B) - \gamma_r T)]/(1-\gamma_r/\gamma_n)$, in Fig. 3. The unusual feature of these results is that the temperature of the onset of the transition to the vortex state is independent of the magnetic field. Strong fluctuations are known to produce such an effect in the cuprate superconductors, but there is no obvious reason to expect them in $Na_{0.3}CoO_2 \cdot 1.3H_2O$.

Two mechanisms for the residual DOS were recognized in early specific-heat measurements on poor quality samples of the cuprates. There were samples with essentially the same T_c but different values of $\Delta C_e(T_c)$ that decreased with increasing γ_r. These results were interpreted in terms of different volume fractions of the *same* normal and superconducting phases, with normal regions occurring on the scale of the coherence length, which is of the order of a lattice parameter in the cuprates. Penetration-depth measurements suggested a similar interpretation, which in that context was described as the "Swiss cheese" model. There were also series of samples for which increases in γ_r were correlated with decreases in T_c, and these results were interpreted in terms of pair breaking. For these samples there were generally indications of the presence of paramagnetic centers, e.g., *substantial* low-T "upturns" in $C(0)/T$, and in-field Schottky anomalies. The reductions in T_c were relatively small in comparison with the corresponding effect in classical gapless super-

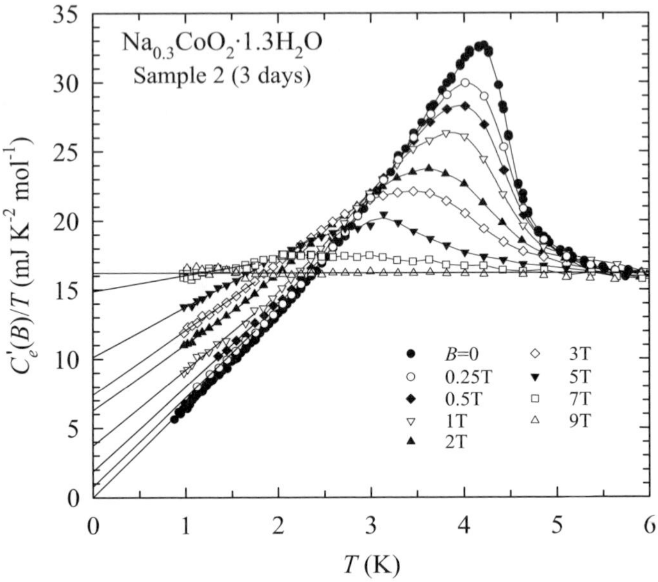

Fig. 3 Vortex-state electron specific heat for Sample 2

conductors, but this was explained theoretically as a consequence of resonant scattering by the pair-breaking scattering centers. As shown in Fig. 2, the superconductivity is *not* the same in Samples 2 and 3, and the Swiss cheese model, at least as it was understood for the cuprates, is not applicable. The longer coherence length in $Na_{0.3}CoO_2 \cdot 1.3H_2O$ also makes that model less plausible. The specific-heat results do not completely rule out the presence of normal material, but they are more consistent with pair breaking as the origin of the residual DOS. The same sample-dependent residual DOS measured by γ_r is seen in measurements of the spin-lattice relaxation time (T_1). Those results have generally been attributed to pair breaking (see, e.g. [13, 14]), and they present a compelling case for that mechanism: It is possible to fit the time dependence of the relaxation at 1.7 K [14] as a superposition of contributions from normal and superconducting regions, but the same mixture of normal and superconducting regions would not fit the data at other temperatures. This shows that the *same* nuclei see both the residual DOS and T_c, and they are not in separate phases. For pair breaking, the separation of C_{es} into contributions from states filled by pair breaking and from states associated with the superconducting-state condensate of Cooper pairs is not strictly correct, but there are reasons to expect it to be a reasonable approximation: Empirically, both C_{es} (as shown in Figs. 1–3) and $1/T_1$ (see, e.g. [13, 14]) show such a separation, and calculations of the effect of pair breaking [15] give good agreement with the T_1 data [14]. In addition, for classic gapless superconductors, e.g., Th – Gd alloys [16], C_{es} is, to a good approximation, the sum of $\gamma_r T$ and BCS-like terms. With these results as justification, we take $C_{es} - \gamma_r T = (1 - \gamma_r/\gamma_n)C'_{es}$ as the contribution of the superconducting condensate to C_{es}.

For Sample 2 C'_{es} shows substantial, unusual deviations from BCS theory that are strikingly similar to those for MgB_2 (Fig. 4). At the time of the discovery of its superconductivity, the deviations for MgB_2 were unprecedented, but it is now well established that they are a result of additive contributions to C_{es} from two electron bands, one with a small superconducting-state energy gap that produces the positive deviations at low T, and one with a large gap that produces the strong-coupling character of C_{es} near T_c, the positive curvature in Fig. 4. A a two-gap fit to C_{es} gives the parameters characteristic of the two contributions in good agreement with theory and other experimental results [17]. (More recently, Sr_2RuO_4 has shown somewhat similar deviations from BCS theory [18], but they are also associated with contributions from bands with different gaps [18, 19]). The unusual deviations of C_{es} from BCS theory and the similarity to MgB_2, as well as penetration-depth measurements [20], show the presence of two gaps in $Na_{0.3}CoO_2 \cdot 1.3H_2O$.

The C'_{es} data are well represented by two-gap fits, similar to that for MgB_2 [17], in which the 0-K gap amplitude, $\Delta(0)$, is taken as an adjustable parameter represented by $\alpha \equiv \Delta(0)/k_B T_c$, which is 1.764 in the weak-coupling limit (for a single-gap superconductor). For each band, $i = 1, 2$, the fits give

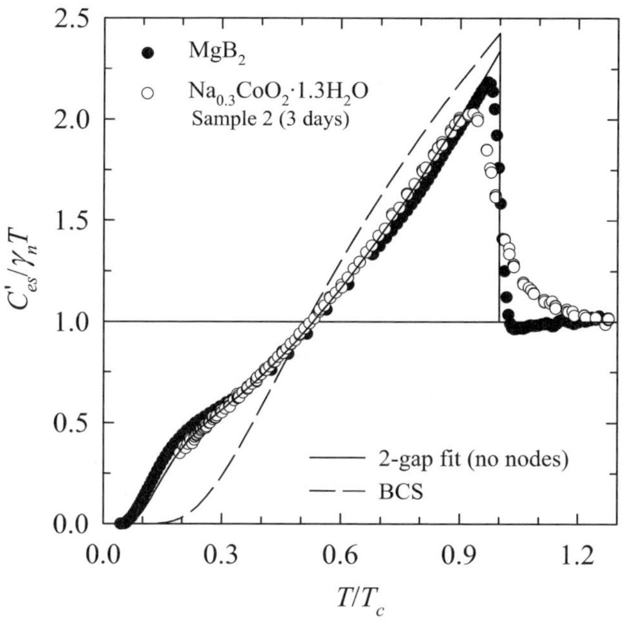

Fig. 4 The superconducting-state electron specific heat for Sample 2, compared with MgB_2 and BCS theory. The *solid curve* is a 2-gap fit without nodes

α_i and the fractional contribution of the normal-state DOS to the superconducting condensate γ_{is}/γ_n. For Sample 2, two 2-gap fits that represent the data within the experimental uncertainty are shown in Fig. 5, together with the contribution of each band to C'_{es}. One, the dotted curve in Fig. 5, gives $\alpha_1 = 2.15$, $\alpha_2 = 1.00$, $\gamma_{1s}/\gamma_n = 0.32$, and $\gamma_{2s}/\gamma_n = 0.27$. It extrapolates the T^2 dependence of the experimental data to 0 K and corresponds to the presence of nodes in the small gap. The other, the solid curve in Fig. 5 (and the solid curves in Figs. 2 and 4), gives $\alpha_1 = 2.20$, $\alpha_2 = 0.70$, $\gamma_{1s}/\gamma_n = 0.32$, and $\gamma_{2s}/\gamma_n = 0.27$. It extrapolates to 0 K exponentially in $1/T$, and corresponds to the absence of nodes. For Sample 3, the deviations from BCS theory are less obvious but they still suggest two gaps [11]: The low-T deviations associated with the small gap are not evident in Fig. 2, but the high values and positive curvature of C'_{es}/T near T_c are clear indications of strong coupling ($\alpha > 1.764$) and the large gap. A two-gap fit without nodes, shown in Fig. 2, gives $\alpha_1 = 2.30$, $\alpha_2 = 1.10$, $\gamma_{1s}/\gamma_n = 0.24$, and $\gamma_{2s}/\gamma_n = 0.06$.

Correlations among sample age, the degree of pair breaking, and the changes in C'_{es} show that the sample-to-sample differences are all related to sample age. The DOS associated with pair breaking, measured by the fraction γ_r/γ_n, increases with sample age: for Sample 2 (3 days), $\gamma_r/\gamma_n = 0.41$; for

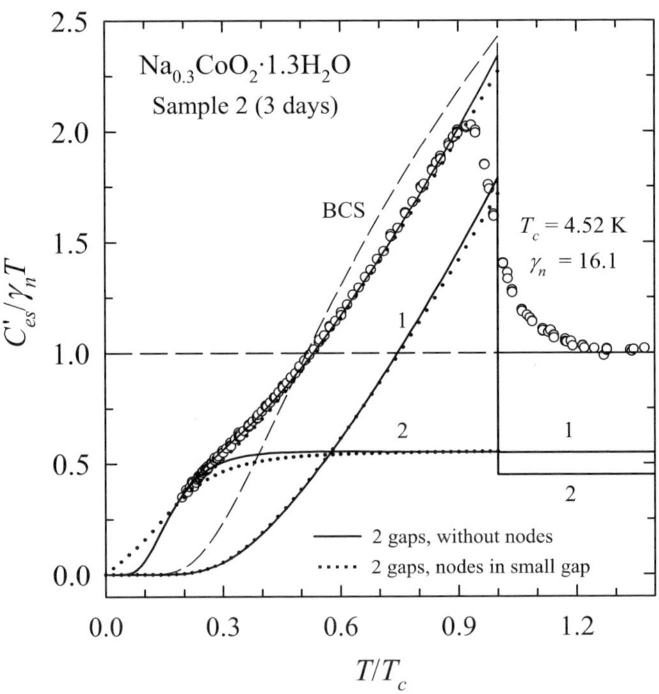

Fig. 5 Superconducting-state electron specific heat for Sample 2 with two 2-gap fits, and the two components of each fit, distinguished by the numbers 1 and 2

Sample 3 (5 days), $\gamma_r/\gamma_n = 0.70$; for Sample 1 (40 days), there is no superconductivity, and in that sense the pair breaking is complete. For the superconducting samples the DOS associated with the superconducting condensate is measured by the complementary fractions, 0.59 and 0.30, for Samples 2 and 3, respectively. The decrease, 0.29, is the sum of the decrease in γ_{is}/γ_n in each band, 0.08 in the large-gap band, and 0.21 in the small-gap band. The pair breaking acts in both bands, but, not surprisingly, it acts preferentially in the small-gap band, in which the pairing is weaker. This shift in the relative contributions to the superconducting condensate is the difference in the superconductivity. For Sample 1 the pair breaking is even stronger; the superconducting-state electron pairing has been weakened to the degree that it has given way to a competing order.

Comparisons with other changes that occur on a similar time scale support and extend this interpretation of the specific-heat results: Several studies have shown that samples, which are initially not superconducting, become superconducting, T_c passes through a maximum, ~ 4.5 K, and superconductivity disappears, all on a time scale on the order of days [7, 21]. Changes in lattice parameters and an increase in the concentration (δ) of O vacancies progress steadily, beginning before superconductivity appears, and continuing after its disappearance [7]. Identification of the O vacancies as the pair-breaking scattering centers would account for the specific heat of all three samples: Sample 3 would have a higher value of δ and a higher level of pair breaking than Sample 2, but essentially the same T_c, with the decrease in T_c that might be expected to accompany the increased pair breaking compensated by the competing effects that must be present to give the maximum in T_c. Sample 1, with a still higher δ, would not be superconducting. The correlations among δ, γ_r, and T_c suggest that *any* sample with a high T_c will also have a substantial γ_r. Unlike the cuprates and heavy-Fermion superconductors, improvements in sample quality may not reduce γ_r in $Na_{0.3}CoO_2 \cdot 1.3H_2O$.

For Sample 2, C_{es} can be fitted with or without the low-T T^2 term that would indicate the presence of line nodes in the small gap. For Sample 3, there is no evidence of the T^2 term, but, since the pair breaking that has increased the $\gamma_r T$ term would also decrease the T^2 term (Scalapino DJ, private communication), the existence of nodes cannot be ruled out. While the existence of *nodes* is left uncertain, the pair breaking without magnetic centers requires a sign-changing *component* of the OP, and rules out any kind of s-wave pairing, including "extended s-wave", which could give nodes. A doped t–J model on a triangular lattice [4] gave $d_{x2-y2} + id_{xy}$ pairing, which has no nodes, and is consistent with the specific heat. Of the four combinations of pairing symmetries and scattering mechanisms considered by Bang et al. [16], d_{x2-y2} or $d_{x2-y2} + id_{xy}$ pairing, with unitary or Born-limit scattering, $d_{x2-y2} + id_{xy}$ pairing with unitary scattering gave the best fit to T_1 data [14], and the calculated DOS is, qualitatively at least, consistent with the specific heat. These considerations suggest $d_{x2-y2} + id_{xy}$ pairing as a plausible candidate.

In summary, the specific heat shows the presence of two energy gaps in the superconducting state, which implies contributions to the DOS from two bands with different superconducting-state energy gaps. In this respect, $Na_{0.3}CoO_2 \cdot 1.3H_2O$ is similar to MgB_2. Non-magnetic pair breaking that increases with sample age produces the residual DOS. It acts preferentially in the band with the smaller energy gap to produce the sample-to-sample differences in the superconducting-state condensate. The results do not determine unequivocally the presence or absence of nodes in the energy gap. However, they do require a sign-changing component of the OP, and in that respect there is some similarity to the cuprates.

Acknowledgements The work at LBNL was supported by the Director, Office of Basic Energy Sciences, Materials Sciences Division of the U. S. DOE under Contract No. DE–AC03–76SF00098; at Princeton, by NSF grant DMR-0213706 and a DOE-BES grant DE-FG02-98-ER45706. N. O. was supported, in part, by the DAAD.

References

1. Bednorz JG, Müller KA (1986) Z Phys B 64:189
2. Schilling A et al. (1993) Nature (London) 363:56
3. Takada K et al. (2003) Nature 422:53
4. Wang Q-H et al. (2004) Phys Rev B 69:92504
5. Johannes MD et al. (2004) Phys Rev Lett 93:097005–1
6. Mazin II, Johannes ME (2005) Nature Physics 1:91
7. Barnes PW et al. (2005) Phys Rev B 72:134515
8. Nagamatsu J et al. (2001) Nature (London) 410:63
9. Schaak RE et al. (2003) Nature 424:527
10. Lee D-H et al., to be published
11. Oeschler N et al., to be published
12. Yang HD et al. (2005) Phys Rev B 71:20504
13. Ishida K et al. (2003) J Phys Soc Jpn 72:3041
14. Fujimoto T et al. (2004) Phys Rev Lett 92:47004
15. Bang Y, Graf MJ, Balatsky AV (2003) Phys Rev B 68:212504
16. Finnemore DK et al. (1965) Phys Rev 137:A550
17. Fisher RA et al. (2003) Physica C 385:180
18. Deguchi K et al. (2004) Phys Rev Lett 92:047002–1
19. Zhitomirsky ME, Rice TM (2001) Phys Rev Lett 87:057001–1
20. Yuan HQ et al. (2005) Bull Am Phys Soc 50:929
21. Ohta H et al. (2005) J Phys Soc Jpn 74:3147

Schneider T (2007) In: Bussmann-Holder A, Keller H (eds)
High T_c Superconductors and Related Transition Metal Oxides.
Springer-Verlag Berlin Heidelberg, 269–276
DOI 10.1007/978-3-540-71023-3

Quantum Superconductor-Metal Transition in Al, C doped MgB_2 and Overdoped Cuprates?

T. Schneider

Physik-Institut der Universität Zürich, Winterthurerstrasse 190, CH-8057 Zürich, Switzerland
toni.schneider@swissonline.ch

Abstract We consider the realistic case of a superconductor with a nonzero density of elastic scatterers, so that the normal state conductivity is finite. The quantum superconductor-metal (QSM) transition can then be tuned by varying either the attractive electron-electron interaction, the quenched disorder, or the applied magnetic field. We explore the consistency of the associated scaling relations, $T_c \propto \lambda(0)^{-1} \propto \Delta(0) \propto \xi(0)^{-1} \propto H_{c2}(0)^{1/2}$ and $T_c(H) \propto \lambda(0,\mathrm{H})^{-1} \propto \Delta(0,\mathrm{H}) \propto (H_{c2}(0) - H)^{1/2}$, valid for all dimensions $D > 2$, with experimental data, in Al, C doped MgB_2 and overdoped cuprates.

Keywords Al and C doped MgB_2 · Overdoped cuprates · Quantum superconductor-metal transition

Understanding the phenomenon of superconductivity, now observed in quite disparate systems, such as simple elements, fullerenes, molecular metals, cuprates, borides, etc., involves searching for universal relations between superconducting properties across different materials, which might provide hints towards a unique classification. In spite of the great impact of the BCS theory [1], the discovery of superconductivity in the cuprates in 1986 [2] made it clear that the BCS relations between the critical amplitudes of the gap (Δ_0), the correlation length (ξ_0), the magnetic penetration depth (λ_0), the upper critical field (H_{c20}) and the transition temperature T_c, namely [3]

$$T_c \propto \lambda_0^{-1} \propto \Delta_0^{-1} \propto \xi_0^{-1} \propto H_{c20}^{1/2} , \tag{1}$$

do not apply. Here, $\lambda(T) = \lambda_0 t^{1/2}$, $\Delta(T) = \Delta_0 t^{1/2} \simeq 1.76\Delta(0)t^{1/2}$, $\xi(T) = \xi_0 t^{-1/2}$, and $H_{c2} = H_{c20}t$, close to the superconductor metal transition, with $t = 1 - T/T_c$ and $2\Delta(0)/(k_B T_c) \simeq 3.52$. Furthermore, there are empirical relations between T_c and the zero-temperature superfluid density, $\rho_s(0)$, related to the zero-temperature magnetic field penetration depth $\lambda(0)$ in terms of $\rho_s(0) \propto \lambda^{-2}(0)$. In various families of underdoped cuprate superconductors there is the empirical relation $T_c \propto \rho_s(0) \propto \lambda^{-2}(0)$, first identified by Uemura et al. [4, 5], while in molecular superconductors, $T_c \propto \lambda^{-3}(0)$, appears to apply [6]. Both scaling forms appear to have no counterpart in the BCS scenario and even in more conventional superconductors, including $Mg_{1-x}Al_xB_2$, $Mg(C_xB_{1-x})_2$, and MgB_{2+x}, such relationships remain to be explored.

According to the theory of quantum critical phenomena a power law relation between T_c and $\rho_s(0) \propto \lambda^{-2}(0)$ is expected whenever there is a critical line $T_c(x)$ with a critical endpoint $x = x_c$ [7–9]. Here the transition temperature vanishes and a quantum phase transition occurs. x denotes the tuning parameter of the transition. A variety of underdoped cuprate superconductors exhibit such a critical line, ending at the quantum superconductor to insulator (QSI) transition, where the materials become essentially two dimensional [10]. If the finite temperature behavior in this regime is controlled by the 3D-xy critical point, T_c and $\rho_s(0)$ scale as [7–9]

$$T_c \propto \rho_s(0)^{z/(D+z-2)} . \tag{2}$$

z denotes the dynamic critical exponent of the quantum transition in D dimensions. For $D = 2$ we recover the empirical Uemura relation, $T_c \propto \rho_s(0)$ [4, 5], irrespective of the value of z.

There is also considerable evidence for a critical line $T_c(x)$ in more conventional superconductors, including $Mg_{1-x}Al_xB_2$, $Mg(C_xB_{1-x})_2$, $MgB_{2-x}Be_x$, and NbB_{2+x}, where superconductivity disappears at some critical value $x = x_c$ [11–14], whereupon a quantum superconductor to metal (QSM) transition is expected to occur. To illustrate this behavior we depicted in Fig. 1 the data for T_c vs. the nominal concentration x for $Mg_{1-x}Al_xB_2$ and $Mg(C_xB_{1-x})_2$ taken from Postorino et al. [11] and Gonnelli et al. [12]. Since in these nearly isotropic materials the anisotropy does not change substantially upon substitution the QSM transition occurs in $D = 3$. Furthermore, there is considerable

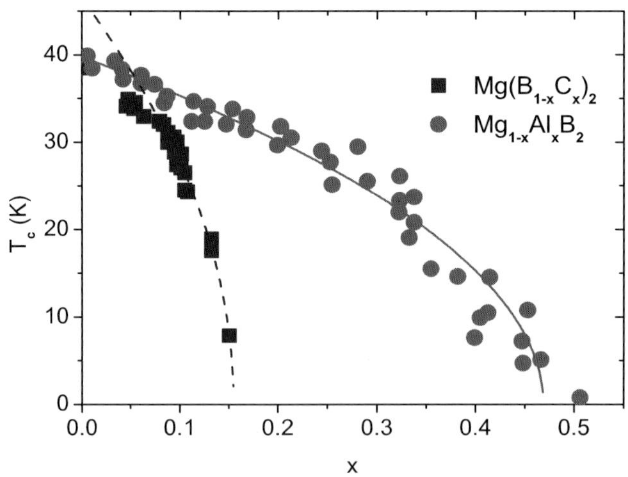

Fig. 1 T_c vs. the nominal concentration x for $Mg_{1-x}Al_xB_2$ and $Mg(C_xB_{1-x})_2$ taken from Postorino et al. [11] and Gonnelli et al. [12]

evidence that on increasing the Al or C content the homogeneity and the crystallographic order decrease even in segregation-free samples [15–17].

For this reason we consider the realistic case of a superconductor with a nonzero density of elastic scatterers, so that the normal state conductivity is finite. In the absence of an applied magnetic field the QSM transition can then be tuned by varying either the attractive electron-electron interaction, or the quenched disorder. In a theoretical description, quenched disorder can occur on a microscopic level, e.g., due to randomly distributed scattering centers. For such systems it was shown that the upper critical dimension D_c^+, above which the critical behavior is governed by a simple Gaussian fixed point (FP), is lower than that of the corresponding classical or finite temperature transition, namely $D_c^+ = 2$ [18, 19]. For $D > 2$ the transition is then governed by a Gaussian FP with unusual properties. Since the mean-field/Gaussian theory yields the exact critical behavior at $T = 0$, all relations between observables that are derived at finite temperature within BCS theory are valid. Accordingly, the zero temperature counterpart of the scaling relation (1) reads

$$T_c \propto \lambda(0)^{-1} \propto \Delta(0) \propto \xi(0)^{-1} \propto H_{c2}(0)^{1/2} , \quad (3)$$

while T_c and the dimensionless distance from the critical point δ ($\delta < 0$ in the disordered phase) are related by [18]

$$T_c \propto \exp(-1/|\delta|) . \quad (4)$$

Hyperscaling is violated by the usual mechanism that is operative above an upper critical dimension. Indeed, this QSM transition occurs in $D = 3$, while the upper critical dimension of the QSM transition is $D_c^+ = 2$ [18, 19]. For this reason the scaling relation (Eq. 2), involving hyperscaling, does not apply in the present case where $z = 2$ [18].

We are now prepared to confront the scaling predictions for a disorder tuned QSM transition with experiment. In Fig. 2 we show T_c vs. zero-temperature muon-spin depolarization rate $\sigma(0) \propto \rho_s(0) \propto \lambda^{-2}(0)$ for $Mg_{1-x}Al_xB_2$ taken from Serventi et al. [20]. Although the data is rather sparse, in particular close to the QSM transition, the reduction of the superfluid density $\sigma(0) \propto \rho_s(0) \propto \lambda^{-2}(0)$ with decreasing T_c is clearly observed, consistent with the flow to the QSM transition where T_c and $\sigma(0)$ scale as $T_c \propto 1/\lambda(0) \propto \sigma(0)^{1/2}$ (Eq. 3).

To substantiate this supposition further, we consider the T_c dependence of the gap $\Delta(0)$. In Fig. 3 we depicted the experimental data of Daghero et al. [17] for the two gap superconductor $Mg_{1-x}Al_xB_2$. Although the data does not extend very close to the QSM transition, the flow to $T_c \propto \Delta(0)$ (Eq. 3) can be anticipated. According to Fig. 4, showing $\Delta_\sigma(0)$ and $\Delta_\pi(0)$ vs. T_c for $Mg(B_{1-x}C_x)_2$ single crystals taken from Gonnelli et al. [15], the flow to the QSM transition (Eq. 3) is apparent in the σ-gap as well, while the π-gap, nearly constant down to $T_c = 19$ K, appears to merge the σ-gap below this transition temperature [15].

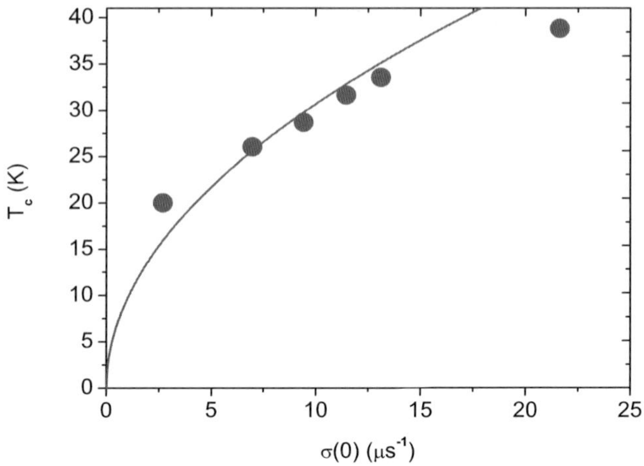

Fig. 2 T_c vs. zero-temperature muon-spin depolarization rate $\sigma(0) \propto \rho_s(0) \propto \lambda^{-2}(0)$ for $Mg_{1-x}Al_xB_2$ derived from Serventi et al. [20]. The solid line is Eq. 3 in terms of $T_c = 9.7\sigma_{sc}(0)^{1/2}$

Next we turn to the T_c dependence of the upper critical fields $H_{c2}^{ab}(0)$ and $H_{c2}^{c}(0)$. Although the experimental data for $Mg_{1-x}Al_xB_2$ single crystals taken from Klein et al. [16] and Kim et al. [21] shown in Fig. 5 is rather sparse, consistency towards QSM scaling behavior $H_{c2}^{ab,c}(0) \propto T_c^2$ (Eq. 3), indicated by the solid and dashed lines, can be anticipated. From these lines we deduce for the zero temperature anisotropy the estimate $\gamma(0) = (H_{c2}^{c}(0)/H_{c2}^{ab}(0))^{1/2} =$

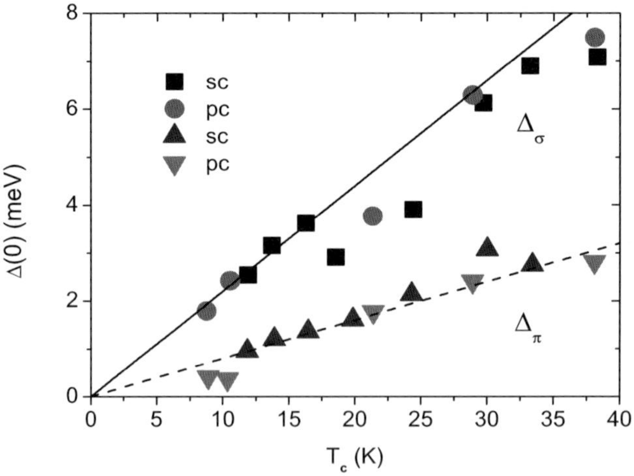

Fig. 3 $\Delta_\sigma(0)$ and $\Delta_\pi(0)$ vs. T_c for $Mg_{1-x}Al_xB_2$ single crystals (sc) and polycrystals (pc) taken from Daghero et al. [17]. The *solid* and *dashed lines* are (Eq. 3) in terms of $\Delta_\sigma(0) = 0.22T_c$ and $\Delta_\sigma(0) = 0.08T_c$

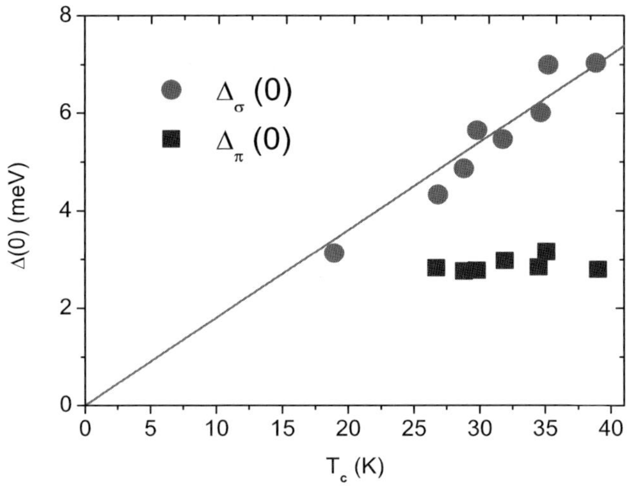

Fig. 4 $\Delta_\sigma(0)$ and $\Delta_\pi(0)$ vs. T_c for $Mg(B_{1-x}C_x)_2$ single crystals taken from Gonnelli et al. [15]. The *solid line* is (Eq. 3) in terms of $\Delta_\sigma(0) = 0.18 T_c$

$\xi_{ab}(0)/\xi_c(0) \simeq 1.9$. Because $H_{c2}^{ab}(0) \propto \xi_c(0)^{-2} \propto T_c^2$ and $H_{c2}^{c}(0) \propto \xi_{ab}(0)^{-2} \propto T_c^2$ the reduction of the upper critical fields mirrors the increase of the correlation lengths as the QSM transition is approached.

Finally we consider the magnetic field tuned QSM transition for fixed x. Noting that the correlation length and the magnetic field scale as $\xi(0)^2(H_{c2}(0,x) - H) \propto \Phi_0$, together with Eq. 4, the zero temperature gap

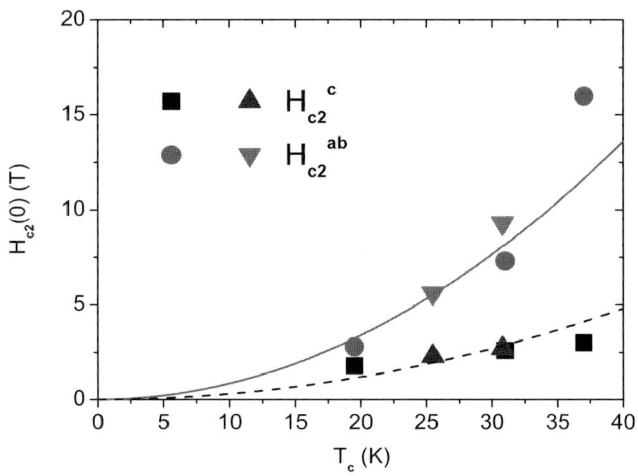

Fig. 5 $H_{c2}^{ab}(0)$ and $H_{c2}^{c}(0)$ vs. T_c for $Mg_{1-x}Al_xB_2$ single crystals taken from Klein et al. (■, •) [16] and Kim et al. (▲, ▼) [7]. The *solid* and *dashed lines* are Eq. 3 in terms of $H_{c2}^{ab}(0) = 0.0085 T_c^2$ and $H_{c2}^{c}(0) = 0.003 T_c^2$

scales then close to the QSM transition as

$$\Delta(0,x) \propto (H_{c2}(0,x) - H)^{1/2} . \tag{5}$$

Here the critical line $T_c(x,H)$ ends at $H_{c2}(T=0,x)$. In Fig. 6 we depicted $\Delta_\pi(T=6.5\,\text{K})$ vs. H applied along the c-axis of a $Mg_{1-x}Al_xB_2$ single crystals with $x=0.2$ ($T_c \simeq 24\,\text{K}$) taken from Giubileo et al. [22]. Although Eq. 5 represents the asymptotic behavior we observe remarkable agreement with the local tunneling data over the entire magnetic field range. Clearly, the occurrence of the magnetic field tuned QSM transition is not restricted to the gap. From Eqs. 3 and 5 we deduce

$$\begin{aligned} T_c(x,H) &\propto \lambda(0,x,H)^{-1} \propto \Delta(0,x,H) \\ &\propto (H_{c2}(0,x) - H)^{1/2} , \end{aligned} \tag{6}$$

which remains to be tested experimentally. The corresponding schematic phase diagram is shown in Fig. 7. As the substituent concentration or the magnetic field is increased T_c is suppressed and driven all the way to zero, where along the line $H_{c2}(T=0,x)$ the QSM transition, characterized by the scaling relations (Eq. 3) and (Eq. 6) occurs.

These scaling relations also imply that close to the QSM transition the isotope and pressure effects on these observables are not independent of one another. From Eqs. 3 and 6 we deduce for the relative changes upon isotope substitution or applied pressure the relations

$$\frac{\Delta T_c}{T_c} = \frac{\Delta a}{a} - \frac{\Delta\lambda(0)}{\lambda(0)} = \frac{\Delta b}{b} + \frac{\Delta\Delta(0)}{\Delta(0)} = \frac{\Delta c}{c} + \frac{\Delta H_{c2}(0)}{2H_{c2}(0)} , \tag{7}$$

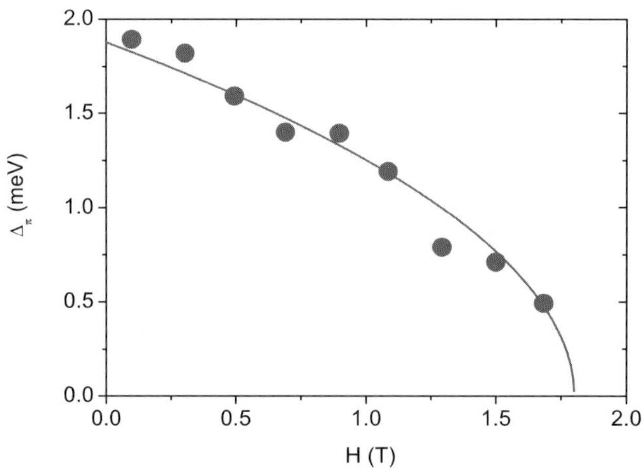

Fig. 6 $\Delta_\pi(T=6.5\,\text{K}, x=0.2)$ vs. H applied along the c-axis of $Mg_{1-x}Al_xB_2$ single crystals taken from Giubileo et al. [22]. The *solid line* is Eq. 5 in terms of $\Delta_\pi(T=6.5\,\text{K}, x=0.2) = 1.4(H^c_{c2}(0,x=0.2) - H)^{1/2}$ with $H^c_{c2}(0,x=0.2) = 1.8T$

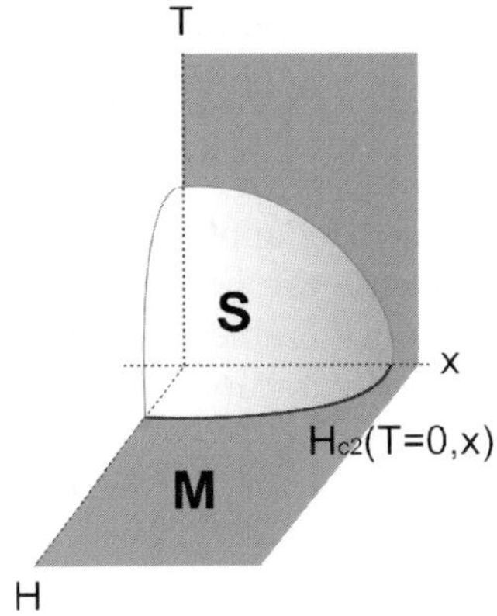

Fig. 7 Schematic phase diagram. There is the surface of finite temperature superconductor (S) to metal (M) transitions ending at the critical line $H_{c2}(T=0,x)$ of quantum superconductor-metal (QSM) transitions

where $T_c = a/\lambda(0) = b\Delta(0) = cH_{c2}(0)^{1/2}$ and

$$\frac{\Delta T_c(x,H)}{T_c(x,H)} = \frac{\Delta d}{d} - \frac{\Delta\lambda(0,x,H)}{\lambda(0,x,H)} = \frac{\Delta e}{e} + \frac{\Delta\Delta(0,x,H)}{\Delta(0,x,H)} \qquad (8)$$
$$= \frac{\Delta f}{f} + \frac{\Delta H_{c2}(0,x)}{2H_{c2}(0,x)},$$

where $T_c(x,H) = d/\lambda(0) = e\Delta(0,x,H) = fH_{c2}(0,x)^{1/2}$. a to f are non-universal coefficients.

We sketched, following Kirkpatrick and Belitz [18] the scaling relations of a quantum superconductor to metal (QSM) transition for nearly isotropic three dimensional systems, considering the realistic case of a superconductor with a nonzero density of elastic scatterers, so that the normal state conductivity is finite. The QSM transition can then be tuned by varying either the attractive electron-electron interaction, the quenched disorder, or the applied magnetic field. We have shown that $Mg_{1-x}Al_xB_2$ and $Mg(B_{1-x}C_x)_2$, where increasing Al or C enhances the disorder even in segregation-free samples [15–17], are potential candidates to observe this QSM transition, characterized by the scaling relations Eqs. 3, 6, 7, and 8. Indeed, as a whole, the spare experimental data points to this QSM transition, but more extended experimental data are needed to confirm this characteristic scaling relation unambiguously. Indeed, based on band structure calculations and the Eliashberg theory, it was argued that the observed decrease of T_c of Al and C doped MgB_2 samples can be understood mainly in terms of a band filling effect due to the electron doping by Al and C [23, 24]. Finally we note that NbB_{2+x} ([14, 25] and Khasanov et al., unpublished results), $Nb_{1-x}B_2$ [26], $MgB_{2-x}Be_x$ [13] are additional potential candidates, as well as overdoped cuprates [27–29]. In particular, evidence for $1/\lambda(0)^2 \propto T_c^2$ emerges for NbB_{2+x} from the muon-spin rotation study of Khasanov et al., unpublished results. Moreover, based on our analysis, a plot of T_c vs. $1/\lambda(0)^2$ of cuprate superconductors should rise more steeply in the underdoped limit ($T_c \propto 1/\lambda(0)^2$) than in the overdoped limit ($T_c \propto 1/\lambda(0)$). Various experiments appear to support this behavior qualitatively [4, 5, 27–29] but more data are necessary to confirm it quantitatively. On the other hand, there is considerable experimental evidence [30] that for overdoped cuprates the zero temperature gap is proportional to T_c (Eq. 3).

Acknowledgements The author is grateful to T.R. Kirkpatrick, D. Belitz, R. Khasanov and H. Keller for very useful comments and suggestions on the subject matter.

References

1. Bardeen J, Cooper LN, Schrieffer JR (1957) Phys Rev 108:1175
2. Bednorz JG, Müller KA (1986) Z Phys B 64:189
3. Ketterson JB, Song SN (1999) Superconductivity. Cambridge University Press, Cambridge
4. Uemura YJ et al. (1989) Phys Rev Lett 62:2317
5. Uemura YJ et al. (1991) Phys Rev Lett 66:2665
6. Pratt FL, Blundell SJ (2005) Phys Rev Lett 94:097006
7. Kim K, Weichman PB (1991) Phys Rev B 43:13583
8. Schneider T, Singer JM (2000) Phase Transition Approach to High Temperature Superconductivity. Imperial College Press, London
9. Schneider T (2004) In: Bennemann KH, Ketterson JB (eds) The Physics of Superconductors. Springer, Berlin
10. Schneider T (2003) Physica B 326:289
11. Postorino P et al. (2001) Phys Rev B 65:020507
12. Gonnelli RS et al. (2006) J of Physics and Chemistry of Solids 67:360
13. Ahn JS, Kim Y-J, Kim M-S, Lee S-I, Choi EJ (2002) Phys Rev B 65:172503
14. Escamilla R, Huerta L (2006) Supercond Sci Technol 19:623
15. Gonnelli RS et al. (2005) Phys Rev B 71:060503(R)
16. Klein T et al. (2006) Phys Rev B 73:224528
17. Daghero D et al., cond-mat/0608029
18. Kirkpatrick TR, Belitz D (1997) Phys Rev Lett 79:3042
19. Lubo Zhou, Kirkpatrick TR (2005) Phys Rev B 72:024514
20. Serventi S et al. (2004) Phys Rev Lett 93:217003
21. Kim H-J et al. (2006) Phys Rev B 73:064520
22. Giubileo F et al., cond-mat/0604354
23. de la Peña O, Aguayo A, de Coss R (2002) Phys Rev B 66:012511
24. Kortus J, Dolgov OV, Kremer RK (2005) Phys Rev Lett 94:027002
25. Takagiwa H et al. (2004) J Phys Soc Jpn 73:2631
26. Yamamoto A, Takao C, Masui T, Izumi M, Tajima S (2002) Physica C 383:197
27. Niedermayer C et al. (1993) Phys Rev Lett 71:1764
28. Uemura YJ et al. (1993) Nature 364:605
29. Bernhard C et al. (1995) Phys Rev B 52:10488
30. Peets DC et al., cond-mat/0609250

Scott JF (2007) In: Bussmann-Holder A, Keller H (eds)
High T_c Superconductors and Related Transition Metal Oxides.
Springer-Verlag Berlin Heidelberg, 277–285
DOI 10.1007/978-3-540-71023-3

Ferroelectric-on-Superconductor Devices

J. F. Scott

Centre for Ferroics, Earth Sciences Department, University of Cambridge,
Cambridge CB2 3EQ, UK
jsco99@esc.cam.ac.uk

Abstract A brief review of ferroelectric-on-superconductor devices is given, emphasizing strip lines for agile-frequency microwave devices and phased-array radar.

Introduction

The first ferroelectric-on-superconductor device was reported by the present author and his earlier Colorado coauthors in 1993–1994 [1–3]. Following that work a number of efforts have been reported in the USA and abroad, often led by Felix Miranda's group at NASA, O. Vendik's group in Leningrad/St. Petersburg, or Boikov's lab in Gothenberg. At present the commercial impact has not yet been large, owing to the dielectric loss, which in thin-film form renders the figure of merit somewhat disappointing. The phase shift per applied Volt is quite good, but the losses remain unacceptably high for many applications. Nevertheless, the present paper tries to give a brief account of developments over the past decade.

In practically all the devices considered the superconductor acts as a ground plane. Their operation below T_c is basically to make the sheet resistance of this ground plane negligible. Therefore liquid nitrogen operation is possible using high-T_c superconductors. Because ferroelectric thin films are employed, the dielectric loss is often dependent upon film thickness. This is probably due to damaged interfacial layers near the electrodes, as first shown in this context for BST by Kaiser et al. [4–6] at the National Bureau of Standards (now NIST); similarly, Zafar et al. [4–6] at IBM showed that the frequency dependence C(f) in BST films could be explained by assuming a 1.0 nF capacitance in series with the bulk capacitance of such films.

Following our initial work [1–3] with barium strontium titanate (BST) and pure strontium titanate ($SrTiO_3$) on yttrium barium copper oxide (YBCO) and on the thallium high-temperature superconductors (HTSCs) originally invented by Prof. Hermann in our group, there were some immediate applications by other groups to similar prototype devices [4–8]. A particularly outstanding study of fabrication and characterization of such devices was the report by Galt et al. [9, 10].

Early Prototypes

One of the first efforts was by Boikov et al. on sapphire [4–6]. Heterostructures with a thin (300 nm) layer of BST or $KTaO_3$ (KTO) ferroelectric layer between two YBCO superconducting films were laser deposited on Si-on-sapphire substrates (SOS). To avoid chemical interaction between Si and YBCO, a double yttrium-doped ZrO_2/CeO_2 (9.5% Y_2O_3) buffer layer was grown epitaxially on the (001) Si surface. This five-layer structure was the basis of several voltage controlled, microwave integrated devices.

The (001) planes of both BST and KTO were normal to the c-axis of the YBCO in the trilayers. The superconducting T_c of the bottom and top YBCO layers were 88–90 K. The loss ($\tan\delta$) in the BST layers at 50–100 K was in the range of 0.7–2.0% at 100 kHz.

This was followed soon by efforts on the Tl-based high-T_c materials [8]. Epitaxial $Tl_2Ba_2CaCu_2O_8$ thin films with excellent electrical transport characteristics were grown in a two-step process involving metal-organic chemical vapor deposition (MOCVD) of a BaCaCuO(F) thin film followed by a post-anneal in the presence of Tl_2O vapor. Vapor pressure characteristics of the recently developed liquid metal-organic precursors $Ba(hfa)_2$.mep, (hfa = hexafluoroacetylacetonate, mep = methylethylpentaglyme); $Ca(hfa)_2$.tet (tet = tetraglyme); and the solid precursor $Cu(dpm)_2$ (dpm = dipivaloylmethanate) were characterized. These precursors were employed in a low pressure horizontal, hot-wall, film growth reactor for growth of BaCaCuO(F) thin films on (110) $LaAlO_3$ substrates. The resulting films were shown to be epitaxial with the c-axis normal to the substrate surface. The best films exhibited T_c = 105 K, transport-measured J(c) = 1.2×10^5 A/cm^2 at 77 K, and surface resistances as low as 0.4 mΩ (40 K, 10 GHz).

By 1997 significant efforts were being made to optimize strip-line materials and architecture [9, 10]. Boikov's group measured the voltage tunable phase shift in coplanar waveguides based on HTSC/ferroelectric thin film structures. The transmission lines were fabricated from YBCO/BST double layers deposited on $LaAlO_3$ substrates. It was shown that a significant improvement of the tunability can be achieved by proper choice of the ferroelectric material and the waveguide dimensions. BST with $x = 0.05$ was chosen because its Curie temperature is close to 77 K. A phase shift of 254 degrees tuned by 35 V dc bias was demonstrated at 20 GHz and 75 K on a device with both YBCO and BST as epitaxial films.

At about this time careful studies of strontium titanate and BST films were underway, and the use of a resonator technique had been employed to measure the relative permittivity of single crystal squares of strontium titanate in the temperature range 20–300 K at microwave frequencies; it was found [11] that the method of metallization of the strontium titanate crucially affects the effective measured dielectric constant, with an air gap as small as 0.5 λ (visible light) giving a 50% error in the inferred value of ε at 20 K; as the gap size increases the corresponding error increases dramatically. This has important implications when using strontium titanate, or indeed any high-permittivity crystals, at cryogenic temperatures.

By 1998 other ferroelectric materials were being tried out on superconductors, and Regnier et al. [12] investigated lead scandium tantalate. A breakthrough also came in 1998 with two papers by Li et al. [13, 14] achieving very low losses in strontium titanate thin films, nearly as low as in single crystals. At the same time Merkulov et al. demonstrated via Raman spectroscopy how the metal layer influenced the dielectric properties of the strontium titanate [15], and a more general review appeared on the use of BST thin films for capacitors in random-access memories (DRAMs) [16].

By 1999 the Gothenberg group under Boikov was making progress: Using PLD deposition of BST on YBCO, they deposited [17] three kinds of ferroelectric-on-superconductor device: A trilayer capacitor; a planar interdigital capacitor; and a coplanar waveguide. A high voltage tunability was demonstrated (up to 40%) at a loss level as low as 1%. A compact YBCO/BST coplanar waveguide with a gap as narrow as 18 microns was tested as an electrically tunable phase-shifter, and field-induced phase shifts of more than 180 degrees were obtained with 35 V dc bias at 20 GHz. Shortly thereafter there followed a series of six papers from Felix Miranda's group. [18–23] the first of these described the performance of a K-band conductor/ferroelectric thin film/dielectric tunable diplexer that was developed as a critical part of a discriminator-locked tunable oscillator. This diplexer consisted of two K-band (18 to 22 GHz), two-pole YBCO/$SrTiO_3$/$LaAlO_3$ bandpass filters coupled to the microwave signal input port through a 50 W power splitter. When tested independently at 77 K and varying bias of ± 400 V, these filters exhibited a pass-band frequency shift greater than 1.7 GHz, while maintaining a non-de-embeded insertion loss near 2.0 dB. Electromagnetic modeling results for the diplexer indicate that varying the dielectric constant of the $SrTiO_3$ from 3000 (77 K value) to 500 (300 K value) shifts the diplexer's "cross-over" frequency from 19 to 21 GHz, with only minor variations in insertion loss. Measurements at 298 K of this diplexer circuit using a gold/$SrTiO_3$/$LaAlO_3$ structure resulted in non-deembeded insertion losses of 3.3 dB and 2.5 dB at the filters' maxima at 20.65 and 22.43 GHz, respectively, with a cross-over frequency at 21.07 GHz.

The second contribution described the correlation of electric field and critical design parameters such as the insertion loss, frequency tunability, return loss, and bandwidth of conductor/ferroelectric/dielectric microstrip tunable K-band microwave filter. This work consisted of (BST) ferroelectric thin film based tunable micro-strip filters for room temperature applications. Two new parameters which simplified the evaluation of ferroelectric thin films for tunable microwave filters were defined. The first of these, called the sensitivity parameter, is defined as the incremental change in center frequency with incremental change in maximum applied electric field – E(Peak) – in the filter.

The other, the loss parameter, is defined as the incremental or decremental change in insertion loss of the filter with incremental change in maximum applied electric field. At room temperature, the Au/BST/$LaAlO_3$ micro-strip filters exhibited a sensitivity parameter value between 15 and 5 MHz/cm/kV. The loss parameter varied for different bias configurations used for electrically tuning the filter, from 0.05 to 0.01 dB/cm/kV at room temperature. The third paper gave a detailed comparison between the performance of the YBCO/STO/LAO filters with that of gold/STO/LAO counterparts.

In the fourth and fifth papers of this series, the group reported for the first time the performance of a Au/BST/MgO two-layered microstrip voltage-controlled Lange coupler (VCLC) designed for Ku- and K-band frequencies at room temperature. Tight coupling of 3 dB or higher was obtained over a frequency range of 14–19 GHz. At K-band frequencies the coupling was voltage-controllable using the nonlinear de electric field dependence of the relative dielectric constant of BST.

Parenthetically we note that the physics behind such devices relies on the voltage (field) dependence of their capacitance C(E). In an RLC circuit, changing the capacitance with electric field results in changing the frequency. Unfortunately, in a thin film the capacitance varies with voltage V or field E for two different reasons: In the expression for the capacitance of the dielectric

$$C(E) = \varepsilon(E)A/d(E)\,. \tag{1}$$

The dielectric constant $\varepsilon(E)$ depends upon applied field in a way that can be calculated from the Devonshire-Landau free energy. This is the only contribution considered in work by Waser, Kingon et al. [24, 25] However, as can be seen in Eq. 1, the effective thickness $d(E)$ is also a function of field E and is related to the depletion width. This $d(E)$ is the only field dependence of $C(E)$ considered by most researchers who have come into the field from the semiconductor chip area [26, 27]. In reality, both $\varepsilon(E)$ and $d(E)$ must be considered carefully. This has not been done yet in the published literature and is not trivial, because in general it requires knowledge of the field distribution in a partially depleted semiconductor.

The present author has confirmed [28, 29] that for strontium titanate at 14 GHz and 77 K $C(V)$ varies as

$$C(V) = aV^{3/2}\,. \tag{2}$$

This result was first predicted theoretically by Tagantsev and Gurevich [30–33]. Reference [28, 29] also measured carefully the loss tangent versus temperature T and film thickness d over a wide temperature range (> 100 K) and showed that it satisfied the theoretical predictions [30–33]

$$\begin{aligned}\tan = &\left\{\tan d_s + \tan d + \varepsilon_s' d\left(1 + \tan^2 d_s\right) / \left[\varepsilon' d_s(1 + \tan^2 d)\right]\right\} / \\ &\left\{1 + \varepsilon_s' d(1 + \tan^2 d_s)/[1 + \tan^2 d]\right\}\end{aligned} \tag{3}$$

with only one completely adjustable parameter – a defective surface layer $d_s = 50$ nm thick; here in Eq. 3 the subscripts refer to the damaged surface layer and parameters without subscripts, to the interior of the film, so that tan d means the loss tangent of the undamaged film. The other parameters in Eq. 3 – tand d and tan d_s – are both functions of T but can be constrained by values of the loss tangent very far in temperature from the Curie temperature, and for very thick or very thin films.

The dependence of loss upon film thickness was fitted over from $d = 0.01$ microns to 0.5 microns with a single parameter ratio d_S/d of a little less than 0.01 characterizing the loss tangent from 1% to 7% over this spread of thickness. Details are given in [28, 29]. Thus, in summary, for BST thin films, [28, 29] established that the dependence of $C(V, T, d)$ upon voltage satisfied the theory of Tagantsev and Gurevich, and the dependence upon temperature T and film thickness d satisfied the theory of Neumann and Hoffmann [30–33]. This is not completely satisfactory, because the former theory is intrinsic and the latter extrinsic; but perhaps the functional dependence of the Gurevich-Tagantsev theory is correct even when the magnitudes are renormalized due to extrinsic contributions.

The coupling level was improved from – 11.6 to – 3.7 dB at 20 GHz with an applied de electric field of 16 kV/cm. The introduction of the ferroelectric tuning layer enhances the bandwidth of the VCLC in comparison with a Lange coupler with no ferroelectric layer. This work demonstrates another advantageous application for ferroelectric thin films in passive microwave components. An overview was also presented of the progress that Miranda's group had achieved toward integration of this technology into wireless and satellite communication systems.

Meanwhile by 2000 Hermann's group in Colorado had continued to make progress on PLD of BST films [28, 29]. Using interdigitalized capacitors, lower loss tangents (0.2% at 300 K) were observed for highly oriented BST films. Other groups in Beijing, Karlsruhe, and New York entered this line of research [30–33]. Rather good diagrams of X-band devices from the Russian group are shown by Petrov et al. [34] and Kozyrev et al. [35] The former paper illustrates a planar capacitor structure with an alumina substrate and Cu electrode; a strontium titanate film is applied to the substrate over a CeO_2 buffer layer. This capacitor is then mounted onto a micro-strip resonator. The latter paper details a YBCO-on-strontium titanate structure with Au contact pads on a lanthanum aluminate substrate. It has strip widths varying from 4 to 30 microns and gap widths of 18 to 75 microns, respectively.

The Last Five Years: 2002–2007

Miranda's group gave a review in 2002 [36]. That paper dealt with the status of thin film ferroelectric-based frequency and phase-agile microwave devices for communication applications. Results were discussed in terms of device parameters such as RF losses, tunability, phase shift, frequency of operation, ease of fabrication, and projected cost. Although the presentation emphasized a NASA perspective, the information provided offered a perspective of the attributes of this technology for other government agencies' electronically steerable electronics requirements, and for commercial communication applications. At about the same time, more fundamental physics questions were being addressed by the present author [37, 38]. One of those is the question of effective masses in thin-film ferroelectric tunneling devices. The band mass m* for heavy electrons in the *d*-like conduction band of $SrTiO_3$ is about 6 m_e. However, tunneling experiments typically give values near unity ([39, 40] and Bibes M, private communication). Although this has never been explained, it seems likely that the tunneling involves only light electrons.

In 2002 improvements in tunability and loss factors for BST films by doping them with TiO_2 was reported from Los Alamos [41]. And a very fine review of the general field was given by Speller at Oxford [42], plus a short up-date from Miranda's group [43]. Speller summarized what he described as the first large scale commercial application of high temperature superconducting (HTSC) thin films – the area of passive microwave devices. He concluded that the thallium-based superconductors are particularly attractive for these applications because they have high transition temperatures (up to 127 K) and are relatively stable against atmospheric attack. $Tl_2Ba_2CaCu_2O_8$ films (TBCCO) with high critical current density and low surface resistance had already been fabricated by growing biaxially aligned films epitaxially on single crystal substrates. However, the processing of Tl based HTSC thin films is particularly difficult due to the complexity of the crystal structure and the volatility and toxicity of thallium. Most Tl HTSC films were fabricated using a two-step process involving the deposition of an amorphous precursor followed by an ex situ thalliation anneal. The relative stability of the phases in the TBCCO system depends on the partial pressure of oxygen and thallous oxide, the processing temperature and the initial stoichiometry of the precursor. Unfortunately, it had not been possible to grow "epitaxial" Tl HTSC films directly on substrates such as MgO and sapphire which have dielectric properties almost ideal for high frequency applications, because the lattice mismatch is too great and/or chemical reaction occurs. But he pointed out that buffer layers can be deposited on MgO or sapphire substrates to reduce the lattice mismatch and/or prevent chemical interaction with the HTSC film. In my opinion this review was very helpful in focusing attention onto the thallium compounds and away from YBCO.

Present Situation and Application to Phased–Array Radar

At present the funding for this line of work has waned, and commercial applications are waiting for some further breakthrough. The related area of applications – phased-array radar – has had some attention, including a collection of papers in Integrated Ferroelectrics. Early work in Leningrad and the USA looked very promising.

Phased-array radar devices produce an angular scanning of the horizon without mechanical rotation of the antenna. This can be achieved by a voltage-dependent phase shift in the antenna elements. A good diagram of such a device is given by Barnes [44, 45]. Such shifts can be generated in bulk ferroelectrics, but only at very high voltages (kV), which entails large sizes and weights. It would be desirable to produce phased array radar devices with ferroelectric thin films, since this could entail very low voltages (5 V) compatible with silicon chip technology, and hence embodiments with small size and weight. Thin-film ferroelectrics exhibit a large decrease in dielectric constant with application of modest voltages, typically 25–60% at 5 V. This suggested that they could be used as the active phase-shift element in phased-array radar, a project studied carefully in the 1990s by researchers at Grumman and TRW in the USA [46–52], by US Army laboratories [53, 54], and by Vendik's group in Leningrad [35]. Unfortunately the dielectric loss tangent in these films remains too large for acceptable insertion losses in such devices, so nothing has been made commercially [36]. There is hope that this loss is not intrinsic, since it is orders of magnitude smaller in bulk. Therefore significant efforts to improve the quality of thin films has been underway [13, 14], with perhaps the best results obtained at Penn State University.

In the past year Miranda's group has reported large-scale production tests of BST microwave phase shifters [55], fabricated via self-assembled monolayers [56]. Thus, this review ends on a note of optimism.

References

1. Hermann AM, Yandrofski RM, Scott JF et al. (1994) J Superconduct 7:463–469
2. Yandrofski R et al. (1995) US Patent 5 472 935
3. Hermann AM et al. (1993) Bull Am Phys Soc 38:689
4. Kaiser DL, Vaudin MD, Rotter LD et al. (1999) J Mater Res 14:4657–4666
5. Boikov YA, Ivanov ZG, Claeson T (1995) Inst Phys Conf Series 148:1123–1126
6. Boikov YA, Ivanov ZG, Vasiliev AL et al. (1996) Superconduct Sci Technol 9:A178–A181
7. Hinds BJ, Mc RJNeely, Studebaker DB et al. (1997) J Mat Res 12:1214–1236
8. Chakalov RA, Ivanov ZG, Boikov YA et al. (1997) Inst Phys Conf Series 158:327–330
9. Galt D, Price JC, Beall JA, Ono RH (1993) Appl Phys Lett 63:3078–3080
10. (1993) IEEE MTT-S Microwave Symp. Digest, p 1421
11. Lacey D, Gallop JC, Davis LE (1998) Meas Sci Technol 9:536–539

12. Regnier S, Caranoni C, Marfaing J (1998) J Physique IV 8:45–49
13. Li HC, Si WD, West AD et al. (1998) Appl Phys Lett 73:190–192
14. Li HC, Si WD, West AD et al. (1998) Appl Phys Lett 73:464–466
15. Merkulov VI. Fox JR, Li HC et al. (1998) Appl Phys Lett 72:3291–3293
16. Scott JF (1998) Ann Rev Mat Sci 28:79–100
17. Chakalov RA, Ivanov ZG, Boikov YA et al. (1998) Physica C 308:279–288
18. Miranda FA, Subramanyam G, Van Keuls FW et al. (1999) IEEE Trans Appl Superconduct 9:3581–3584
19. Subramanyam G, Van Keuls FW, Miranda FA et al. (1999) Integ Ferroelec 24:273–285
20. Subramanyam G, Van Keuls FW, Miranda FA et al. (2000) IEEE Trans Microwave Theory Techniques 48:525–530
21. Subramanyam G, Van Keuls FW, Miranda FA et al. (2000) IEEE Microwave Guided Wave Lett 10:136–138
22. Miranda FA, Subramanyam G, Van Keuls FW et al. (2000) IEEE Trans Microwave Theory Techniques 48:1181–1189
23. Subramanyam G, Van FW Keuls, Miranda FA (2000) Integ Ferroelec 29:81–93
24. Outzourhit A et al. (1995) Integ Ferroelec 8:227–234
25. Dietz GW et al. (1997) J Appl Phys 82:2359–2364
26. Miller SL, Nasby RD, Schwank JR, Rodgers MS, Dressendorfer PV (1990) J Appl Phys 68:6463–6471
27. Hwang CS et al. (1998) J Appl Phys 83:3703–3713
28. Scott JF et al. (1995) Integ Ferroelec 6:189–196
29. Hermann AM, Veeraraghavan B, Balzar D et al. (2000) Integ Ferroelec 29:161–173
30. Lin L (2000) Mat Sci Eng Rpts 29:153–181; Aidamand R, Schneider R (2001) Thin Solid Films 384:1–8
31. Xu H, Gao EH, Ma QY (2001) IEEE Trans Appl Superconduct 11:353–356
32. Gurevich VL, Tagantsev AK (1986) Zh Eksp Teor Fiz 91:245–261
33. Neumann R, Hoffmann G (1993) Ferroelec 134:202–208
34. Petrov PK, Carlsson EF, Larson P, Friesel M, Ivanov ZG (1998) J Appl Phys 84:3134–3140
35. Kozyrev AB et al. (1998) J Appl Phys 84:3326–3332
36. Miranda FA, Van Keuls FW, Romanofsky RR et al. (2002) Integ Ferroelec 42:131–149
37. Scott JF (2002) Integ Ferroelec 42:1–14
38. Scott JF, Dawber M (2001) J Physique IV 11:9–14
39. Stolichnov I, Tagantsev AK (1998) J Appl Phys 84:3216–3225
40. Baniecki JD, Laibowitz RB, Shaw TM, Parks C, Lian J, Xu H, Ma QY (2001) J Appl Phys 89:2873–2885
41. Jia QX, Park BH, Gibbons BJ et al. (2002) Appl Phys Lett 81:114–116
42. Speller SC (2003) Mat Sci Technol 19:269–282
43. Subramanyam G, Van Keuls FW, Miranda FA et al. (2003) Mater Chem Phys 79:147–150
44. Barnes FS et al. (1995) Integ Ferroelec 8:171
45. Scott JF, Ferroelectric Memories (2000) Springer, Heidelberg, 2000, p 181
46. Jack L (1991) US Patent 5 070 241
47. Babbitt RW, Koscica TE, Drach WC (1993) US Patent 5 212 463
48. Collier DC (1994) Integ Ferroelec 4:113–119
49. Kain A, Pham T, Pettie-Hall C et al. (1995) Integ Ferroelec 8:45–51
50. Jackson CM et al. (1994) Integ Ferroelec 4:121–128
51. Babbitt RW, Koscica TE, Drach WC (1992) Microwave J 35:63–71
52. Babbitt RW, Koscica TE, Drach WC (1995) Integ Ferroelec 8:65–76

53. Sengupta LC, Stowell S, Ngo E et al. (1995) Integ Ferroelec 8:77–88
54. Wilber WD, Tidrow SC, Tauber A et al. (1998) Integ Ferroelec 19:149–155
55. Van Keuls FW, Varaljay NC, Mueller CH et al. (2005) Integ Ferroelec 77:51–62
56. Morales-Cruz AL, Van Keuls FW, Miranda FA et al. (2005) Appl Surf Sci 252:981–995

Shengelaya A, Kochelaev BI et al. (2007) In: Bussmann-Holder A, Keller H (eds)
High T_c Superconductors and Related Transition Metal Oxides.
Springer-Verlag Berlin Heidelberg, 287–302
DOI 10.1007/978-3-540-71023-3

Electronic Phase Separation and Unusual Isotope Effects in $La_{2-x}Sr_xCuO_4$ Observed by Electron Paramagnetic Resonance

A. Shengelaya[1] (✉) · B. I. Kochelaev[2] · K. Conder[3] · H. Keller[4]

[1]Physics Institute of Tbilisi State University, Chavchavadze 3, GE-0128 Tbilisi, Georgia
salex@physik.unizh.ch

[2]Department of Physics, Kazan State University, 420008 Kazan, Russia

[3]Laboratory for Developments and Methods, Paul Scherrer Institut,
CH-5232 Villigen PSI, Switzerland

[4]Physik-Institut der Universität Zürich, Winterthurerstrasse 190, CH-8057 Zürich,
Switzerland

Abstract We review the results of our recent studies of $La_{2-x}Sr_xCuO_4$ cuprate superconductor using electron paramagnetic resonance (EPR). It is shown that the EPR of Mn^{2+} doped into $La_{2-x}Sr_xCuO_4$ (LSCO) provides an unique microscopic information concerning the magnetic, electronic and lattice properties of cuprates. The main conclusions followed from these experiments are discussed. Most attention is given to important and highly debated questions, such as the role of the lattice, electron–phonon interaction, polaron formation and microscopic electronic phase separation in cuprates.

Introduction

Electron Paramagnetic Resonance (EPR) is a powerfull tool in solid state physics, which allows to study crystal electric fields, electron–phonon interactions, static and dynamic magnetic correlations on a *microscopic* level. The remarkable discovery of the high-T_c superconductivity by Bednorz and Müller [1] created a great interest to the EPR study of doped layered cuprates and their parent compounds. The main reason is that the basic superconducting events occur in the CuO_2 planes. Moreover, the search for high-T_c superconductivity by Bednorz and Müller was guided by the Jahn–Teller polaron model, associated with the Cu^{2+} ion in the oxygen octahedron [1]. The Cu^{2+} ion in this plane has an electronic configuration d^9 with a singlet orbital $d(x^2 - y^2)$ ground state. The Cu^{2+} ion was used as an EPR probe in many dielectric crystals and gives a very good EPR signal [2, 3]. However, many attempts to observe the bulk EPR signal of the Cu spin-system in cuprates gave a negative result. The nature of the EPR silence of cuprate superconductors and their parent compounds was a subject of intensive theoretical and experimental investigations. The reason for the EPR silence is most probably due to the extremaly broad linewidth, which was estimated in our work [4]. Because of the EPR silence the investigation of high-T_c superconductors has, from a magnetic resonance point of view, till recently been dominated by nuclear magnetic resonance (NMR). The observation of EPR would be of great interest, because the time-domain of observation of EPR is two to three orders of magnitude shorter than that of NMR. In addition the sensitivity of the EPR is by 6 orders of magnitude higher than that of NMR which is very important for the elucidation of a subtle phenomena taking place in the key structural fragments of the superconducting compounds CuO_2 planes. Another approach in the application of EPR to high-T_c superconductors is to dope these compounds with a small amount of some paramagnetic ions which are used to probe the intrinsic behavior. One of the best candidates is Mn, which in the 2+ valent state gives a well defined signal and substitutes for the Cu^{2+} in the CuO_2 plane. Kochelaev et al. have intensively studied the EPR of Mn^{2+} doped $La_{2-x}Sr_xCuO_4$ [5]. They found that the Mn ions are strongly coupled to the collective motion of the Cu spins (the so called bottleneck regime). In order to explain the bottleneck condition let us consider a schematic block diagram of the relaxation paths between Mn and Cu spin systems and the lattice shown in Fig. 1. It is known that the Larmor frequencies of Mn and Cu ions are very close and the isotropic exchange between them is very strong, being of the same order of magnitude as the isotropic exchange betwe en the Cu ions. In this situation a collective motion of the transverse homogeneous magnetizations of the Mn and Cu ions appears, if the relaxation rate between them $\Gamma_{s\sigma}$ is faster than the relaxation of each of them to the lattice Γ_{sL} and $\Gamma_{\sigma L}$: $\Gamma_{s\sigma} \gg \Gamma_{sL}, \Gamma_{\sigma L}$. This is so-called bottleneck effect.

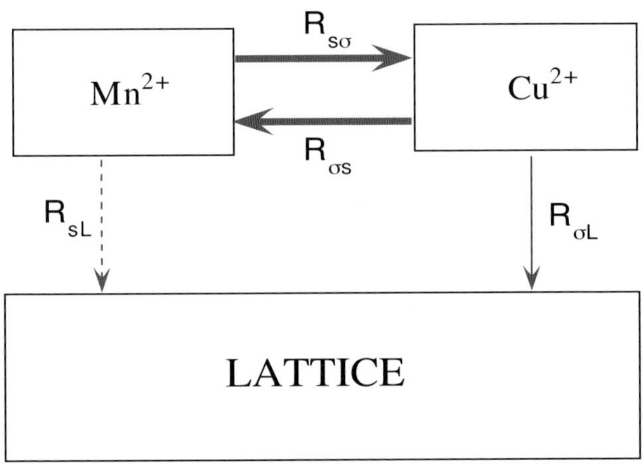

Fig. 1 Block diagram showing the energy flow paths for the Mn and Cu spin systems and the lattice. The relaxation rate R_{ab} represents relaxation from subsystem a to subsystem b. The thickness of the arrows is a measure of the magnitude of the particular relaxation rate R_{ab}

In the bottleneck regime the effective relaxation rate Γ_{eff} is controlled mainly by the individual relaxation rates of Mn and Cu ions to the lattice Γ_{sL} and $\Gamma_{\sigma\text{L}}$ weighted by the corresponding static spin susceptibilities:

$$\Gamma_{\text{eff}} = \frac{\chi_{\text{s}}^0 \Gamma_{\text{sL}} + \chi_\sigma^0 \Gamma_{\sigma\text{L}}}{\chi_{\text{s}} + \chi_\sigma} + \frac{\langle(\Delta\omega)^2\rangle}{\Gamma_{\text{s}\sigma}} . \quad (1)$$

The last term takes into account a partial opening of the bottleneck at low temperatures where $\Gamma_{\text{s}\sigma} \longrightarrow 0$. $\langle(\Delta\omega)^2\rangle$ is the mean square of the local fields distribution at the Mn sites. χ_{s}^0, χ_σ^0 and χ_{s}, χ_σ are the bare and renormalized spin susceptibilities of the Mn and Cu spin-systems, respectively. The latter are:

$$\chi_{\text{s},\sigma} = \chi_{\text{s},\sigma}^0 \frac{1 + \lambda\chi_{\text{s},\sigma}^0}{1 - \lambda^2 \chi_{\text{s}}^0 \chi_\sigma^0} , \quad \chi_{\text{s}}^0 = yN\frac{(g_{\text{s}}\mu_{\text{B}})^2}{3k_{\text{B}}T} , \quad \lambda = \frac{2zJ_{\text{s}\sigma}}{Ng_{\text{s}}g_\sigma\mu_{\text{B}}^2} , \quad (2)$$

where g_{s} and g_σ are the g-factors, y is the concentration of Mn ions, z is the number of their neighboring Cu ions. The effective g-factor and the intensity of the collective EPR signal are:

$$g_{\text{eff}} = \frac{\chi_{\text{s}} g_{\text{s}} + \chi_\sigma g_\sigma}{\chi_{\text{s}} + \chi_\sigma} , \quad I_{\text{eff}} \propto \chi_{\text{s}} + \chi_\sigma . \quad (3)$$

It is worth to point out that in a deep bottleneck regime and the strong coupling between two subsystems, $\Gamma_{\text{s}\sigma}$ disappears from the effective relaxation rate (Eq. 1), since the last term can be neglected. This is a consequence of a commutation of the total spin of the two subsystems with the isotropic exchange coupling between them. In addition the relaxation rate Γ_{sL} is much smaller than $\Gamma_{\sigma\text{L}}$ and can be neglected, since the ground state of the Mn^{2+} ion is an orbital S-state and the spin-orbit interaction of Mn is rather small. Also, for our Mn concentrations $\chi_{\text{s}}^0 \gg \chi_\sigma^0$. Then in the deep bottleneck regime the EPR linewidth is controlled mainly by the strongly reduced relaxation rate of the Cu magnetic moment to the lattice $\Gamma_{\text{eff}} \propto (\chi_\sigma^0/\chi_{\text{s}})\Gamma_{\sigma\text{L}}$, while the EPR intensity is defined by the Mn susceptibility $I_{\text{eff}} \propto \chi_{\text{s}}$. Based on the above analysis we come to the important conclusion that due to the bottleneck condition we are able to access $\Gamma_{\sigma\text{L}}$ despite the fact that the EPR of Cu ions are not observable in cuprates. We decided to take advantage of the bottleneck regime in Mn^{2+} doped $La_{2-x}Sr_xCuO_4$ to study the copper spin-lattice relaxation and search for possible oxygen isotope effects on the EPR signal [4].

EPR in Moderately Doped $La_{2-x}Sr_xCuO_4$

We studied polycrystalline samples of $La_{2-x}Sr_xCu_{1-y}Mn_yO_4$ with $0.06 \leq x \leq 0.20$ and $0.01 \leq y \leq 0.04$. Oxygen isotope substitution has been made at 800 °C for 30 h in an oxygen pressure of ~ 1.0 bar. Oxygen isotope enrichment of the samples were determined using thermogravimetry. The ^{18}O samples have about 85% ^{18}O and 15% ^{16}O. It is important to note that the EPR spectra for the samples with different oxygen isotopes were taken with exactly the same spectrometer conditions. Special care was taken also to perform the experiments on samples with the same mass, mounted in identical sample tubes, etc. The reproducibility of the measured EPR signals was checked several times. Thus the EPR signals obtained for samples with different oxygen isotopes can be directly compared. We observed an EPR signal in all examined samples. The lineshape of the signal is Lorentzian and symmetric throughout the whole temperature range. The resonance field corresponds to $g \sim 2$, a value very close to the g-factor for the Mn^{2+} ion. Figure 2a shows typical EPR spectra for $x = 0.06$, $y = 0.02$ sample with different oxygen isotopes ^{16}O and ^{18}O. From Fig. 2a one can see a difference between the EPR signals of the two isotope samples. Analysis of the spectra showed that the integral intensities of the EPR signals in two isotope samples are the same, but the linewidths are different. The linewidth for the ^{18}O sample is larger than for the ^{16}O sample, giving rise to different amplitudes for the EPR signal for two isotope samples. We performed measurements on back-exchanged samples in order to check whether the observed isotope effect is intrinsic. The corresponding results are presented in Fig. 2b. It is seen that the isotope effect is completely reversible. We studied also the samples with different Mn concentration ($y = 0.01$, 0.04) in order to clarify the role of Mn doping in the observed isotope effect. We found that the absolute value of linewidth changes with Mn concentration, however the isotope effect itself is independent on Mn content. This shows that Mn^{2+} serves as an EPR probe and the observed isotope effect is intrinsic to $La_{2-x}Sr_xCuO_4$. Taking this into account we studied samples with a fixed Mn concentration of $y = 0.02$ and different Sr doping.

Figure 3 shows the temperature dependence of the linewidth for the $x = 0.06$ sample with different oxygen isotopes. With decreasing temperature the linewidth decreases, passes through a minimum at a temperature T_{min} and increases again on further cooling. From Fig. 3 one can see that the observed isotope effect on the EPR linewidth is temperature dependent. The isotope effect is large at low temperatures and gradually disappears above T_{min} at high temperatures.

We found that the isotope effect decreases with increasing Sr concentration. EPR spectra and the temperature dependence of the EPR linewidth for the more doped sample $x = 0.10$ with different oxygen isotopes are shown in Figs. 4 and 5, respectively. One can see that the isotope effect is smaller than for the $x = 0.06$ sample, but the qualitative behavior is the same. We also measured optimally doped and overdoped samples. In the optimally doped sample $x = 0.15$ the isotope effect still exists at low temperatures, but it is very small and completely disappears in the overdoped sample $x = 0.20$. To see what happens in samples without Sr doping, we performed the EPR in parent compound La_2CuO_4 with different oxygen isotopes. No isotope effect was

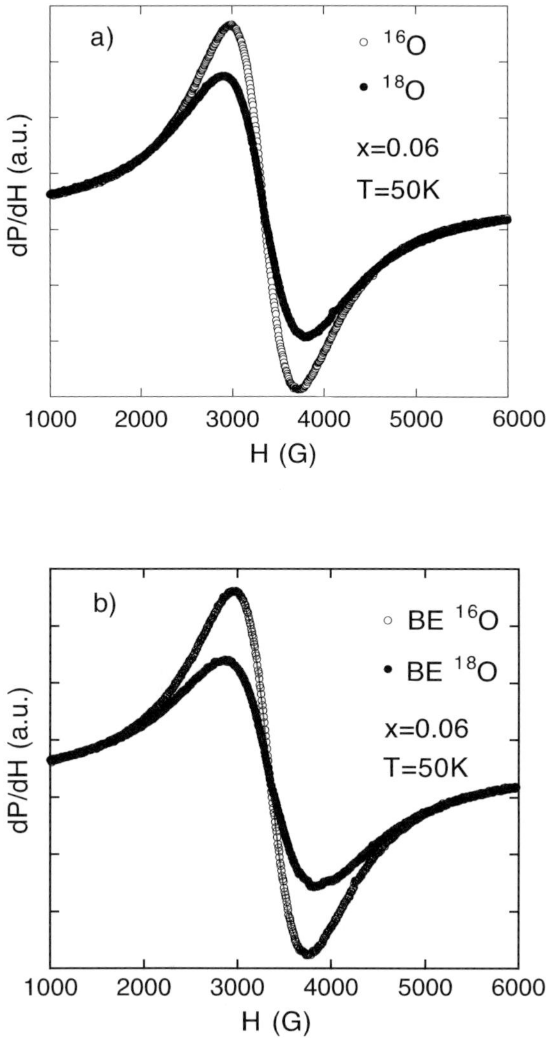

Fig. 2 **a** EPR signal of ^{16}O and ^{18}O samples of $La_{1.94}Sr_{0.06}Cu_{0.98}Mn_{0.02}O_4$ measured at $T =$ 50 K under identical experimental conditions. The fits to a Lorentzian line shape are represented by *solid lines*. **b** EPR signal of ^{16}O and ^{18}O samples of $La_{1.94}Sr_{0.06}Cu_{0.98}Mn_{0.02}O_4$ after isotope back-exchange from ^{18}O to ^{16}O (BE ^{16}O) and from ^{16}O to ^{18}O (BE ^{18}O). One can see that the isotope effect is reversible

found for these samples. This result indicates that the holes doped to CuO_2 planes are crucial to observe the isotope effect on the EPR signal.

The observed isotope effect indicates an important role of the lattice motion in the relaxation of the Cu magnetization. We propose that an interaction between the Cu^{2+} spin-system and the lattice motion in general is the same as

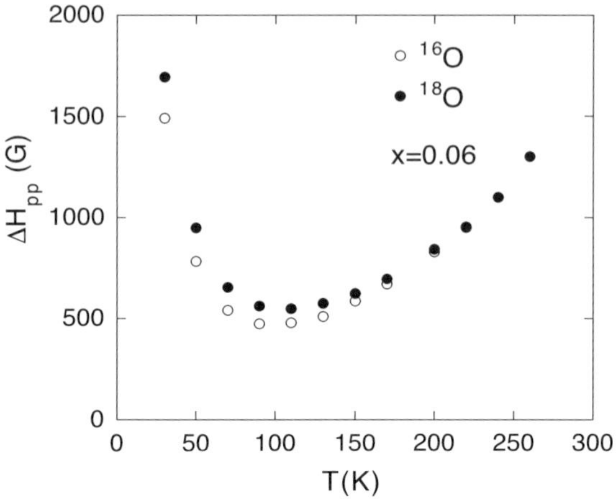

Fig. 3 Temperature dependence of the peak-to-peak EPR linewidth ΔH_{pp} for ^{16}O and ^{18}O samples of $La_{1.94}Sr_{0.06}Cu_{0.98}Mn_{0.02}O_4$

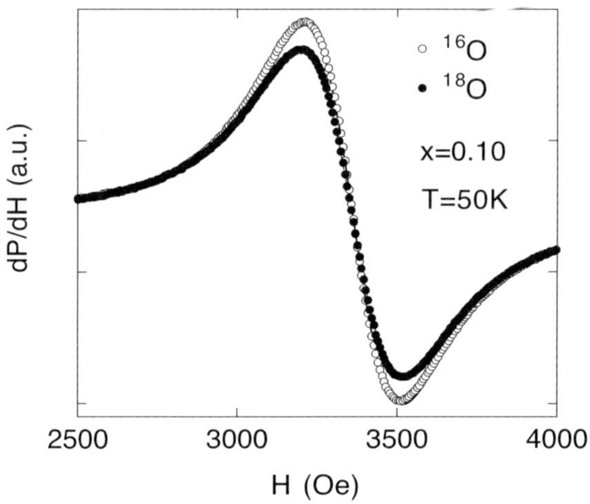

Fig. 4 EPR signal of ^{16}O and ^{18}O samples of $La_{1.90}Sr_{0.10}Cu_{0.98}Mn_{0.02}O_4$ measured at T = 50 K under identical experimental conditions

in insulators, i.e. due to the modulation of the crystal electric field by the lattice distortions and the spin-orbit coupling of the Cu ions. This mechanism usually leads to a rather slow spin-lattice relaxation rate, since the Kramers doublet is not sensitive to the electric field because of time reversal symmetry. Nonvanishing matrix elements of the spin-lattice coupling appear only due to the external magnetic field. This gives an additional very small factor $g\beta H/\Delta$, where Δ is the crystal field splitting of the orbital states. In our case the eigenstates of the Cu spin-system are defined mainly by the very large isotropic Cu – Cu exchange, instead of the Zeeman interaction. Roughly speaking, the

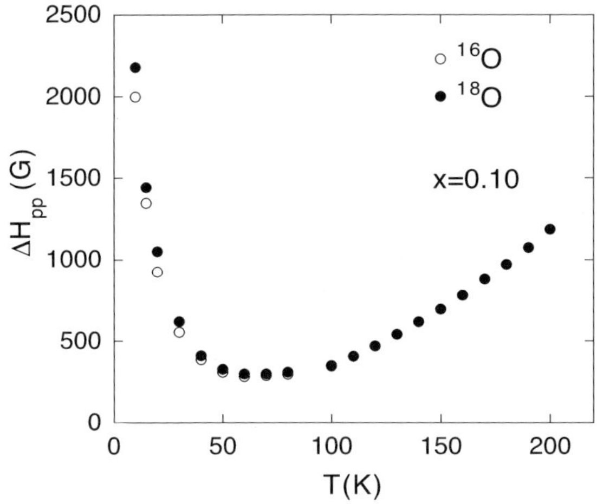

Fig. 5 Temperature dependence of the peak-to-peak EPR linewidth ΔH_{pp} for ^{16}O and ^{18}O samples of $La_{1.90}Sr_{0.10}Cu_{0.98}Mn_{0.02}O_4$

Cu – Cu exchange coupling $J = 1500$ K will play the role of the magnetic field splitting $g\beta H$. The spin-orbit interaction couples the ground state of the Cu^{2+} ion $d(x^2 - y^2)$ to the excited states $d(xz)$ and $d(yz)$ only and the relations between the matrix elements are dictated for the Kramers doublet by the time reversal symmetry. Taking this into account it was found that the Hamiltonian of the spin-lattice interaction has the following form [6]:

$$H_{\sigma \mathrm{L}} = \frac{J\Lambda}{8a} \sum_{<ij>} \left[\left(\sigma_y^i \sigma_z^j - \sigma_z^i \sigma_y^j \right) \left(Q_4^i - Q_4^j \right) + \left(\sigma_x^i \sigma_z^j - \sigma_z^i \sigma_x^j \right) \left(Q_5^i - Q_5^j \right) \right] , \tag{4}$$

$$\Lambda = \frac{3\lambda G}{\Delta^2} .$$

Here Q_4^i and Q_5^i are the normal coordinates of the oxygen octahedron CuO_6, which describe its distortions due to the tunneling motion of the apical oxygen between the four potential minima without the rigid rotation of the octahedra as a whole. The constants λ and G correspond to the spin-orbit and orbit-lattice coupling, Δ is the crystal field splitting between the ground and excited orbital energy levels, $< ij >$ means the sum over neighboring Cu sites in the x–y plane, a is the lattice constant.

Using this Hamiltonian we calculated the temperature dependence of the EPR linewidth and compared it with the experimental data [4]. The solid lines in Fig. 6 represent the best fit curves using our theoretical model. One can see that there is a good agreement between theory and experiment. Furthermore, this novel theoretical approach allows an estimation of the intrinsic EPR relaxation time, and provides an explanation for the long-standing problem of

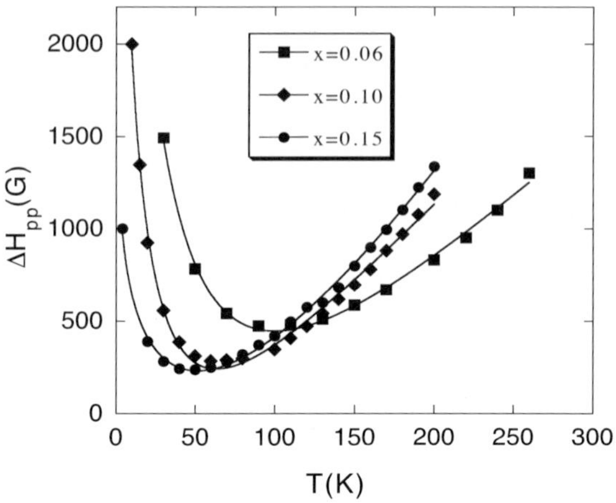

Fig. 6 Temperature dependence of the EPR linewidth ΔH_{pp} for $La_{2-x}Sr_xCu_{0.98}Mn_{0.02}O_4$ samples with $x = 0.06$, 0.10 and 0.15. The *solid lines* represent the best fit using the theoretical model described in the text

EPR silence in high-T_c superconductors. Our estimations show that the intrinsic EPR linewidth from copper spins in these compounds would be of the order of 10^4 G. This value is too large for observing an EPR signal at the usual frequencies. From the analysis above it turns out that the spin-lattice interaction in the Cu orbital ground state $d(x^2 - y^2)$ involves only the Q_4 and Q_5 modes related to the tilts and rotations of the oxygen octahedra. These modes are strongly anharmonic being responsible for a structural phase transition from a high-temperature tetragonal to a low-temperature orthorhombic and tetragonal phases. As a result, besides vibrations near the minima of a potential energy $V(Q_4, Q_5)$, there is a quantum tunneling between the minima. The latter leads to a stronger modulation of the crystal field at the Cu ion causing a more effective spin-lattice relaxation. Since the tunneling frequency between the minima depends exponentially on the oxygen mass, it gives a large isotope effect of the Cu electron spin-lattice relaxation rate. This explains the large isotope effect on the EPR linewidth observed experimentally [4, 7]. The oxygen isotope effect on the EPR linewidth indicates an intimate connection between the lattice vibrations and the magnetism and provides evidence for the polaronic charge carriers in cuprate superconductors.

EPR in Lightly Doped $La_{2-x}Sr_xCuO_4$

The generic phase diagram in hole-doped cuprates is by now well established. At a critical concentration of doping $x_{c1} \approx 0.06$, superconductivity sets in at $T = 0$, and ends at a higher doping level $x_{c2} \approx 0.25$ [8]. Both are the critical endpoints of the superconducting phase-transition line [9]. At the former, a transition from an insulating to the superconducting state has been assumed untill very recently [9]. However, using finite-size scaling for the susceptibility of a series of concentrations $x < x_{c1}$, the following was inferred: The material consists of antiferromagnetic (AF) domains of variable size, separated by metallic domain walls [10]. More recently Ando et al. corroborated this early finding by measuring the in-plane resistivity anisotropy in untwinned single crystals of $La_{2-x}Sr_xCuO_4$ (LSCO) and $YBa_2Cu_3O_{7-\delta}$ in the lightly doped region, interpreting their results in terms of metallic stripes present [11]. Our EPR study of the lightly doped sample $La_{2-x}Sr_xCu_{0.98}Mn_{0.02}O_4$ with $x = 0.03$ revealed an evidence of the electronic phase separation [7]. Besides the broad EPR line described in previous section, we observed at low temperatures an additional narrow EPR line at the same resonance magnetic field (Fig. 7).

One can see from Fig. 7 that the narrow line shows practically no isotope effect, whereas the broad line exhibits a huge isotope effect. A number of experiments on high-T_c superconductors suggest the possible existence of two quasiparticles: a heavy Jahn–Teller (JT) type polaron and a light fermion [12]. In the context of the two-carrier paradigm, the narrow line in the EPR spectra can be attributed to centers with nearly undistorted environment in regions with highly mobile carriers, whereas the broad one is due to the carriers with a distorted environment and slow polaronic carriers. This can naturally explain why two signals have so different linewidths as well as their different isotope dependence. Therefore the observation of two components in EPR spectra of the $x = 0.03$ sample indicates a microscopic electronic phase separation in the $La_{2-x}Sr_xCuO_4$ compound.

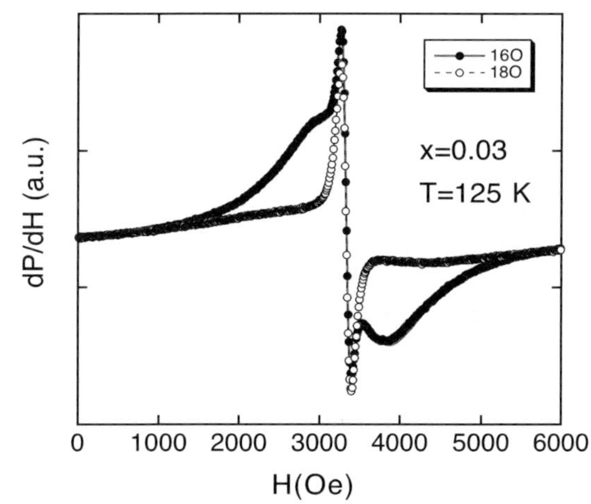

Fig. 7 EPR signal of ^{16}O and ^{18}O samples of $La_{1.97}Sr_{0.03}Cu_{0.98}Mn_{0.02}O_4$ measured at $T =$ 125 K under identical experimental conditions. The *solid lines* represent the best fits using a sum of two Lorentzian components: a narrow and a broad one

More recently we performed a detailed investigation of the narrow EPR line in lightly doped $La_{2-x}Sr_xCuO_4$ with $0 \leq x \leq 0.06$ [13]. An important information was obtained from the temperature dependence of the EPR intensity shown in Fig. 8. One can see from Fig. 8 that the two components observed in EPR spectra follow a completely different temperature dependence. The intensity of the broad line has a maximum and strongly decreases with decreasing temperature. On the other hand, the intensity of the narrow line is negligible at high temperatures and starts to increase almost exponentially below ~ 150 K. We note that the temperature below which the intensity of the broad line decreases shifts to lower temperatures with increasing doping. However, the shape of the $I(T)$ dependence for the narrow line is practically doping-independent and only slightly shifts towards higher temperatures with increased doping. A similar tendency is observed also for the

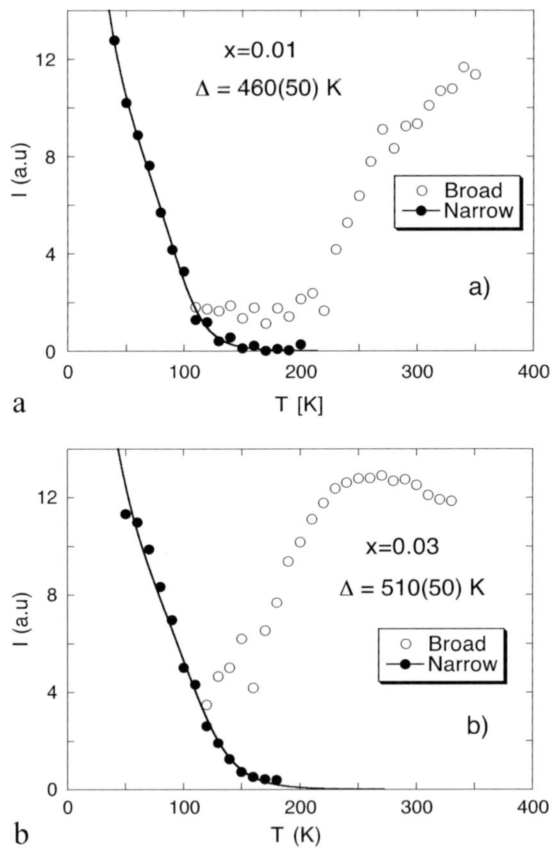

Fig. 8 Temperature dependence of the narrow and broad EPR signal intensity in $La_{2-x}Sr_xCu_{0.98}Mn_{0.02}O_4$ with different Sr dopings: **a** $x = 0.01$; **b** $x = 0.03$. The *solid lines* represent fits using the model described in the text

temperature dependence of the EPR linewidth. The linewidth of the broad line and its temperature dependence are strongly doping-dependent, whereas the linewidth of the narrow line is very similar for samples with different Sr doping (Fig. 9).

It is important to point out that the observed two-component EPR spectra are an intrinsic property of the lightly doped LSCO and are not due to conventional chemical phase separation. We examined our samples using X-ray diffraction, and detected no impurity phases. Moreover, the temperature dependence of the relative intensities of the two EPR signals rules out macroscopic inhomogeneities and points towards a microscopic electronic phase separation. The qualitatively different behavior of the broad and narrow EPR signals indicates that they belong to distinct regions in the sample. First we notice that the broad line vanishes at low temperatures. This can be explained by taking into account the AF order present in samples with very low Sr concentration [10]. It is expected that upon approaching the AF ordering temperature, a strong shift of the resonance frequency and an increase of the relaxation rate of the Cu spin system will occur. This will break the bottleneck regime of the Mn^{2+} ions, and as a consequence the EPR signal becomes unobservable [4].

In contrast to the broad line, the narrow signal appears at low temperatures and its intensity increases with decreasing temperature. This indicates that the narrow signal is due to the regions where the AF order is supressed. It is known that the AF order is destroyed by the doped holes, and above $x = 0.06$ AF fluctuations are much less pronounced [14]. Therefore, it is natural to relate the narrow line to regions with locally high carrier concentration and high mobility. This assumption is strongly supported by the absence of an oxygen isotope effect on the linewidth of the narrow line as well. It was

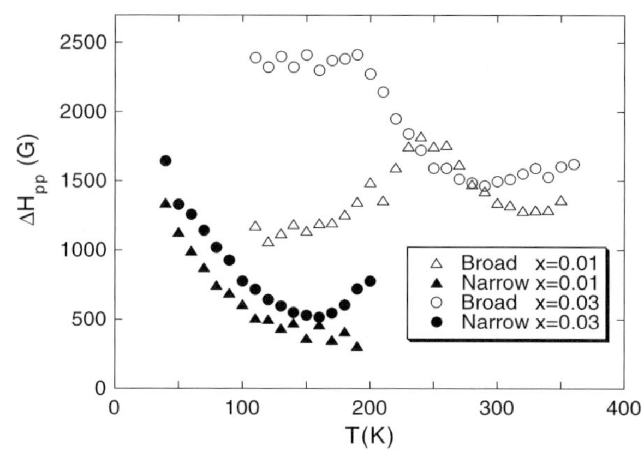

Fig. 9 Temperature dependence of the peak-to-peak linewidth ΔH_{pp} for the narrow and broad EPR lines in $La_{2-x}Sr_xCu_{0.98}Mn_{0.02}O_4$ with $x = 0.01$ and 0.03

shown previously that the isotope effect on the linewidth decreases at high charge-carrier concentrations close to the optimum doping [4]. We obtain another important indication from the temperature dependence of the EPR intensity. Because we relate the narrow line to hole-rich regions, an exponential increase of its intensity at low temperatures indicates an energy gap for the existence of these regions. In the following we will argue that this phase separation is assisted by the electron–phonon coupling. More precisely, the latter induces anisotropic interactions between the holes via the phonon exchange,resulting in the creation of extended nano-scale hole-rich regions.

An interaction between holes via the phonon exchange can be written in the form [15]:

$$H_{\text{pol-ph}} = G \sum_{n\alpha} P^{+}_{n\alpha} P_{n\alpha} \epsilon_{\alpha\alpha} , \tag{5}$$

where $P^{+}_{n\alpha}$ is a creation operator of one polaron, $\epsilon_{\alpha\alpha}$ is a deformation tensor, G is a coupling constant. It was shown that this interaction reduces to usual elastic forces if we neglect the retardation effects and optical modes. Following Aminov and Kochelaev [15], Orbach and Tachiki [16] we can find an interaction due to an exchange by phonons between two holes oriented along the axes α and β and separated by the space vector $\boldsymbol{R} = \boldsymbol{R}_{n\alpha} - \boldsymbol{R}_{m\beta}$

$$H_{\text{int}} = F\left(\boldsymbol{R}_{n\alpha} - \boldsymbol{R}_{m\beta}\right) P^{+}_{n\alpha} P_{n\alpha} P^{+}_{m\beta} P_{m\beta} ; \tag{6}$$

$$F_{xx}(\boldsymbol{R}) = \frac{1}{8\pi\rho C_{\text{l}}^2} \frac{G^2}{R^3} \left\{2\left(1 - 3\gamma_x^2\right) + \eta\left[12\gamma_x^2 - 15\gamma_x^4 - 1\right]\right\} ,$$

$$F_{xy}(\boldsymbol{R}) = \frac{1}{8\pi\rho C_{\text{l}}^2} \frac{G^2}{R^3} \left(2 - 15\gamma_x^2\gamma_y^2\right) .$$

Here $C_{\text{l}}, C_{\text{t}}$ are longitudinal and transversal sound velocities; $\gamma_x = x/R$, $\gamma_y = y/R$; $\eta = (C_{\text{l}}^2 - C_{\text{t}}^2)/C_{\text{t}}^2$. This interaction is highly anisotropic being attractive for some orientations and repulsive for others [17]. The attraction between holes may result in a bipolaron formation when holes approach each other closely enough.The bipolaron formation can be a starting point for the creation of hole-rich regions by attracting of additional holes. Because of the highly anisotropic elastic forces these regions are expected to have the form of stripes. Therefore the bipolaron formation energy Δ can be considered as an energy gap for the formation of hole-rich regions.

In applying the above model to the interpretation of our EPR results we have to keep in mind that the spin dynamics of the coupled Mn-Cu system experiences a strong bottleneck regime, as discussed in previous section. In the bottleneck regime the intensity of the joint EPR signal, being proportional to the sum of spin susceptibilities $I \sim \chi_{\text{Mn}} + \chi_{\text{Cu}}$, is determined mainly by the Mn susceptibility, since $\chi_{\text{Mn}} \gg \chi_{\text{Cu}}$ for our Mn concentration and temperature range. This results in a Curie–Weiss temperature dependence of the EPR signal.

Taking into account this remark we conclude that the EPR intensity of the narrow line is proportional to the volume of the sample occuppied by the hole-rich regions because the Mn ions are randomly distributed in the sample. We expect that the volume in question is proportional to the number of bipolarons, which can be estimated in a way proposed by Mihailovic and Kabanov [18]. If the density of states is determined by $N(E) \sim E^{\alpha}$, the number of bipolarons is

$$N_{\mathrm{bipol}} = \left(\sqrt{z^2 + x} - z\right)^2 , \quad z = KT^{\alpha+1} \exp\left(-\frac{\Delta}{T}\right) , \tag{7}$$

where Δ is the bipolaron formation energy, x is the level of hole doping, and K is a temperature- and doping-independent parameter related to the free polaron density of states. The EPR intensity from the hole-rich regions will be proportional to the product of the Curie–Weiss susceptibility of the bottlenecked Mn-Cu system and the number of the bipolarons

$$I_{\mathrm{narrow}} \sim \frac{C}{T-\theta} N_{\mathrm{bipol}} , \tag{8}$$

where C is the Curie constant and θ is the Curie–Weiss temperature. The experimental points for the narrow-line intensity were fitted for the two-dimensional system ($\alpha = 0$), and we used the value $\theta = -8$ K, which was found from measurements of the static magnetic susceptibility (an attempt to vary θ yielded about the same value). The values of C and θ are determined mainly by the concentration and magnetic moment of the Mn ions and their coupling with the Cu ions. Since these parameters are expected to be doping independent (or weakly dependent), they were found by fitting for the sample $x = 0.01$ and then were kept constant for other concentrations leaving the only free parameter the energy gap Δ.

The results of the fit are shown in Fig. 8. For the bipolaron formation energy we obtained $\Delta = 460(50)$ K, which is practically doping-independent. This value agrees very well with the value of Δ obtained from the analysis of inelastic neutron-scattering and Raman data in cuprate superconductors [18]. Mihailovic and Kabanov identified the pairs as intersite Jahn–Teller pairs which may be called bipolarons [19]. Recently Kochelaev et al. performed theoretical calculations of the polaron interactions via the phonon field using the extended Hubbard model [17]. They estimated the bipolaron formation energy and obtained values of $100\ \mathrm{K} \le \Delta \le 730$ K, depending on the value of the Coulomb repulsion between holes on neighboring copper and oxygen sites V_{pd}, $0 \le V_{\mathrm{pd}} \le 1.2$ eV. This means that the experimental value of Δ can be understood in terms of the elastic interactions between the polarons. We note also that recently Khomskii and Kugel emphasized an important role of the elastic interactions for stripe formation in doped manganites [20].

It is interesting to compare our results with other experiments performed in lightly doped LSCO. Recently Ando et al. measured the in-plane anisotropy of the resistivity ρ_b/ρ_a in single crystals of LSCO with $x = 0.02–0.04$ [11]. They found that at high temperatures the anisotropy is small, which is consistent with the weak orthorhombicity present. However, ρ_b/ρ_a grows rapidly with decreasing temperature below ~ 150 K. This provides macroscopic evidence that electrons self-organize into an anisotropic state because there is no other external source to cause the in-plane anisotropy in $La_{2-x}Sr_xCuO_4$. With EPR being a real-space and local probe it is difficult to determine the shape of the hole-rich regions. However, we noticed that the temperature dependence of the narrow EPR line intensity is very similar to that of ρ_b/ρ_a obtained by Ando et al. (see Fig. 2d in [11]). To make this similarity clear, we plotted $I_{narrow}(T)$ and $\rho_b/\rho_a(T)$ on the same graph (Fig. 10). It is remarkable that both quantities show very similar temperature dependences. It means that our *microscopic* EPR measurements and the *macroscopic* resistivity measurements by Ando et al. provide evidence of the same phenomenon: the formation of hole-rich metallic stripes in lightly doped LSCO well below $x_{c1} = 0.06$. This conclusion is also supported by a recent angle-resolved photoemission study of LSCO which showed the existence of metallic quasiparticles near the nodal direction below $x = 0.06$ [21].

As was discussed in previous section, for doping levels $x \geq 0.06$ only a single EPR line is observed and the temperature dependence of the signal intensity recovers an usual Curie–Weiss behavior. On the other hand there is still a substantial isotope effect on the EPR line. To understand the change of the EPR spectra at $x = 0.06$, one should first comment on the observability of the phase separation in our EPR experiments. The main difference of the EPR signals from the hole-rich and hole-poor regions is the spin relaxation rate of the Cu spin system, which results in different EPR linewidths. One would expect these local differences of the relaxation rate to be averaged out by the spin

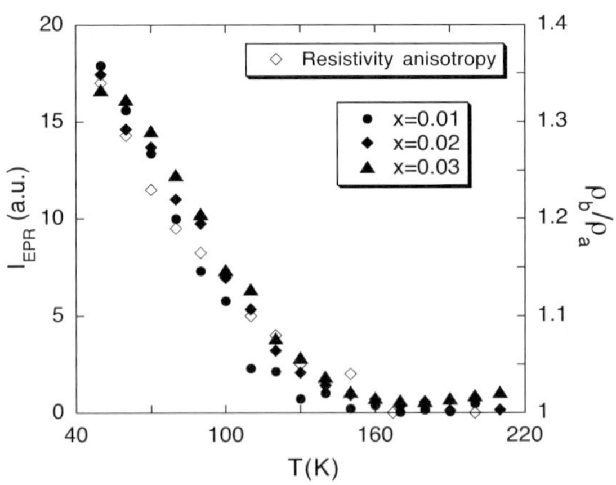

Fig. 10 Temperature dependence of the narrow EPR line intensities in $La_{2-x}Sr_xCu_{0.98}Mn_{0.02}O_4$ and of the resistivity anisotropy ratio in $La_{1.97}Sr_{0.03}CuO_4$ obtained in [11]

diffusion. The spin diffusion in the CuO_2 plane is expected to be very fast because of the huge exchange integral between the Cu ions. A rough estimate shows that during the Larmor period a local spin temperature can be transported over 100 Cu – Cu distances. It means that all the different nanoscale regions will relax to the lattice with a single relaxation rate, and we cannot distinguish them with EPR. However, the AF order which appears below T_N in the hole-poor regions in lightly doped LSCO freezes the process of spin diffusion, and this is the reason we can see different EPR lines from the two types of regions. From this we expect that with increasing doping, where magnetic order gets suppressed, spin diffusion will become faster, extended, and we can no longer distinguish different regions with EPR. This is most probably what happens in samples with $x \geq 0.06$, where only a single EPR line is observed [4]. This does not mean that the phase separation in hole-rich and hole-poor regions does not exist at $x \geq 0.06$, but that the spin diffusion averages out the EPR response from these regions. In fact, ARPES measurements showed the presence of two quasiparticles in the whole doping range [22, 23], indicating that the electronic phase separation exists also at higher doping levels. Also, recent Raman and infrared measurements provided evidence of one-dimensional conductivity in LSCO with $x = 0.10$ [24].

Summary

It was shown that the EPR of Mn^{2+} doped into $La_{2-x}Sr_xCuO_4$ (LSCO) offers a microscopic and often unique tool to study the magnetic, electronic and lattice properties of cuprates. The main results obtained can be summarized as follows:

1. A large oxygen isotope effect (OIE) was observed on EPR linewidth in Mn-doped LSCO. The isotope effect is large in underdoped region and decreases with increasing Sr concentration.
2. A new model of the Cu spin relaxation is proposed for cuprates. This model allows an estimation of the intrinsic EPR linewidth in cuprates, and provides an explanation for the long-standing problem of EPR silence in these materials. Furtherome, the observed OIE on the EPR linewidth can be quantitatively accounted for by this model.
3. EPR measurements in lightly doped LSCO revealed the presence of two resonance signals: a narrow and a broad one. Their behavior indicates that the narrow signal is due to hole-rich metallic regions and the broad signal due to hole-poor AF regions. The narrow-line intensity is small at high temperatures and increases exponentially below ~ 150 K. The activation energy inferred, $\Delta = 460(50)$ K, is nearly the same as that deduced from other experiments for the formation of bipolarons, pointing to the origin of the metallic regions present.

Obtained results demonstrate that the classical EPR technique which was successfully used by K. A. Müller for many decades to study perovskite materials, still contributes at the forefront in condensed matter physics, such as high temperature superconductivity.

Acknowledgements We gratefully acknowledge Karl Alex Müller for initiating this work and for continues collaboration and support for many years. His expertise and knowledge of EPR were crucial for the implementation of this project. This work was supported by the Swiss National Science Foundation, and in part by the SCOPES grant No. IB7420-110784.

References

1. Bednorz JG, Müller KA (1986) Z Phys B – Condensed Matter 64:189
2. Altshuler SA, Kozyrev BM (1972) Electron paramagnetic resonance of transition elements compounds, 2nd edn. Nauka, Moscow
3. Abragam A, Bleaney B (1970) Electron Paramagnetic Resonance of transition ions. Clarendon Press, Oxford
4. Shengelaya A, Keller H, Müller KA, Kochelaev BI, Conder K (2001) Phys Rev B 63:144513
5. Kochelaev BI, Kan L, Elschner B, Elschner S (1994) Phys Rev B 49:13106
6. Kochelaev BI (1999) J Supercond 12:53
7. Shengelaya A, Keller H, Müller KA, Kochelaev BI, Conder K (2000) J Supercond 13:955
8. Tallon JF et al. (1995) Phys Rev B 51:12911
9. Schneider T (2003) In: Bennemann KH, Ketterson JB (eds) The Physics of Conventional and Unconventional Superconductors. Springer, Berlin
10. Cho JH et al. (1993) Phys Rev Lett 70:222
11. Ando Y et al. (2002) Phys Rev Lett 88:137005
12. Mihailovic D, Müller KA (1997) In: Kaldis E et al. (eds) High-T_c Superconductivity: Ten Years after the Discovery. Kluwer Academic Publishers, p 243
13. Shengelaya A, Bruun M, Kochelaev BI, Safina A, Conder K, Müller KA (2004) Phys Rev Lett 93:017001
14. Niedermayer C et al. (1998) Phys Rev Lett 80:3843
15. Aminov LK, Kochelaev BI (1962) Zh Eksp Theor Fiz 42:1303
16. Orbach R, Tachiki M (1967) Phys Rev 158:524
17. Kochelaev BI et al. (2003) Mod Phys Lett B 17:415
18. Kabanov VV, Mihailovic D (2002) Phys Rev B 65:212508
19. Mihailovic D, Kabanov VV (2001) Phys Rev B 63:054505
20. Khomskii DI, Kugel KI (2001) Europhys Lett 55:208
21. Yoshida T et al. (2003) Phys Rev Lett 91:027001
22. Lanzara A et al. (2001) Nature 412:510
23. Zhou XJ et al. (2003) Nature 423:398
24. Venturini F et al. (2002) Phys Rev B 66:R060502

Slichter CP (2007) In: Bussmann-Holder A, Keller H (eds)
High T_c Superconductors and Related Transition Metal Oxides.
Springer-Verlag Berlin Heidelberg, 303–310
DOI 10.1007/978-3-540-71023-3

Some Science History with a Mutual Connection

C. P. Slichter

Department of Physics, University of Illinois, Urbana, Illinois, 61801-3080, USA
cslichte@uiuc.edu

Abstract An account is given of some science history involving the author with several people whom Alex Müller mentions in his Nobel Biography since they played important roles in his life. Topics are the measurement of the spin susceptibility of metals first derived theoretically by Wolfgang Pauli, the discovery of the self-trapped hole by Werner Känzig, plus relationships with Heine Gränicher and John Armstrong.

Introduction

I am honored to be invited to submit a contribution on the happy occasion of the 80th birthday of Alex Müller. He has had a profound influence on science and on scientists, and in particular on my life since my entire research activities for the last twenty years spring directly from the discoveries that he made together with George Bednorz that culminated in the award to them of the Nobel Prize in Physics. He has written a delightful autobiography for the Swedish Academy of Sciences. In it, he tells of some of the scientists who were influential in his own life. Several of them have also been very important in my life. I take this opportunity to tell a bit of this "science history".

In his autobiography, Alex Müller says that Werner Känzig, his instructor for advanced laboratory courses, convinced him to continue as a physicist at a time when he was thinking of changing fields. He describes the important influence of Wolfgang Pauli, his instructor in his advanced theoretical courses. He reports that Heine Gränicher (who was Känzig's first PhD student), suggested that Müller study the newly synthesized material $SrTiO_3$ by electron spin resonance (ESR), launching Müller on the study of such materials. Müller's PhD thesis was based on his discovery of the ESR signal from Fe^{3+} impurities in this system [1]. He descrfies how later, in the early 1980's when he became an IBM Fellow, Dr. John Armstrong, Director of Research at the IBM, invited him to spend a sabbatical at the Yorktown Heights IBM Laboratory to think about superconductivity. Shortly after his return to the Rüschlikon Laboratory, Bednorz and Müller made their famous discovery of the cuprate high temperature superconductors.

I likewise started out in the field of electron spin resonance. My undergraduate advisor had been J.H. Van Vleck. He had long been interested in magnetism, so when Purcell, Pound, and Torrey had demonstrated nuclear magnetic resonance at Harvard, Van Vleck was anxious that magnetic resonance of electrons be instituted to study paramagnetic materials. In the spring of 1947 Van Vleck proposed to me that I ask Professor E.M. Purcell to be my thesis advisor for an electron spin resonance study of paramagnetic salts. At that time, Purcell's other students were Nicholas Bloembergen and George Pake. They were working on nuclear magnetic resonance. For my thesis I built a 3 centimeter electron spin resonance apparatus with which I measured spectra of materials like iron alum. I also made measurements of electron spin-lattice relaxation times by the saturation method, using a pulsed high power magnetron left over from wartime radar countermeasure development activities.

Since both Alex Müller and I worked in the field of electron spin resonance, we knew each other's work, but it was not however until the 1980s when we were both members of the IBM Science Advisory Committee that we really got to know each other at the personal level. At that time I learned that my friend Werner Känzig had been important in Alex's life, and that Heine Gränicher, who was my colleague at Illinois for a few short months before family illness forced him to return to Switzerland, had suggested the thesis topic to Alex Müller's. Had Gränicher stayed at Illinois, I am sure my students and I would have undertaken studies of materials like $SrTiO_3$ and I would have gotten to know Alex Müller well in the early 1960s. Alex Müller also told me that when he was a student at the ETH, there was a Journal Club in the Physics Department where students were assigned journal articles to read and describe in an oral report. He said that he was assigned the study and presentation of the paper that my students and I had written presenting the first experimental measurement of the electron spin contribution to the magnetic susceptibility of metals. The theory of the electron spin contribution was first derived by Wolfgang Pauli. Alex Müller told me that the audience for his report included Pauli himself. I was thrilled that our experiment had come to Pauli's attention.

Often when one publishes an article, one omits historical aspects, things such as how one came to have the idea for the work, or how it came to pass, such things as the role of chance as well as the key role of other scientists. These are often some of the most interesting parts of the story. So I take this occasion to give below a bit of this sort of history for several activities in my life that are connected to some of the people mentioned by Alex Müller in his Nobel autobiography.

The Conduction Electron Resonance in Metals and the Measurement of the Pauli Spin Susceptibility

The measurement my students Bob Schumacher, Tom Carver, and I made of the Pauli Spin Susceptibility was first published in 1954 [2, 3]. The essential concept of the measurement is illustrated by Fig. 1. It shows the magnetic res-

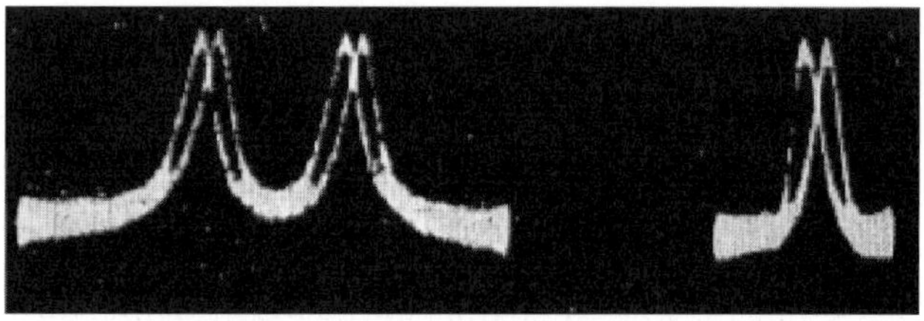

Fig. 1 Oscilloscope pictures of the resonance absorption in lithium metal vs. magnetic field at about 17 MHz. *Left*: Electron spin resonance. Zero field at the center of the sweep. The four peaks arise from a small phase shift between the horizontal and vertical axes. Sweep amplitude about 40 gauss. *Right*: ^{7}Li nuclear resonance at about 10 000 gauss. Sweep amplitude approximately 3 gauss [2, 3]

onance of conduction electrons of Li metal on the left and the ^{7}Li NMR signal of the same sample on the right, using the same NMR apparatus for both resonances operating at 17 MHz, the only change between the signals being the strength of the magnetic field to produce resonance. The experiment enables one to compare the unknown static magnetic susceptibility of the conduction electrons to that of the nuclei in the same sample. The nuclear static spin susceptibility is known since it can be calculated exactly from Curie's law. The area under the absorption signal is related to the static magnetic susceptibility are related through the Kramer–Kronig relation

$$\chi_0 = \frac{2}{\pi} \int_0^\infty \frac{\chi''(\omega)\,\mathrm{d}\omega}{\omega} \approx \frac{2}{\pi} \frac{\chi''_{\mathrm{max}} \Delta\omega}{\omega_0}, \tag{1}$$

where χ_0 is the static susceptibility, χ'' the imaginary part of the magnetic susceptibility, ω_0, the Larmor frequency, and $\Delta\omega$ the line width. Thus, the ratio of the areas under the absorption curves gives the ratio of the static magnetic susceptibilities. All instrumental variables such as sample volume, coil *Q*, amplifier gain cancel out.

I had first encountered this expression when starting work on my thesis in the summer of 1947, designing and building the microwave apparatus for electron spin resonance. Purcell pointed out that I could use it to estimate how big the signals might be since we knew the electron static spin susceptibility from Curie's law.

I was not thinking about conduction electron spin resonance at that time. I was thinking of the ESR of Fe^{3+} in iron alum. But one day Purcell came into the lab and asked whether I saw any signs of the resonance of the conduction electrons in the walls of my sample cavity. Such a thought had not occurred to me, and I wondered why I had not seen such a thing. The next day, Purcell came in to tell me that he thought the problem was lifetime broadening: the conduction electrons, moving at the Fermi velocity, are only acted on by the microwave alternating fields for the very short time that they remain within the skin depth of the metal surface.

It was Overhauser who made me think once again about conduction electron spin resonance when he came to the University of Illinois in 1951 as a post doc. For his PhD thesis at Berkeley with Charles Kittel, he had calculated the spin-lattice relaxation time of conduction electrons and suggested that my students and I try to see it since no one had seen it as yet. Moreover, he had discovered the idea for which he became famous, "the Overhauser effect": saturating the conduction electron spin resonance will enormously enhance the nuclear polarization [4, 5].

Overhauser's calculations suggested that the relaxation time should be long, so the conduction electron spin resonance should be narrow and correspondingly intense. Mindful of Purcell's remarks about lifetime broadening, we tried to find it in metal powder samples of materials like Cu and Al, using metal powder samples in an NMR apparatus where the skin depth was large compared to the particle size. We were unsuccessful[1] [6–8]. Then Griswold, Kip, and Kittel found it using microwaves in Li and Na [9]. We quickly then were able to find the ESR of conduction electrons in Li and Na using NMR apparatus. The lines were much broader (several gauss) than the original estimates, but still gave strong signals. We were then able to demonstrate the correctness of Overhauser's prediction of nuclear polarization of nuclei [10, 11].

It was David Pines, who was a post doc working with John Bardeen, who called our attention to the fact that no one had ever measured the electron spin contribution to the magnetic susceptibility of metals, made famous by thetheoretical formula of Pauli that was in all the textbooks on condensed matter physics. The problem is that conventional measurements of magnetic susceptibility do not give just the contribution of the spins. Rather, there are also orbital currents that contribute. Pines lamented this fact since he had just developed a new method of calculating the many body corrections to Pauli's formula and wished that they could be measured. Eq. 1 was something I knew well from my time with Purcell, so I told David that we knew how to do it. My students Bob Schumacher, Tom Carver, and I immediately set to work to do the measurement. The complete idea of measuring the ratio of the electron spin susceptibility to the nuclear static susceptibility came to me naturally since we had been doing electron spin resonance in an NMR rig and had also being doing electron-nuclear double resonance to demonstrate Overhauser's idea for dynamic polarization of nuclei. Our new experiments led to the papers of reference 2. As one sees, their existence is the product of many minds, many workers, many suggestions, a succession of experiments, and of the enormous good fortune that people talked to one another. For me, it was then

[1] It was many years before these resonances were found, using the clever spin transmission method of Carver and of Schumacher

a very special thrill to learn from Alex Müller that he knew of our experiment when a student, and that he had described our measurement to an audience that contained Pauli himself!

Alex Müller began studies of conduction electron spin resonance while at the Battelle Institution in Geneva shortly after finishing his PhD thesis. He investigated neutron-irradiated single crystals of graphite [12]. He has of course collaborated on electron spin resonance studies of the cuprate superconductors. They pose a special problem since no signal is found in the pure cuprate material. One must dope them with small amounts of foreign atoms such as Gd or Mn whose resonance one can detect, using them as a probe of their surroundings much as one employs nuclear spins as a probe when doing NMR. A particularly interesting paper is his study of Mn-doped LSCO [13] presenting evidence of bipolarons and studying the isotope effect.

Werner Känzig and the Discovery of the Self-trapped Hole

In the mid 1950's, Werner Känzig came to the University of Illinois as a postdoctoral research associate of Professor Robert Maurer. Unknown to me, Känzig set up an electron spin resonance apparatus in Professor Maurer's labs with which to search for an impurity center found in alkali halide crystals called the V_1 center. Such centers were revealed through optical spectroscopy where they gave rise to strong optical absorption typically in the visible (hence their common name "color centers"). Their makeup was a mystery. I first learned of Werner's work with Maurer when Werner appeared in my lab with an electron spin resonance spectrum taken at liquid nitrogen temperature that he had just discovered in a sample of KCl that he had X-rayed at that temperature. The spectrum was very complex, consisting of many lines and it's meaning was a mystery. The spectrum is shown is Fig. 2 [14].

Känzig [15] believed that it was the spectrum of an entity called the V_1 center, since his procedure for generating the spectrum was the one used to produced V_1 centers. Although there are many lines, one sees what appears to be a family of lines equally spaced in frequency with intensities 123432. Such a spectrum had recently been seen in experiments at Bell labs by Fletcher and colleagues [16, 17] in their studies of impurity states in Si. The spectrum arose from pairs of As impurities that were strongly exchange coupled, and could be explained from the fact that the As nuclei have spin 3/2 [18]. Känzig's strong 1234321 spectrum showed that it arose from a Cl_2^- molecule in which the unpaired electron spin spent equal time on two ^{35}Cl atoms. Other lines arose from ^{37}Cl pairs and from ^{35}Cl – ^{37}Cl pairs. He called it a self-trapped hole since the existence of resolved hyperfine structure showed that the spectrum arose from an entity that was not moving through the lattice, and the g-factor showed that there was a missing electron. I had just added a new student to my group, Theodore Castner. He joined Känzig in working on what we now labeled the

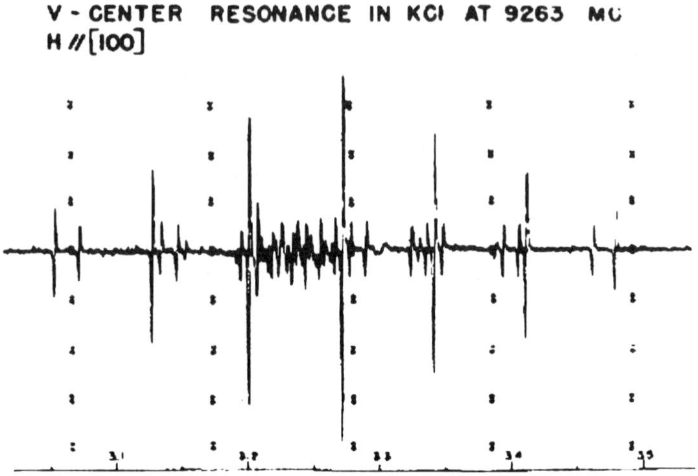

Fig. 2 The electron spin resonance of the V_K center in KCl with the applied static magnetic field oriented along the crystal [100] direction, showing the strong 1 234 321 spectra from pairs of the Cl atoms with the abundant isotope ^{35}Cl. ([14] with permission of the authors)

V_K center (after it's discoverer) in developing a very detailed understanding of all aspects of the ESR spectrum. This experiment opened the way for numerous other studies of hole centers. Känzig went on at the GE Research Laboratory to make many more important contributions to the ESR study of impurity centers. Eventually, he returned to the ETH where my family and I had the great pleasure of visiting him and his family when we visited Switzerland.

The IBM Research Laboratories

In 1961 I was invited to give the Loeb Lectures as a visiting Professor of Physics at Harvard. I gave a one-semester course in magnetic resonance. To commemorate this occasion, I turned the lectures into a textbook on magnetic resonance. One of the students in my class was a very talented graduate student named John Armstrong who was doing his thesis with Professor George Benedek. I tried very hard, but without success, to get John to join my lab as a post doc, but he stayed to work with Bloembergen[2]. Next, he joined the IBM Research laboratory at Yorktown heights eventually, rising in the ranks to become Director of Research and then Vice President for Science and Technology. It was in that capacity that, in the early 1980s, he invited Alex Müller, then an IBM Fellow, to spend a sabbatical at Yorktown Heights to

[2] There is now a professorship at Harvard, the John A. and Elizabeth S. Armstrong Professorship given by John Armstrong and his wife Elizabeth

study superconductivity. And it was also at about that time that Alex Müller and I shared membership on the IBM Scientific Advisory Committee. This Committee gave us the first chance to really get to know each other.

I remember vividly how exciting it was for all of us on the Committee when two scientists who had been working under his supervision, Binnig and Rohrer, invented a device they called a scanning tunneling microscope. The Advisory Committee saw the instrument first hand on a visit to Rüschlikon. It was even more exciting when Binnig and Rohrer were awarded the Nobel Prize in 1986. We were very proud of the role of our colleague Alex Müller. The next year, however, was even more spectacular. Our colleague Alex Müller together with George Bednorz discovered a whole new class of superconductors with vastly higher transition temperatures than ever obtained before!

The American Physical Society March Meeting took place in New York City that year. The session on superconductivity was mobbed. It has been called the Woodstock of Science after the famous gigantic gathering of rock music fans in Woodstock New York in 1969. Everyone was amazed at the discovery since so many scientists had tried in vain to raise the superconducting transition temperature. After so many failed attempts, the funding agencies looked askance at proposals to do so! I wanted to hear from Alex exactly how he had gotten the idea to study the cuprates, and hoped to get him to myself for just a few minutes. Of course, everyone wanted to talk to him. I found him in the hallway and persuaded him to sneak off to a basement restaurant in the hotel for a glass of wine in the evening just before the session of invited talks at which Alex answered questions to a gigantic audience about his amazing new discovery. The very next year, Alex Müller and George Bednorz shared the Nobel Prize!

I feel very fortunate to have had the chance to get to know some of the individuals who played important roles in the life of Alex Müller. Above all, I am supremely fortunate to know Alex Müller himself. His 80th Birthday calls up many wonderful memories for me. I dedicate this article to him.

References

1. Müller KA (1958) Helv Phys Acta 31:173
2. Schumacher RT, Carver TR, Slichter CP (1954) Phys Rev 95:1089
3. Schumacher RT, Slichter CP (1956) Phys Rev 101:58
4. Overhauser AW (1953) Phys Rev 91:476
5. Overhauser AW (1953) Phys Rev 92:411
6. Lewis RB, Carver TR (1964) Phys Rev Lett 25:693
7. Vander Ven NS, Schumacher RT (1964) Phys Rev Lett 12:695
8. Schultz S, Latham C (1965) Phys Rev Lett, p 15148
9. Griswold TW, Kip AF, Kittel C (1952) Phys Rev 88:951
10. Carver TR, Slichter CP (1953) Phys Rev 92:212

11. Carver TR, Slichter CP (1956) Phys Rev 102:975
12. Müller KA (1961) Phys Rev 123:1550
13. Shengalaya A, Keller H, Müller KA, Kochalaev BI, Conder K (2001) Phys Rev B63: 144513
14. Castner TG, Känzig W (1957) J Phys Chem Solids 3:178
15. Känzig W (1955) Phys Rev 99:1890
16. Fletcher RC et al. (1954) Phys Rev 94:1392
17. Fletcher RC et al. (1954) Phys Rev 95:844
18. Charles Slichter P (1955) Phys Rev 99:479

Takashige M (2007) In: Bussmann-Holder A, Keller H (eds)
High T_c Superconductors and Related Transition Metal Oxides.
Springer-Verlag Berlin Heidelberg, 311–314
DOI 10.1007/978-3-540-71023-3

Reminiscences of Collaboration in 1986

M. Takashige

Department of Electronics, Iwaki Meisei University, Chuohdai Iino 5-5-1, Iwaki, 970-8551 Fukushima, Japan
masa@iwakimu.ac.jp

To celebrate Prof. Alex Müller's 80th birthday, it is my great honor and pleasure to have the opportunity to contribute to this special book. I was in IBM Zurich Research Laboratory as a visiting scientist from Institute for Solid State Physics, the University of Tokyo (at that time), attached to Alex Müller's group in that memorable 1986. This article is reminiscences during my stay in IBM Laboratory, Rüschlikon, Switzerland.

The purpose of my stay was collaboration with Alex Müller and Dr. Erich Courtens about low temperature properties of ferroelectrics. At that time my research interests were in disordered systems such as the amorphous state of the perovskite $PbTiO_3$, the dipole glass of RDP-ADP. Our common interest is, for example, how does the quantum paraelectric feature of $SrTiO_3$ change, when the system is brought into the amorphous state or in the fine grained state crystallized from its amorphous state? For this purpose, low temperature dielectric measurements were to be planned.

It was February 3, 1986, when we arrived in Zurich-Kloten airport, where Dr. Georg Bednorz came to pick up our family. Since then, I shared Georg's office next to Alex Müller's room in IBM laboratory for one year.

For the first three months, under Georg's support, I was doing experiments such as preparation of amorphous films of $SrTiO_3$ using electron sputtering. I did not know that Alex and Georg were studying superconductivity at all. Later I heard that Georg already observed a resistivity drop of a sample of BaLaCuO, just one week before our arrival. Soon I realized that they were dealing with at least some conducting (black) samples including Cu-ions, since Georg often performed chemical co-precipitation to prepare mixtures of starting materials for BaLaCuO from aqueous solutions with blue color indicating the presence of a copper ion clearly. Co-precipitation is a showy method and it was not easy to perform it in secret, since we always spent time together both in the office and the laboratory.

It was April 21, 1986, when I understood the situation more exactly. In this morning, Georg gave me a preprint entitled "Possible High T_c Superconductivity in the BaLaCuO System" [1], submitted to Z. Phys. I never forget this day, since the day was half holiday of the laboratory due to traditional celebration of Zurich, called Sechselauten (six ringing festival) with burning of snowman festival to mark the end of winter. When the bells of St. Peter chime at six o'clock ring, a wooden snowman filled with firecrackers is lit. The faster the snowman is burnt out, it is said, the better is the summer which comes and also the better is the harvest. Our family went to the city of Zurich to see this festival in the evening. The snowman was burnt down rather faster, which means good harvest in the year, but I never imagined that harvest of our research would be much bigger.

After observation of my reaction for a few weeks, Alex asked me to join the research in superconductivity. I agreed willingly. At that point I was not able to have conviction about the occurrence in superconductivity, however, as an experimental scientist I thought that the most right way is to try to check by my own hand. To add one more, I was very glad that Alex and Georg trusted me and invited me to collaborate in such a serious problem. In this way, our collaboration started in May, 1986. It was continued until January, 1987.

I learned several points from the preprint [1]. When I read the background part of this paper described about strong electron phonon interaction of the perovskite and formation of Jahn–Teller polaron, I thought that it was really the concept made by Alex Müller who studied motion of octahedra in perovskite for a long time. Without the idea of Jahn–Teller polaron, nobody arrives at materials like BaLaCuO including Cu^{2+} and Cu^{3+} mixed valency state.

BaLaCuO was originally the material, which aimed at the composition of $(Ba, La)CuO_3$ type perovskite referred to C. Michel et al. [2]. However, unlike a way of sample preparation in [2], Georg's co-precipitation and setting of lower sintering temperature, which was an extremely important decision to determine fate, led to the samples with coexisting three phases. One is a 2-dimensinal layered perovskite with K_2NiF_4 structure, the second is a 3-dimensinal perovskite first aimed at and the third is an unknown phase (later determined to be CuO phase). Thus, our next work would be to answer "which is the superconductive phase?" To confirm Meissner–Ochsenfeld effect, the susceptometer was to be installed in summer

While awaiting the installation, using X-ray diffraction, electrical resistivity measurement and chemical analysis by systematically changing the composition, we tried to identify the superconducting phase. Before midsummer, my skeptical feeling about superconductivity of this system completely vanished and we had strong indication that the phase which is responsible for superconductivity was $La_2CuO_4 : Ba$. However, final judgement had to depend on the measurement of dc susceptibility.

In the middle of September, the susceptometer was installed later than the schedule of one month or more. We had no time, since already the first paper had been open literature [1]. Immediately we tried to measure several samples. The result was what we expected. The presence of the Meissner-Ochsenfeld effect was confirmed for the samples showing the resistivity drop. We identified the phase responsible for superconductivity a $La_2CuO_4 : Ba$, having K_2NiF_4 structure.

These results were summarized in the paper, entitled “Susceptibility Measurements Support High-T_c Superconductivity in the BaLaCuO System” [3] and submitted to Euro. Phys. Lett. in the middle of October. Related to this, there was very wonderful recollection. When Alex, Georg and I were sitting together to make final correction of the above draft, there was the announcement that the 1986 Nobel Prize in Physics had been awarded to our colleagues Gerd Binnig and Heinrich Rohrer for their design of the scanning tunneling microscope (STM). I had not imagined yet then that the Nobel Prize of the next year would come to Alex and Georg.

When the work was settled temporarily, Alex and Mrs. Inge Müller invited our family to their mountain villa. I had a special feeling, something like that I was able to play one role and very much relaxed. Alex drove us many places and we enjoyed the nature of Switzerland in early autumn, especially Engadine. During my stay in IBM Laboratory, I learned much from Alex. He talked about a lot of things at tea and lunch time in the IBM restaurant or walking time after lunch, not only physics, the future direction of research, but also the history and culture of Switzerland and nineteenth century European spirit he like.

After confirmation of the diamagnetic susceptibility, we continued to do two works. First, in the magnetic susceptibility measurements, we realized the difference between field cooled (FC) and zero field cooled (ZFC) responses and non-exponential time dependence of ZFC response, which are similar to the behavior of spin glasses. These features were named “superconductive glass” [4] by Alex, and even now it is one of the common features observed in other oxide superconductors. I have been proud of this work to produce such a new concept in the very early stage of high-T_c researches and have deeply respected Alex’s foresight intelligence. Second, further material synthesis such as $La_2CuO_4 : Sr$, $La_2CuO_4 : Ca$ were also made. In $La_2CuO_4 : Sr$, we observed T_c of the 40 K range in the middle of December. However, it was until end of November that we could do research only at our pace.

The first news came from Tokyo in November 28, 1986. My wife Emi accidentally bought a Japanese news paper in the city of Zurich, Asahi Shimbun, International Satellite Edition, where an article appeared about the BaLaCuO system, reporting the effort of Prof. S. Tanaka’s group in the Department of Applied Physics, the University of Tokyo, which in November confirmed the diamagnetic susceptibility in the 30 K range. This news was good news in the sense that our result was supported but I had recognized that international competition already started. Everywhere in the world, scientists started working on BaLaCuO. We had still many unpublished data but it was a pity for me that scientific results already had come to be told earlier by mass communications rather than conventional scientific magazines. After submission of three works [4–6], I left Zurich in January 29, 1987 and our collaboration was finished.

It was very much exciting, pleasant and sometimes hard, and really condensed time. I must say that as a scientist, I was extremely lucky to be there at such a historic moment that happens only once in a life time. Moreover, we learned much from this discovery. For the breakthrough of materials sciences, we always need own philosophy like Alex's Jahn–Teller polaron and own approach like Georg' co-precipitation. I believe that even now we had not yet arrived at high-temperature superconductivity in cuprate oxides, if there was not a pair of these two persons.

Acknowledgements Finally, again I would like to express my sincere thanks to Prof. Alex Müller for giving me opportunity of being visiting scientist of IBM laboratory in 1986 and would like to celebrate heartily his 80th birthday and further development of his researches.

References

1. Bednorz JG, Müller KA (1986) Z Phys B64:189
2. Michel C, Er-Rakho I, Raveau B (1985) Mat Res Bull 20:667
3. Bednorz JG, Takashige M, Müller KA (1987) Europhys Lett 3:379
4. Müller KA, Takashige M, Bednorz JG (1987) Phys Rev Lett 58:1143
5. Bednorz JG, Takashige M, Müller KA (1987) Mat Res Bull 22:819
6. Bednorz JG, Müller KA, Takashige M (1987) Science 236:73

Thomas H (2007) In: Bussmann-Holder A, Keller H (eds)
High T_c Superconductors and Related Transition Metal Oxides.
Springer-Verlag Berlin Heidelberg, 315–319
DOI 10.1007/978-3-540-71023-3

Coupled Order Parameters in Magneto-Ferrolectrics

H. Thomas

Universität Basel, Institut für Physik, Klingelbergstrasse 82, CH-4056 Basel, Switzerland
Harry.Thomas@unibas.ch

Introduction

Recent observations of the simultaneous appearance of non-collinear magnetic order and spontaneous polarization in a special class of multiferroics have revived the interest in the magnetoelectric effect. These phenomena have so far been observed in $TbMnO_3$ [1–4], $DyMnO_3$ [2, 3] and $Tb_{1-x}Dy_xMnO_3$ [5] with an orthorhombically distorted perovskite structure by rotation of the MnO_6 octahedra, and in $Ni_3V_2O_8$ [6], in which the Ni^{2+} ions form a Kagomé staircase. The strong coupling between magnetic order and ferroelectricity gives rise to giant magnetoelectric and magnetocapacitance effects.

I will concentrate in this note on the case of $TbMnO_3$. In this compound, there first appears at $T = 41$ K a collinear modulated sin-wave structure of Mn^{3+} spins,

$$\vec{M}_l = \left\{0, M_b \sin(\vec{q}\cdot\vec{R}_l), 0\right\} \tag{1}$$

with wave vector $\vec{q}$ along the b-axis. At $T = 27$ K, a non-collinear modulated magnetization and a ferroelectric polarization,

$$\Delta\vec{M}_l = \left\{0, 0, M_c \cos(\vec{q}\cdot\vec{R}_l)\right\}\,, \tag{2}$$

$$\vec{P}_l = \{0, 0, P_c\}\,, \tag{3}$$

appear simultaneously. At still lower temperatures, the Tb^{3+} spins begin to contribute.

It was originally suggested that the transition to the magneto-ferroelectric phase is related to an incommensurate-commensurate (lock-in) transition, and that the ferroelectric polarization originates in magnetic interactions through lattice modulations, in analogy to improper ferroelectrics [1, 2]. But later it was found that the wave number continues to vary slowly in the magneto-ferroelectric phase [4, 5]. It was pointed out that the non-collinear part $\Delta\vec{M}$ of the magnetization and the spontaneous polarization $\vec{P}$, both of which break space-inversion symmetry, transform as the same representation of the magnetic group of the collinear sin-wave structure, such that the magnetoelectric coupling term M_cP_c is invariant [4]. The suggestion [7] that the transition is produced by Dzyaloshinskii-Moriya interaction coupled to lattice displacements, as well as a phenomenological approach [8] based on the existence of a Lifshitz invariant, are in agreement with the above conclusion. I propose for this type of transition driven by a genuine magnetoelectric effect the name *magneto-ferroelectric*.

There is a tendency to consider the magneto-ferroelectric transition as a simple magnetic transition, and the spontaneous polarization as an induced effect and a secondary order parameter. This is strictly not correct; with the same right, the magneto-ferroelectric transition could be considered as a ferroelectric transition, and the magnetization change as an induced effect. It is the purpose of this note to present a Landau theory of magneto-ferroelectric transitions in which the ferroelectric moment and the amplitude of the non-collinear part of the modulated magnetization are treated on the same footing. I restrict the discussion to the case of zero magnetic field.

Landau Theory of a Magneto-Ferroelectric

Critical Temperature

In this Section, a Landau theory of magneto-ferroelectric phase transitions is developed in terms of two dynamic variables, the amplitude M_c of the non-collinear part of the modulated magnetization and the spontaneous polarization P_c, treated on equal footing. I introduce the dimensionless quantities

$$m = \frac{M_c}{M_0} \quad \text{and} \quad p = \frac{P_c}{P_0}\,, \tag{4}$$

where M_0 and P_0 are arbitrarily chosen reference amplitudes.

In terms of m and p, the Landau free energy $\mathcal{F}(m,p)$ has the expansion

$$\mathcal{F}(m,p) = \frac{1}{2}\left(a_m m^2 - 2cmp + a_p p^2\right) + \frac{1}{4}\sum_0^4 b_k p^k m^{4-k} + \ldots\,, \tag{5}$$

where a_m is the reciprocal magnetic susceptibility at wave vector $\vec{q}$ for fixed polarization, a_p is the reciprocal electric susceptibility for fixed magnetic moment, and c is the magnetoelectric coupling constant. Either a_m or a_p are assumed to depend on temperature as $a_m(T) \propto T - T_m$ (magnetically driven transition) or $a_p(T) \propto T - T_p$ (ferroelectrically driven transition), where T_m or T_p, respectively, would be the critical temperature in the absence of magnetoelectric coupling. In the presence of magnetoelectric coupling, on the other hand, the critical point T_c is reached when the quadratic part of the Landau potential with determinant $D(T) = a_m(T)a_p(T) - c^2$ ceases to be positive definite. This yields the condition

$$D(T_c) = a_m(T_c)a_p(T_c) - c^2 = 0\,. \tag{6}$$

Thus, a_m and a_p are still positive at $T = T_c$, i.e., the critical temperature T_c is higher than T_m and T_p, respectively. Near $T = T_c$, the determinant D depends on temperature as $D(T) \propto T - T_c$.

Below the critical temperature, both variables m and p vary as $(T_c - T)^{1/2}$, in qualitative agreement with observation.

Response to an Electric Field

The response to a static electric field E is found from

$$\frac{\partial \mathcal{F}}{\partial m} = 0, \qquad \frac{\partial \mathcal{F}}{\partial p} = E\,. \tag{7}$$

This yields the static electrical susceptibility χ_e for $T \geq T_c$

$$\chi_e(T) = \frac{a_m(T)}{D(T)}\,, \tag{8}$$

which diverges for $T \to T_c$ as $(T - T_c)^{-1}$, again in qualitative agreement with observation.

For the calculation of the dynamic susceptibility, we assume overdamped dynamics for both magnetization and electric polarization with damping constants Λ_m and Λ_p, respectively. From the equations of motion

$$\Lambda_m \dot{m} = -\frac{\partial \mathcal{F}}{\partial m}\,, \qquad \Lambda_p \dot{p} = -\frac{\partial \mathcal{F}}{\partial p} + E\,e^{-i\omega t}\,, \tag{9}$$

one obtains the dynamic susceptibility

$$\chi_e(\omega, T) = \frac{a_m(T) - i\Lambda_m\omega}{D(\omega, T)}\,, \tag{10}$$

where

$$D(\omega, T) = \left(a_m(T) - i\Lambda_m\omega\right)\left(a_p(T) - i\Lambda_p\omega\right) - c^2\,. \tag{11}$$

The susceptibility $\chi_e(\omega, T)$ shows a relaxation peak which narrows for $T \to T_c$ in the usual way as $\Delta\omega \propto (T - T_c)$.

The normal-mode frequencies in the field-free case are found as the zeros of $D(\omega, T)$,

$$\omega_{1,2} = -\frac{i}{2}\left[\frac{a_m}{\Lambda_m} + \frac{a_p}{\Lambda_p} \mp \sqrt{\left(\frac{a_m}{\Lambda_m} + \frac{a_p}{\Lambda_p}\right)^2 - 4\frac{D(0, T)}{\Lambda_m\Lambda_p}}\,\right]\,. \tag{12}$$

Proper Order Parameter

So far, the phase transition of a magneto-ferroelectric has been described in terms of two dynamic variables m and p, treated on equal footing. This, however, does *not* mean that the phase transition is of XY-type with a two-component order parameter. In fact, of the two normal-mode frequencies (Eq. 12), only one goes to zero for $T \to T_c$, the other remaining finite. The linear combination of m and p corresponding to the soft mode is the proper order parameter ξ of the transition.

This can also be seen by a transformation to principal axes of the quadratic part of the Landau potential,

$$m = \xi \cos\varphi - \eta \sin\varphi\,, \qquad p = \xi \sin\varphi + \eta \cos\varphi\,, \tag{13}$$

where φ has to be chosen such that

$$\mathcal{F} = \frac{1}{2}\left(\alpha\xi^2 + \beta\eta^2\right) + \text{higher-order terms}\,. \tag{14}$$

From the condition that the coefficient of the term $\xi\eta$ vanishes, one obtains

$$\tan 2\varphi = \frac{2c}{a_\mathrm{p} - a_\mathrm{m}}\,, \tag{15}$$

and the coefficients α and β are given by

$$\alpha = \frac{1}{2}\left(a_\mathrm{m} + a_\mathrm{p} - \sqrt{\left(a_\mathrm{m} + a_\mathrm{p}\right)^2 - 4D}\right)\,, \tag{16}$$

$$\beta = \frac{1}{2}\left(a_\mathrm{m} + a_\mathrm{p} + \sqrt{\left(a_\mathrm{m} + a_\mathrm{p}\right)^2 - 4D}\right)\,. \tag{17}$$

Only the coefficient α goes to zero at $T = T_\mathrm{c}$, whereas the coefficient β stays positive. Thus, the transition is characterized by the single-component order parameter ξ, and the component η describes an induced effect.

Conclusion

A Landau theory of magneto-ferroelectric transitions is presented, which treats the amplitude of the non-collinear part of magnetization M_c and the spontaneous polarization P_c on the same footing. The magnetoelectric coupling increases the critical temperature, and M_c and P_c show the same critical behaviour. The proper order parameter of the transition is a linear combination of M_c and P_c.

References

1. Kimura T et al. (2003) Magnetic Control of Ferroelectric Polarization. Nature (London) 426:55–58
2. Goto T et al. (2004) Ferroelectricity and Giant Magnetocapacitance in Perovskite Rare-Earth Manganites. Phys Rev Lett 92:257201

3. Kimura T et al. (2005) Magnetoelectric Phase Diagrams of Orthorhombic $RMnO_3$ (R = Gd, Tb, and Dy). Phys Rev B 71:224425
4. Kenzelmann M et al. (2005) Magnetic Inversion Symmetry Breaking and Ferroelectricity in $TbMnO_3$. Phys Rev Lett 95:087206
5. Arima T et al. (2006) Collinear to Spiral Spin Transformation without Changing the Modulation Wavelength upon Ferroelectric Transition in $Tb_{1-x}Dy_xMnO_3$. Phys Rev Lett 96:097202
6. Lawes G et al. (2006) Magnetically Driven Ferroelectric Order in $Ni_3V_2O_8$. Phys Rev Lett 96:087205
7. Sergienko IA, Dagotto E (2006) Role of the Dzyaloshinskii-Moriya Interaction in Multiferroic Perovskites. Phys Rev B 73:094434
8. Mostovoy M (2006) Ferroelectricity in Spiral Magnets. Phys Rev Lett 96:067601

Author Index

Printing: Krips bv, Meppel
Binding: Stürtz, Würzburg